AF740820

EW 104

EW Against a New Generation of Threats

For a complete listing of titles in the
Artech House Power Electronic Warfare Library,
turn to the back of this book.

EW 104

EW Against a New Generation of Threats

David L. Adamy

ARTECH HOUSE
BOSTON | LONDON
artechhouse.com

Library of Congress Cataloging-in-Publication Data
A catalog record for this book is available from the U.S. Library of Congress.

British Library Cataloguing in Publication Data
A catalogue record for this book is available from the British Library.

Cover design by John Gomes

ISBN 13: 978-1-60807-869-1

© 2015 ARTECH HOUSE
685 Canton Street
Norwood, MA 02062

All rights reserved. Printed and bound in the United States of America. No part of this book may be reproduced or utilized in any form or by any means, electronic or mechanical, including photocopying, recording, or by any information storage and retrieval system, without permission in writing from the publisher.

All terms mentioned in this book that are known to be trademarks or service marks have been appropriately capitalized. Artech House cannot attest to the accuracy of this information. Use of a term in this book should not be regarded as affecting the validity of any trademark or service mark.

10 9 8 7 6 5 4 3 2 1

This book is dedicated to the young people in uniform who go into harm's way to practice the art and science of electronic warfare. They are the ones who will face the danger from this new generation of threats and do what is necessary to protect the rest of us in this dangerous world.

Contents

Preface

This is the fourth book in the *EW 101* series. It is based on the EW 101 columns in *The Journal of Electronic Defense*. At the time of this writing, the series included 213 columns written over two decades. The first two books, EW 101 and EW 102, covered the basics of electronic warfare (EW). The third book, *EW 103*, focused on communications electronic warfare and was written in response to the situations in the Middle East. There was a new emphasis on EW on the ground and this book was intended to help the ground-forces people who were dealing with hostile communications near the Earth, including the links used to detonate the improvised explosive devices (IEDs) that were causing most of our casualties.

Now the whole EW field is changing. There are new threat radars that are very scary and new types of communication links that are very difficult to handle. Perhaps the scariest aspect of what is happening in EW is that many of the ways that we have been performing EW functions do not do the job any longer. New approaches are required, and this book is intended to help those, in and out of uniform, who are dealing with this new reality, sometimes at risk of life and limb.

Like *EW 101*, *EW 102*, and *EW 103* before it, this is an unclassified book but deals with some subjects that draw information from classified sources. We will not go there. Here is our approach: Use reasonable estimates for antenna gain, effective radiated power (ERP), and so forth. These are probably not the right numbers, and we do not care. If there are no numbers offered in open literature, we will make them up based on reasonable estimates and we will give the logic used in making up each value. Then we will talk about using the equations and we will perform example problems with those estimates plugged in. Different numbers are found in different open sources, so you know some of them will have to be wrong. We will not make judgments on which numbers are right and wrong; we will just pick one number and use that value to work operational problems. The action here is how to use the information. Later, when you are using this information on the job, you can look up the real numbers in your authorized classified sources and plug in the real numbers into the equations that we have discussed.

1

Introduction

The nature of the electronic warfare (EW) field has changed over the last few years and is in a state of accelerating change. The purpose of this book is to deal with those changes from a technical perspective. This book uses threat information from open literature. It is not intended to serve as a threat briefing, but will use reasonable estimates and show what the impact will be on countermeasures.

The important changes in electronic warfare include:

- The recognition of the electromagnetic environment as a distinct battlespace;
- New and extremely dangerous electronically guided weapons;
- New technologies that impact both the accuracy and lethality of weapons.

This book deals with all of these areas to the extent possible with the limitation of using only open-source information. Fortunately, that open-source information is rich in the new technology areas, supporting discussions of their role in new weapons and in the nature and effectiveness of EW measures to counter those weapons.

In the EW vocabulary, we refer to radio emissions associated with threats as "threats." This is not correct; threats are actually things that explode or cause harm in some other way. However, that is the way we talk about such signals. In this book, we will be talking about both radar threats and communication threats. Using this terminology, radar threats are radar signals associated with radar controlled weapons:

- Search and acquisition radars;

- Tracking radars;
- Radio links between radar processors and missiles for guidance and data transfer.

Communication threats include:

- Command and control communication;
- Data links between components of integrated air defense systems;
- Command and data links connecting unmanned aerial vehicles with their control stations;
- Links that fire improvised explosive devices (IEDs);
- Cell phone links when used for military purposes.

Our emphasis in this book is what these signals do and how they impact the effectiveness of weapons and military operations.

We also consider the significant advances in heat-seeking missiles and countermeasures to defeat them.

Briefly stated, we cannot continue to perform EW the way we have been doing it, with great success, for several decades. The world has changed and we must change with it.

This book endeavors to give you some tools to help implement that change.

There are three major thrusts in the rest of this book:

1. There is a discussion of the newly recognized field of electromagnetic warfare (in Chapter 2). This is an additional battlespace that has emerged, in addition to the familiar battlespaces of land, sea, air, and space. As you will see, there are parallels to all of the aspects of the other battlespaces, and EW is an important player. There is a related subject that did not really fit anywhere, but is important. That is the definition of the difference between electronic warfare support (ES) and signal intelligence (SIGINT) in Chapter 11.
2. There are several new technologies and approaches that impact electronically controlled weapons and EW. Each of these areas is covered in its own chapter, including Chapter 5 on digital communication theory, Chapter 8 on digital RF memories (DRFM), and Chapter 10 on radar decoys.
3. There is a discussion of modern threats. Radar threats are covered in two chapters. Chapter 3 is about legacy threats and also includes equa-

tions for the interception and jamming of radar threats. Chapter 4 covers the features of the new generation threats that have been developed. Communications threats are likewise covered in two chapters. Chapter 6 covers legacy threats including the propagation equations for intercept and jamming and also covers emitter location. Chapter 9 is about infrared threats and countermeasures.

2

Spectrum Warfare

The nature of warfare is changing. The realms used to be ground, sea, and air. Then space was added. Now there is a fifth realm: the electromagnetic spectrum. In this chapter, the nature of this new realm of warfare is investigated and related to warfare in the other four realms. This chapter deals with the basic concepts and vocabulary associated with warfare in the electromagnetic (EM) spectrum realm.

2.1 Changes in Warfare

The enhancement in our ability to communicate is making significant changes in the way we conduct warfare. Radio communication started a little over a century ago. Before that, distant communication was only by wire. For practical reasons, military communication was largely by wire until about two generations ago. Ships, aircraft, and ground mobile assets needed to communicate without wires, so much effort went into radio communication. As World War II was starting, radar was developed by most opponents, and radio communication became much more sophisticated.

From the beginning, the use and control of spectrum were issues. When Marconi made his first trans-Atlantic transmissions with a spark gap transmitter, it used so much spectrum that there was room for only one transmission in the world. When tuned transmitters were developed (shortly thereafter), interference between radio links was still a significant problem. The certainty of intercept of radio communication and radar signals and the ability to locate transmitters had significant impact on military operations. Intercept, jamming, emitter location, message security, and transmission security became fundamental to warfare, and are not likely to ever go away.

The basic destructive capabilities employed in warfare have not changed a lot (people who develop these items will probably argue this point). However, the ways that they are employed have changed significantly through use of the EM spectrum (EMS). Now we guide the destructive energy of weapons toward their intended targets using the EMS in various ways. We in the electronic warfare (EW) business also use the EMS to try to prevent those weapons from hitting their intended targets or keeping enemy from knowing where those targets are.

Destructive energy (fast-moving projectiles, significant over-pressure, or heat) is employed to kill enemies or to destroy things that they need to conduct warfare or to sustain their way of life. Sometimes, the destruction of communication capability by an enemy is a goal in itself. Thus, the battlespace, which once had only four dimensions (latitude, longitude, elevation, and time) now has a fifth dimension: frequency (see Figure 2.1).

As a result of increased control of destructive energy, we have moved to more careful focus of the destruction. We want all of the force to go against desired targets. Collateral damage is always a waste of military capability, even by those who do not care about sparing the innocent people who find themselves in the way of warring parties. To those of us who care about avoiding civilian casualties and damage, this focus of weapons is even more pressing.

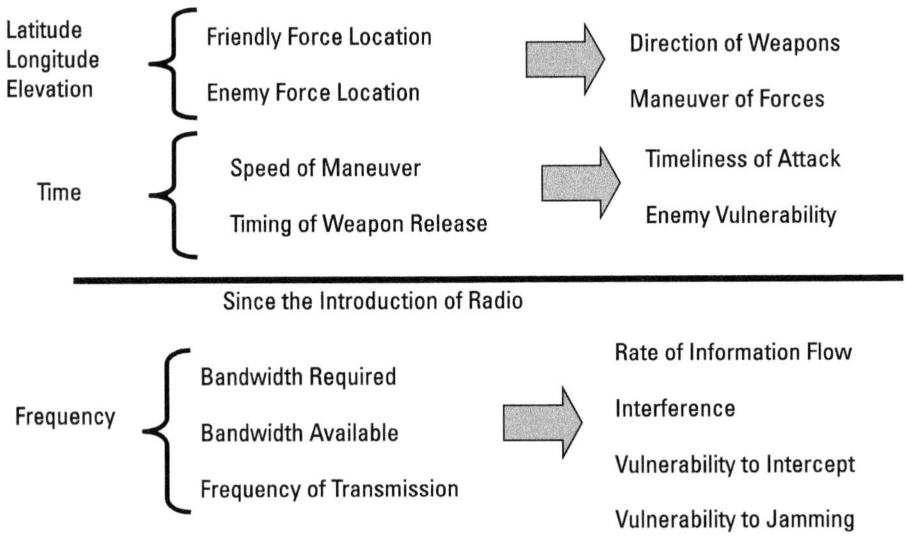

Figure 2.1 Before radio communication, warfare was conducted in four dimensions. Now it has frequency as an additional dimension.

2.2 Some Specific Propagation Related Issues

Range has a significant impact on radio transmission. Depending on the environment, the strength of a receive signal is a function of the square or fourth power of the distance from the transmitter. Therefore, a closer receiver will do a better job of receiving a signal and can also usually locate the transmitter more accurately. If we have multiple receivers, the one closest to a hostile transmitter will have the best information (see Figure 2.2) However, to be useful, that information must get to the place where decisions are made. Thus, those receivers must be part of a network.

Once we depend on inputs from multiple receivers, the network becomes central to our war making ability. We have now entered net-centric warfare.

Then consider the problem of jamming enemy transmissions. Either communication or radar jamming must create adequate jamming to signal ratio. The formulas for both kinds of jamming involve the square (or fourth power) of the range from the jammer to the jammed receiver. If we have a number of jammers geographically spread, we will have the best results if we use the closest jammer. A related problem is jamming our own EMS assets (i.e., fratricide). As shown in Figure 2.3, the jammer closest to the target receiver can jam with the least power, which will reduce the impact of the jamming on friendly communication or radar performance.

Again, those jammers must be part of a network. That network will, of course, be an important target to an enemy. If they can collect information from our network, they will be able to determine much about our tactical inten-

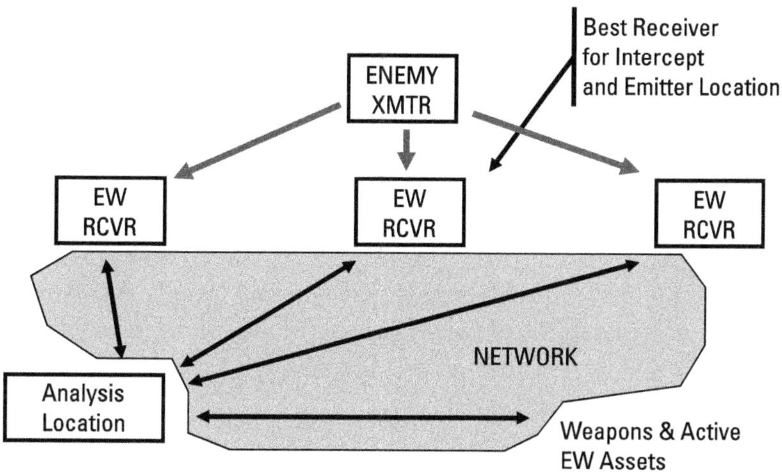

Figure 2.2 Proximity to enemy transmitters has significant impact on intercept and emitter location performance.

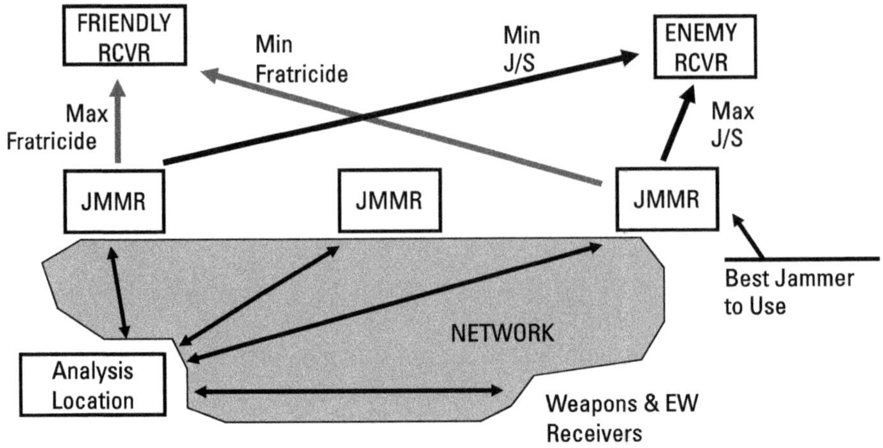

Figure 2.3 Proximity to enemy and friendly receivers has significant impact on jamming effectiveness and fratricide.

tions, and if they can destroy our network, they will diminish or even eliminate our war-making ability.

2.3 Connectivity

Because of our dependence on connectivity in our daily lives and business, an enemy can cause us real damage by attacking the connectivity itself. Consider the economic impact of having our banking system, our rail infrastructure, or our air transportation capability shut down. All of these and many more aspects of our modern economic and military capability are so dependent on connectivity that a radio frequency or cyberattack could cause significant physical damage, loss of military capability, or devastating disruption of economic activity. Before discussing attacks on connectivity in more detail, it will be useful to discuss the nature of connectivity from a technical point of view.

Connectivity can be thought of as any technique for the movement of information from one location or player to another. The medium can be wire, radio propagation, optical propagation, or audio propagation. We must also consider the most basic connectivity; between two people, two devices (e.g., computers), or between devices and people.

2.3.1 The Most Basic Connectivity

In its simplest form, connectivity can be one person talking (or yelling over a distance) to another person or optically transmitting information. Examples of person-to-person optical transmission are writing on a surface for others to read, holding up a sign, code with a steady or flashing light, and use of signaling flags (or perhaps smoke). All are, in fact, used to some extent in almost all of the most sophisticated military and civilian systems. Even when more technical transmission techniques are used, the input of information from humans is by voice or physical input of data from a keyboard or other touch device. Getting the information to another human being can only be done through the senses of hearing, vision, or touch.

All of the simplest techniques share the advantages of simplicity of implementation and robustness. It is very hard to jam this kind of connectivity. It also requires that an enemy be relatively close to intercept transmitted information. That said, security requires diligent measures to prevent an enemy from successfully employing techniques like hidden microphones or cameras or monitoring reflections from lasers bounced off of windows.

However, all of these simple connectivity techniques have the immense disadvantage of short range. Increasing the range of these means of connectivity requires sending a messenger or relaying the information. Both techniques cause significant increase in complexity, reduce security against interception, and reduce the reliability and confidence in the accuracy of the information passed. Thus, it becomes advantageous or even necessary to employ technical transmission paths and techniques to extend the range, perhaps by a few kilometers or perhaps to some significantly different part of the Earth.

2.3.2 Connectivity Requirements

Regardless which connectivity technique is employed, from the simplest to the most complex, the requirements shown in Table 2.1 must be met. First consider the simplest connectivity techniques and the characteristics of the information passed.

2.3.2.1 To or From People

Figure 2.4 shows connectivity with a person:

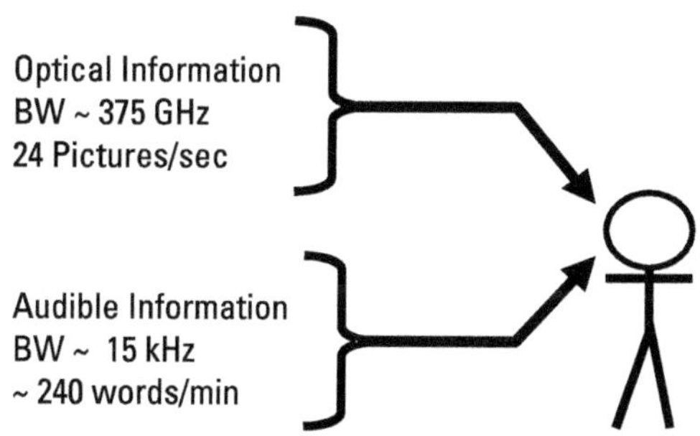

Figure 2.4 Human connectivity is limited by physical bandwidth and data format factors.

Table 2.1
Connectivity Requirements

Requirement	Level
Bandwidth	Adequate to carry highest frequency component of information at required throughput rate
Latency	Short enough to allow an activity loop to operate with required performance
Throughput rate	Adequate to pass information at required speed
Information fidelity	Adequate to allow recovery of required information from a received transmission
Message security	Adequate to protect the information for the duration of its usefulness to an enemy
Transmission security	Adequate to prevent an enemy from detecting a transmission in time to prevent required transmission or to locate a transmitter in time to make an effective attack on it or to determine electronic order of battle in time to effect a military operation
Interference rejection	Adequate to provide required information fidelity in the operating environment
Jamming resistance	Adequate to prevent an enemy with the anticipated jamming capability and geometry from preventing the achievement of adequate information fidelity

- Voice communication: If you have perfect hearing, your ears can handle about 15 kHz, but most information is carried by speech in approximately 4 kHz. Actually, a telephone circuit allows only 300 to 3,400 Hz to carry the voice signal. For us to process received data it must be organized into syllables or words. We can hear and process up to about 240 words per minute.
- Optical communication: Your eyes have much wider bandwidth. If you can see the full rainbow, you can calculate the bandwidth of your eyes from the red to violet spectrum at about 375,000 GHz. However, we see and process whole scenes through our eyes. We can see a new scene 24 times per second. [Note that we see changes in color detail about half that fast and can see light and dark details (luminance) in our peripheral vision faster. A very practical value to consider as the effective bandwidth at which we get visual data might be an analog color television signal which is a little less than 4 MHz wide.]
- Tactile communication: You can probably detect vibration at close to the frequencies you can hear. For example, you can easily detect the vibration of your cell phone at about 1,000 Hz. However, tactile communication is generally limited to alarms that point to more detailed audio or video information. An important exception to this is Braille writing, in which a blind person can receive information through the sensing of patterns of raised dots. There is some discussion in literature of experi-

mental devices which impress graphic images (from video cameras) on the skin of a blind person.

2.3.2.2 Between Machines

Machine to machine or computer to computer connectivity is shown in Figure 2.5. Because computers and other controlled machines are not limited to human connectivity rates, this communication can have much wider bandwidth. Machines can be direct wired to each other, using either parallel or serial interconnectivity, or can be interconnected using a local area network (LAN). The LAN can interconnect machines by digital cable, by RF link, or by optical link. The rates can be from a few hertz to gigahertz.

2.3.3 Long-Range Information Transmission

Now, let's consider the longer range connectivity techniques that move information from one human location to another (or from one computer location to another). We will consider each of the requirements in Table 2.1.

As shown in Figure 2.6, the bandwidth at the point at which the information is input must be adequate to accept that data. However, the bandwidth over which it is transmitted may be different. If the data flow must be continuous, the transmission path must have the full input data bandwidth. However, if the input data is not continuous or has a varying data flow rate, it can be transmitted at a lower rate. Practical systems that perform this way digitize the data and clock it into a register at the sending end of the link. Then the data is

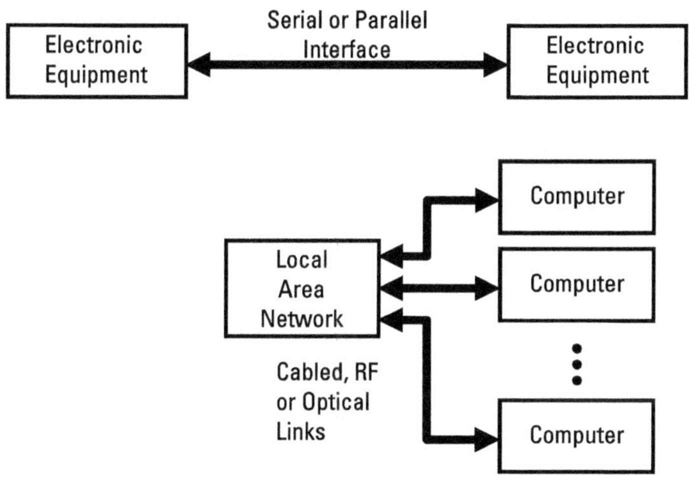

Figure 2.5 Short-range machine connectivity can be direct or through a cabled, RF, or optical LAN.

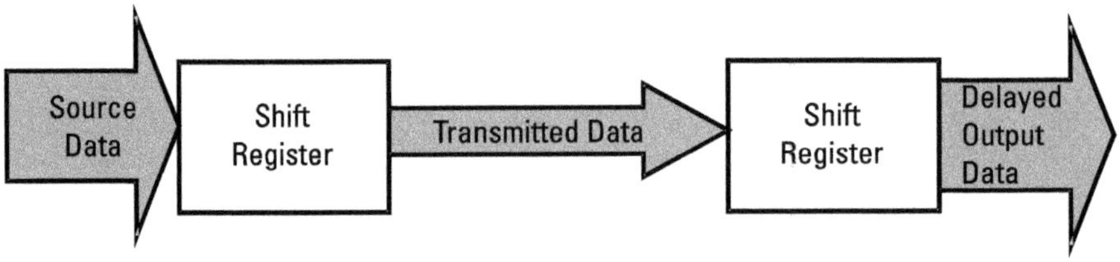

Figure 2.6 High bandwidth, noncontinuous source data can be transmitted at a lower rate and returned to its original format at the receiver, but with latency.

clocked out of the register at a lower rate, which allows a narrower transmission bandwidth. At the receiving end, the data can (if required) be input to another register and clocked out at its original data rate. There are two other factors that impact the required transmission bandwidth: latency and throughput rate.

Latency is the delay in the received data compared to the transmitted data. A good demonstration of latency is a news broadcast involving a local host talking to a reporter who is half a world away. The host asks a question and the reporter is shown standing there not responding for a few seconds before answering. The host's question travels about 85,000 km to and from a satellite at the speed of light, which takes about 2.5 seconds. The reporter's response takes another 2.5 seconds to reach the host's location. The process latency causes the observed 5 seconds of the blank look on the reporter's face. There is additional latency between the host's location and your television set, but you do not notice it because the constant delay allows you to see a continuous flow of data.

Latency becomes critical when the connectivity is inside a process loop. If you are far away trying to manually land an unmanned aerial vehicle, with any significant latency, it would take extraordinary skill to avoid crashing the aircraft by over controlling. The less latency you can tolerate, the less transmission bandwidth reduction you can use. The propagation time versus distance is, of course, also a latency factor.

Throughput rate is the average rate at which information flows. In general, individual pieces of very wideband data can be transmitted over limited bandwidth by spreading them in time. However, if the average rate of information flow is higher than the transmission bandwidth, the latency increases until the process crashes. A simple example of this phenomenon is an individual speaking in a foreign language with limited fluency. The foreign listener typically does not know some of the words used. That person can follow a conversation at some rate, but must mentally review what has been said to pull unknown words out of context. This review process is part of the information path, and thus narrows the effective transmission bandwidth. If the native speaker continues at too high a rate, the listener's review process delay increases the latency until the foreign listener cannot follow the conversation.

In computer-to-computer communication, an analogous process is the storage of wideband data until there is a pause or a period of lower bandwidth data that allows the receiving computer to put the whole data stream back into the proper format to be processed out. The amount of latency allowable depends on the available memory in the receiving computer. When this memory overflows because of excess throughput rate, the process crashes.

Typically, a networked system will get in trouble because of the required throughput rate rather than the peak data rate, which will be discussed later.

2.3.4 Information Fidelity

Earlier we discussed the interaction of bandwidth, latency, and throughput rate. All of these items are also related to information fidelity, which brings up the issue of data compression. When we speak or write, we format information in ways that allow the receiver to receive and process the information in the way the human brain is wired to operate. Language, rules of grammar, sentence structure, punctuation, adjectives, and adverbs all serve to make our meaning clear. They also use up a lot of time and bandwidth. When young people text each other, they poke their cell phones with their thumbs at blinding speed and use abbreviations and grammar that is impenetrable to their elders. What they are doing, from a technical point of view, is encoding for information compression. Because the available bandwidth limits the rate of symbol transmission, the flow of this critically important information is slowed to an unacceptable level by the normal overhead associated with academically acceptable grammar, spelling, and so forth. The encoding is a form of data compression to remove redundancy from the data, thereby allowing the information rate to data rate ratio to increase. The same function is served by digital data compression techniques used for speech and video compression. Figure 2.7 shows the information flow from originator to user including data compression (by any means).

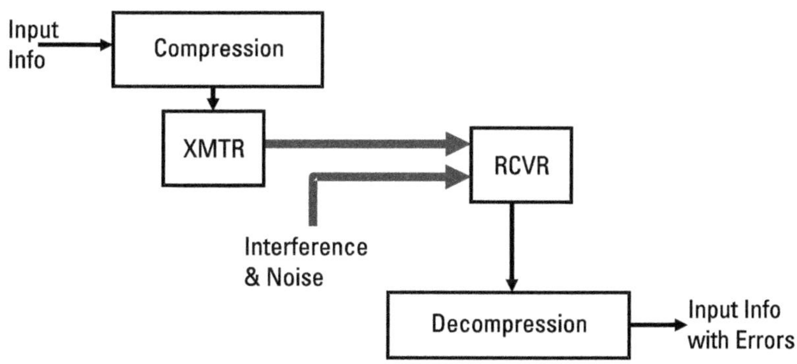

Figure 2.7 Any data compression approach will be subject to errors because of the impact of interference and noise on the decompression.

Note that the signals received by the receiver will also include interfering signals and noise and that the receiver itself generates noise.

The problem, of course, is that any coding used will have some impact on information fidelity. Ideally, the communicator uses a lossless code in which all of the information is preserved through the encoding and decoding process. However, now add the impact of the transmission of the encoded information from the sender to the receiver. First consider digital communication media. As the range increases or interference (intentional or unintentional) occurs, bit errors are created at the point at which the receiver must determine whether a one or zero has been received. Figure 2.8 shows the relationship between the bit error rate and E_b/N_0. E_b/N_0 is the received predetection signal-to-noise ratio (RFSNR) adjusted for the ratio of bit rate to RF bandwidth. To be transmitted, digital data must be carried by a modulation, which requires demodulation to recreate the original digital ones and zeros. Each modulation has a different curve in this figure, but all have about the same shape. In radio transmission, the system is typically designed so that a 10^{-3} to 10^{-7} bit error rate is required. In this range, most modulations provide an error to RFSNR slope of about one order of magnitude of bit error change to 1 dB of change in RFSNR. For transmission within a cable (such as in a telephone network), much higher SNR may be practical, and the slope of this curve steepens.

We will be talking about forward error correction in Chapter 5. Now just consider that error correction and detection codes (EDC) add extra information

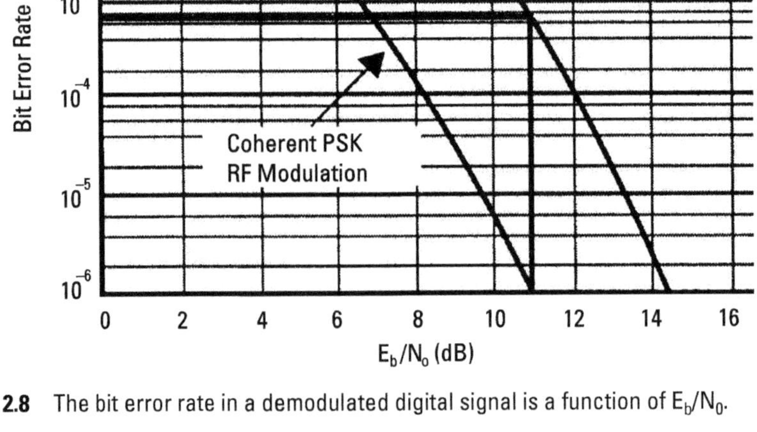

Figure 2.8 The bit error rate in a demodulated digital signal is a function of E_b/N_0.

to transmitted signals to allow some level of errors to be removed at the receiver location.

The point of this discussion is that there will probably be bit errors. These bit errors will degrade the transmitted information by reducing the accuracy of the conversion from the code back to the basic form of the information. For example, when video compression is used, every bit error degrades the reconstructed picture quality.

Note that analogous phenomena occur any time encoding is used, all the way down to the young people texting. One misplaced thumb hit degrades the information fidelity by an amount that is proportional to the power (i.e., the data compression ratio) of the code. This shows the interdependence of the first four rows of Table 2.1.

If the connectivity is over a network that is under attack by an enemy or through a high interference environment, the network and the way it is employed must be robust enough to deliver the necessary information fidelity using the available bandwidth, acceptable latency, and required throughput rate.

Message security is important any time there is a reason to prevent someone else from knowing the information you are sending. This is most obvious for military communication in which an enemy can do your forces great harm by knowing the plans and orders transmitted by command and control communication. By breaking the naval ENIGMA code during the World War II, the allied forces were able to locate (and thus sink) Axis submarines, which changed the whole course of the war. Before the code was broken, ships from Canada to England were sunk twice as fast as they could be built. After the code was broken, submarines were sunk twice as fast as they could be built. Another obvious requirement for message security is the transmission of confidential financial information. Most of us are so afraid of identity theft that we do not transmit credit card numbers or Social Security numbers unless we are confident in the security of the media.

Encryption is the basic way to provide message security. Secure encryption requires that the information be in digital form and that a series of random bits be digitally added to the message ($1 + 1 = 0$ and so forth). At the receiving end, the same random bit series is added to the received message to recover the original message. This does not typically require an increase in the required bandwidth or slow the throughput rate. However, some encryption systems are subject to increases in bit error rates when bit errors are present. One system (many years ago) was carefully measured and it was found that it increased the bit error rate by two orders of magnitude when the encryption was used (i.e., apparently, the decrypter converted one error into 100 errors). According to Figure 2.8, this required two more decibels of received signal power to provide adequate information fidelity.

In Figure 2.9, note that the information flow path starts with compression and then goes to encryption, error correction coding, and transmission. At the receiver, the received information is first subjected to the correction of errors. This is necessary because both the decryption and decompression change the data bits and cause problems related to the number of bit errors present. Note that the EDC also returns the data to its original format. Decryption is after EDC and before decompression because the same code that is encrypted must be decrypted.

A related issue is *authentication* to prevent an enemy from entering your network to insert false information. High-level encryption provides excellent authentication, but proper use of prescribed authentication procedures are also important.

Transmission security requires that an enemy not be able to detect or locate your transmitters. This is quite different from message security in that an enemy may well be able to read the content of your messages under certain circumstances even if you use transmission security measures adequate to provide acceptable protection in expected tactical situations. Transmission security measures include limitation of radiated energy, geometrically narrowing transmission paths, and spectrum spreading. Later in this chapter, we will discuss all of these issues in the context of their impact on the effectiveness of information flow.

Interference rejection and *jamming resistance* are two sides of the same issue. Communication jamming is the process of deliberately creating undesired (interfering) signals in an enemy's receiver to degrade or eliminate the flow

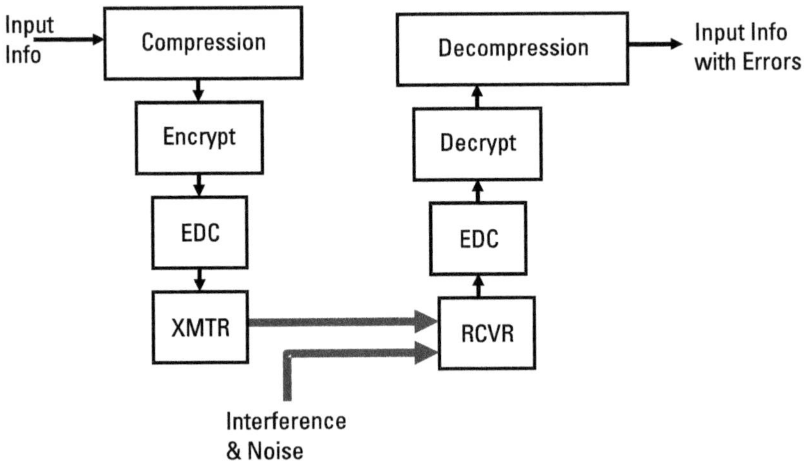

Figure 2.9 The information flow has compression as the first function. EDC is performed between the encryption and decryption, allowing as many errors as possible to be removed before the decryption and finally the decompression functions.

of information. The main difference is that deliberate jamming may be more sophisticated.

Techniques to reduce the impact of interference (either accidental or deliberate) include some related to received signal strength and some related to special modulations. Whichever approaches are used, it is necessary that the network connecting EW assets provide adequate interference protection to allow adequate information fidelity.

2.4 Interference Rejection

Whether intentional or unintentional, interfering signals reduce the fidelity of received information. We will discuss modulation and coding techniques to reduce the impact of interference.

2.4.1 Spreading the Transmitted Spectrum

Spread spectrum techniques will be discussed in detail in Chapter 5. This discussion is focused on the transfer of information versus bandwidth and the nature of the interference environment. The description of low probability of intercept (LPI) is also used to define these signals, but since this deals with only one advantage of the signals, we will talk about them as spread spectrum (SS) signals.

In general, these signals have a much wider transmission spectrum than that required to carry the transmitted information. The despreading of the signal at the receiver recovers the information transmitted while providing a processing gain that increases the ratio of the recovered information to the false outputs from received interference. Note that all of these types of systems trade noise/interference reduction for increased transmission bandwidth requirement. A simple way to think about this is to consider commercial frequency modulated (FM) broadcast signals.

2.4.2 Commercial FM Broadcast

The frequency modulated signal was the first widely used spread spectrum technique. Figure 2.10 shows the modulation. Wideband FM improves signal quality by increasing the signal-to-noise ratio (SNR) and signal-to-interference ratio as a function of the square of the amount by which it spreads the transmission bandwidth. The spreading ratio is called the modulation index. It is the ratio between the maximum frequency offset from the carrier and the highest modulating frequency as shown in Figure 2.11. The cost of this SNR improvement is that transmission requires additional bandwidth. Commercial FM frequency assignments are 100 kHz apart and there must be multiple channel slots be-

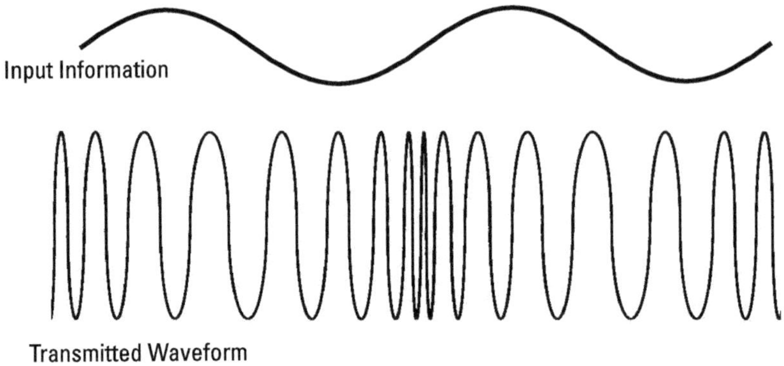

Figure 2.10 An FM signal carries information as variations in the transmitted frequency.

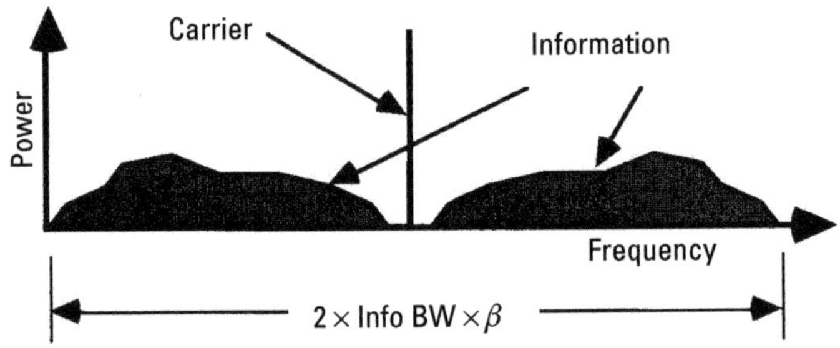

Figure 2.11 The transmitted FM signal carries its information in a bandwidth that is determined by the selected modulation index.

tween occupied channels in a geographic area. With large modulation index, the transmission bandwidth is:

$$BW = 2f_m\beta$$

where BW is the transmitted bandwidth, f_m is the maximum frequency of the information signal, and β is the FM modulation index.

The output signal to noise ratio improvement formula (in decibels) is:

$$SNR = RFSNR + 5 + 20\log\beta$$

where SNR is the output SNR in decibels and RFSNR is the predetection SNR in decibels.

In order to achieve this SNR improvement, the RFSNR must be above a threshold level: either 4 or 12 dB, depending on the type of demodulator used in the receiver. For commercial FM broadcast signals, the maximum modulating frequency is 15 kHz, and the modulation index is 5. With the most common type of demodulator, the RFSNR threshold is 12 dB. Thus, the broadcast bandwidth is 150 kHz (which is 2×15 kHz $\times 5$). With a minimum threshold signal out of the receiving antenna for the most common type of demodulator, the output SNR is 31 dB (which is $12 + 5 + 20 \log 5 = 12 + 5 + 14$). The frequency modulation improved the output SNR by 19 dB.

Note that pre-emphasis (increasing the power of higher modulating frequencies) in the transmitter and de-emphasis (decreasing the power of higher modulating frequencies) in the receiver can allow a few more decibels of SNR improvement, depending on the nature of the information being communicated.

Reduction in interference, either intentional or unintentional, depends on the nature of the interfering signals. If the interference is narrowband, the interference reduction will be similar to the SNR improvement. If the interference is noise-like, for example, noisy power lines, it will be reduced by something like the SNR improvement. However, if the interference is properly modulated jamming or is another similarly modulated FM signal, it will get the same processing gain as the desired signal (i.e., no improvement of performance against interference).

2.4.3 Military Spread Spectrum Signals

Communication in a high interference or hostile environment can profit from the use of special spectrum spreading techniques that are designed to overcome interference. These special modulations include a pseudo-random function that assures that they are unique and sufficiently different from all interfering signals that the desired signal will have a significant processing gain relative to any other received signal.

The pseudo-random function is incorporated into the signal before transmission and all authorized receivers are synchronized so that they can use the same function to despread the received signal allowing recovery of the transmitted information (see Figure 2.12).

There are three types of modulation used in these military spread spectrum systems: frequency hopping, chirp, and direct sequence spread spectrum.

There are also hybrid systems that include multiple spreading modulations. These modulations are discussed in detail in Chapter 5, but our discussion here will focus on their information transfer implications.

There are specific reasons why each type of spread spectrum modulation must carry its information in digital form.

Digital information cannot be directly transmitted. It must be placed on some type of modulation compatible with radio transmission. Digital communication is covered in Chapter 5, but we will go into additional detail here, again with an emphasis on information transfer. Figures 2.13 and 2.14 show

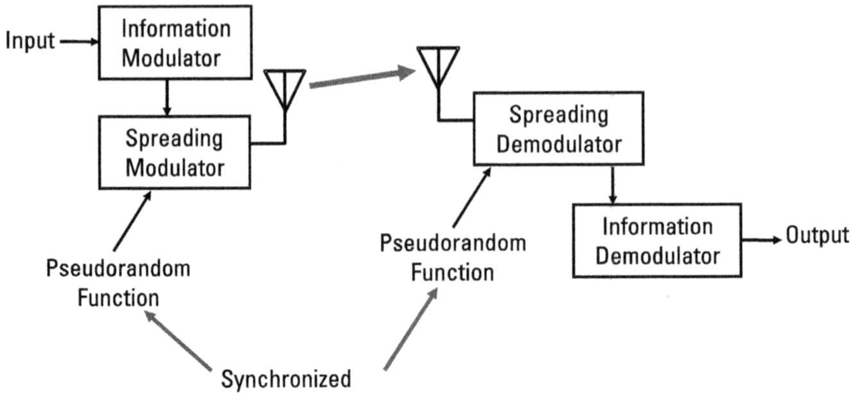

Figure 2.12 LPI communication systems spread their spectrum in response to a pseudorandom function which is synchronized between the transmitter and the receiver.

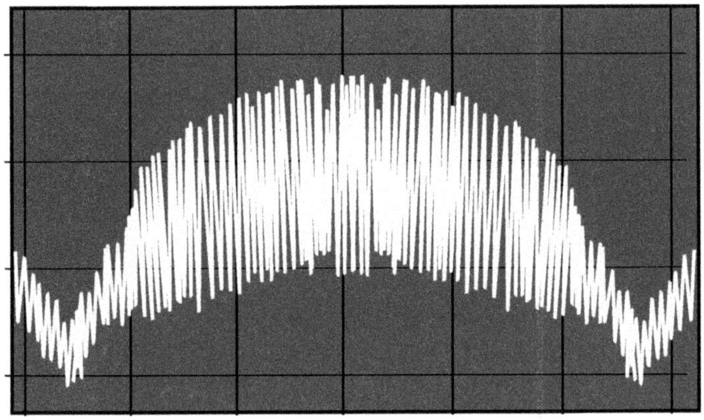

Figure 2.13 A spectrum analyzer display of a digital signal shows a main lobe pattern with clearly defined nulls on either side of the carrier frequency.

the spectrum of a transmitted digital signal as it appears on a spectrum analyzer screen and in diagram form showing the power and frequency dimensions.

You will note that the transmission bandwidth required to carry digital signals is a function of the data clock rate, which is the number of bits per second in the transmitted signal. The required bit rate is a function of the bandwidth of the information carried and the required signal quality. With most digitizing schemes, the Nyquist rate is required. This requirement is that the sample rate be twice the bandwidth (in hertz) of the carried information. The captured signal quality is determined by the number of bits per sample. There are efficient coding approaches that can reduce the required bandwidth. The sample rate (hence the bit rate) can be greater to allow for higher fidelity capture

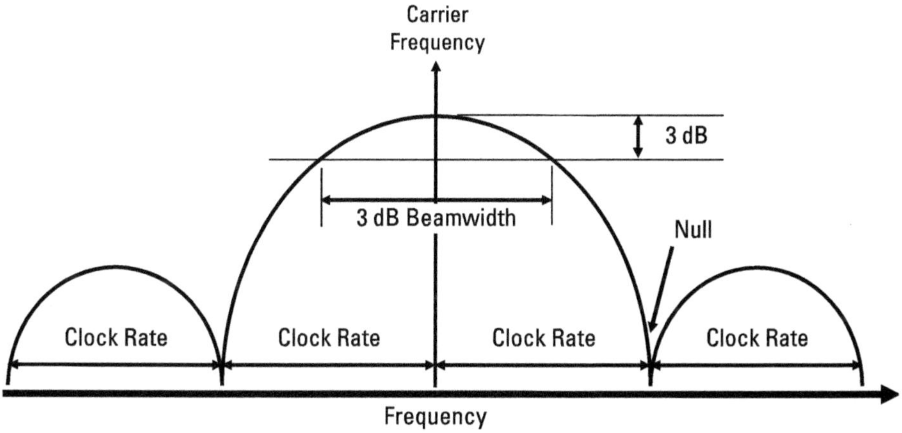

Figure 2.14 The digital signal spectrum includes a main lobe and side lobes with clearly defined nulls spaced at multiples of the clock rate from the carrier frequency.

of the information, and the transmitted signal will in almost all cases require additional bits for addressing, synchronization, and error detection/correction.

2.5 Bandwidth Requirements for Information Transfer

There are several important issues related to the bandwidth used to transport information from one location to another:

- The complexity of the link;
- The location of complex equipment required to generate, store, or use information;
- The vulnerability of links to hostile intercept or transmitter location.

Each of these issues requires tradeoffs in the design of network based military capabilities.

2.5.1 Data Transfer Without a Link

In commercial distributed entertainment and personal computing, all of these trade-offs are being made and are changing rapidly.

Consider electronically delivered movies. First there was the video cassette recorder (VCR), now largely replaced by the digital video disc (DVD). We could purchase or rent videotapes or discs of movies and play them on our own video players. No link delivery of information was required, but we were required to have complex equipment (VCR or DVD player) at the point of use, and the movies had to be physically delivered to the point of use on some media.

An excellent analogy is the loading of threat identification tables into radar warning receivers (RWRs) in the 1970s. The data was stored in the RWR but had to be updated by the physical transportation of updated data sets. Anyone involved with any part of this process is aware of the significant logistical challenges associated with the control, validation, and security of update data and the complexity and maintenance requirements involved.

Figure 2.15 shows the general concept of using transportable media. In EW systems, the transportable media can move collected data from stand-alone systems to a central facility to support operating system and database updates, and the resulting upgrades can be then loaded into the stand-alone systems.

2.5.2 Linked Data Transmission

Now you can have your movies streamed to your personal computer. You can order the movie you want at the time you want it, and the transmitting facility will know (and bill you for) exactly the delivered information. Your receiving equipment can be as complex as a desktop computer or as small and light as a cell phone. Basically, you have no dedicated receiving equipment associated with getting a movie to you. However, you do need to have a rather complex multiple-use device to receive, process, and deliver the information and you need a data link. The greater the bandwidth of your data link, the faster you will get the information and the higher its quality will be. It is normally highly impractical to send video information unless it is compressed, and in general, the quality of the delivered data varies inversely with the amount of compression.

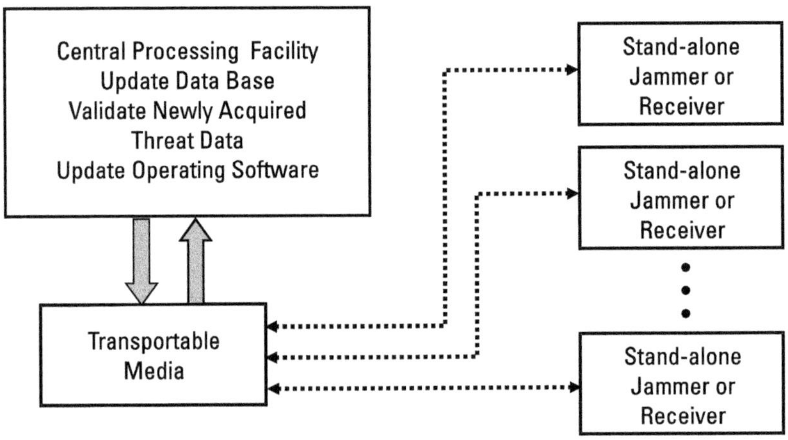

Figure 2.15 Information can be input to or extracted from stand-alone systems using portable media.

2.5.3 Software Location

It is instructive to consider what is happening in the personal computing software business. Originally, you bought and installed software directly onto your personal computer. The software was licensed, but enforcement was difficult. Now, you can rarely activate software without contacting the manufacturer and achieving accountability. The generator of the software knows who has it and can authorize its use by only qualified users. You can also download software with the same controls and security measures. The software manufacturer periodically upgrades the software to all authorized users. This type of software and data distribution is applicable to both commercial and military situations. The level of security and authorization control is generally more rigorous for the military data.

In both cases, the receiving station must have the ability to store all of the software and have enough reconfigurable memory to run the applications. Because there is no real-time interaction required, authorization and data downloading can be accomplished over almost any available link. Narrow links will require significantly more time to transfer data (at a slow rate) than wider links.

Now there is a movement to have the software retained by the manufacturer. The user will access the software over a link, uploading input data and control functions and downloading answers (see Figure 2.16). The benefit is that the user equipment can be significantly less complex, requiring relatively little local memory or computer power. Another benefit is that the manufacturer can perform software maintenance directly; every user will then always have properly upgraded software. What this process is doing is moving capability from the end user to a central location. The result is reduction of complexity at the user location but increase reliance upon links and increased requirement for

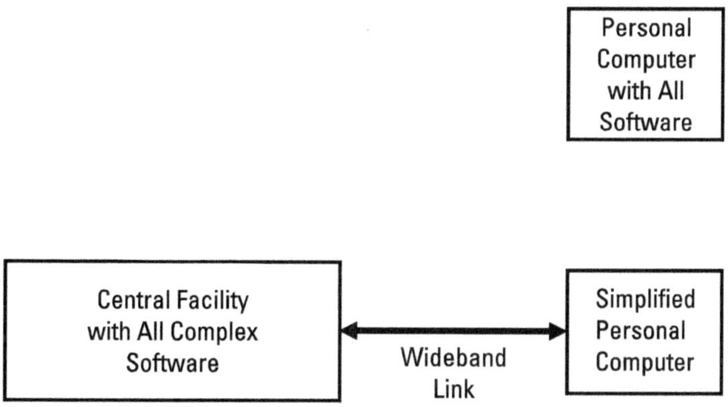

Figure 2.16 Personal computer software can be completely located in the computer or can be held in a central location and accessed as required.

link bandwidth, driven by the real-time (or near-real-time) interaction between the computer and the central facility.

2.6 Distributed Military Capability

Let us generalize to the location of capability in a distributed military system. As shown in Figure 2.17, it is possible to have a great deal of capability at the user location. In EW applications, the user can be an intercept receiver, a jammer, or some other EW equipment. This approach has the advantage of fast access to all system capabilities at the user location without critical real-time dependence on one or more links. Also, multiple user equipment units can operate cooperatively, passing data between themselves as required over relatively narrowband links. There are many user locations, so a great deal of parallel capability will be required. In addition to additional size, weight, power, and cost, there are security concerns. If a piece of user equipment falls into enemy hands, it can be analyzed to determine its capabilities and protected database information may also be extractable.

However, a significant part of the integrated system capability can be implemented at a central location as shown in Figure 2.18. In this case, the total system complexity and maintenance effort is reduced. Further, the user equipment typically goes into harm's way, close to the enemy and is thus more

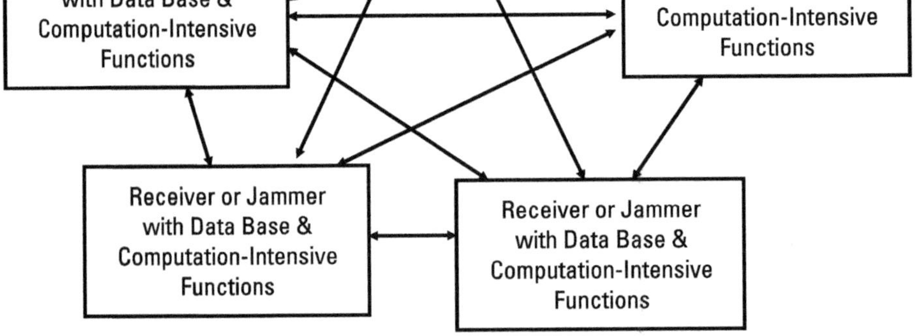

Figure 2.17 A distributed military system can have most of its capability resident in local user devices, allowing narrowband interconnection links.

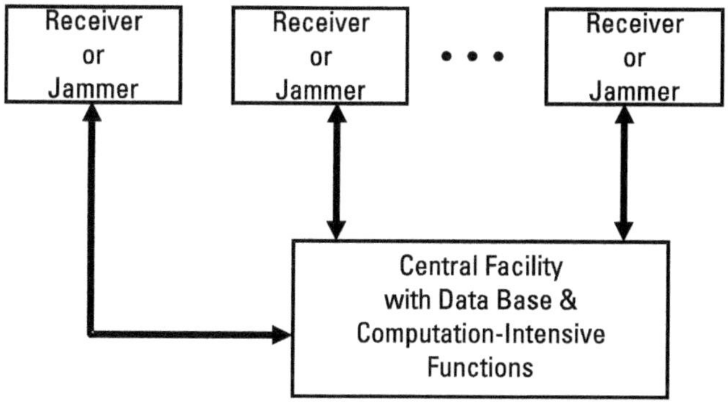

Figure 2.18 The complexity of local user devices can be reduced by accessing a complex central facility over wideband links.

subject to destruction or hostile acquisition than equipment at a (presumably safer) central location.

If databases and computation-intensive processes are held at a central location, there can be no performance without dependable, real-time, wideband communication between the user locations and the central facility. This makes the security and robustness of the data links central to the functionality of the integrated system.

2.6.1 Net-Centric Warfare

When distributed (i.e., net-centric) military operations are implemented, the vulnerability of interconnecting links to jamming and the danger associated with hostile geolocation of transmitters are critical considerations. Both of these problems are reduced by implementing transmission security, which is different from message security.

2.7 Transmission Security Versus Message Security

Message security prevents an enemy from accessing the information carried in a signal by use of encryption. High-quality encryption requires that the signal be digital and adds a pseudo-random bit stream to the signal bit stream as shown in Figure 2.19. For clarity in this discussion, let us call this the encrypting signal. The summed bit stream is itself pseudo-random and makes the message nonrecoverable. In commercial applications, it is often acceptable to use an encrypting signal that repeats after as few as 64 to 256 bits. However, in secure military encryption, the encrypting signal may not repeat for years. (The shorter the encrypting bit stream, the easier it is for an enemy to crack the code.)

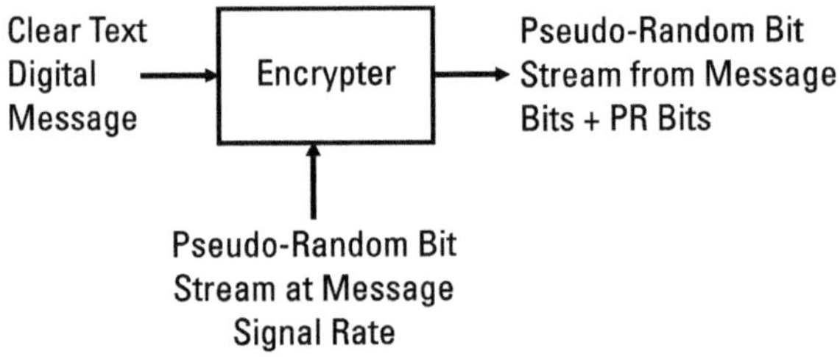

Figure 2.19 Message security is achieved by adding a pseudo-random bit stream to a digitized input message.

At the receiver, the original encrypting bit stream is added to the received bit stream, which returns the signal to its original, nonencrypted form.

However, transmission security involves spreading the spectrum of the transmitted signal in some pseudo-random way that makes it very difficult for an enemy to detect the signal, jam the signal, or locate the transmitter. The three ways to spread the signal are frequency hopping, chirp, and direct sequence spread spectrum. They are discussed (in the context of jamming) in Chapter 5. Here we will consider these techniques from a transmission security point of view. Although there are other operational benefits, the principal benefit of transmission security is to prevent an enemy from locating the transmitter and thus being able to fire on it or use a homing weapon against it. As shown in Figure 2.20, it is most important to provide transmission security for links from high value assets to lower value assets.

A frequency hopped signal switches its full power to a different frequency every few milliseconds (for slow hoppers) or microseconds (for fast hoppers) as shown in Figure 2.21. This makes it fairly easy to detect the presence of the signal, and there are many systems that can sweep for random intercepts that allow the transmitter to be located. This is particularly true of slow hoppers. Thus, frequency hopping is the least desirable technique for protecting the transmitter location.

Chirped signals which employ a wide linear sweep move across a wide frequency range very quickly (see Figure 2.22). Like the frequency hopper, the chirped signal moves its whole signal power to one frequency at a time. However, because it tunes so quickly, a receiver cannot detect the signal unless it has a fairly wide bandwidth. The wide receiver bandwidth reduces receiver sensitivity, but the chirp signal is still fairly easy to detect. Thus, geolocation of the transmitter is fairly straightforward.

Direct sequence spread spectrum signals spread the signal energy over a wide frequency range by adding a secondary digital modulation with a high rate pseudo-random bit stream as shown in Figure 2.23. Note that the bits in the high rate digitization are called chips. The frequency spectrum of a digital signal was described in Section 2.4. The null-to-null bandwidth of the input

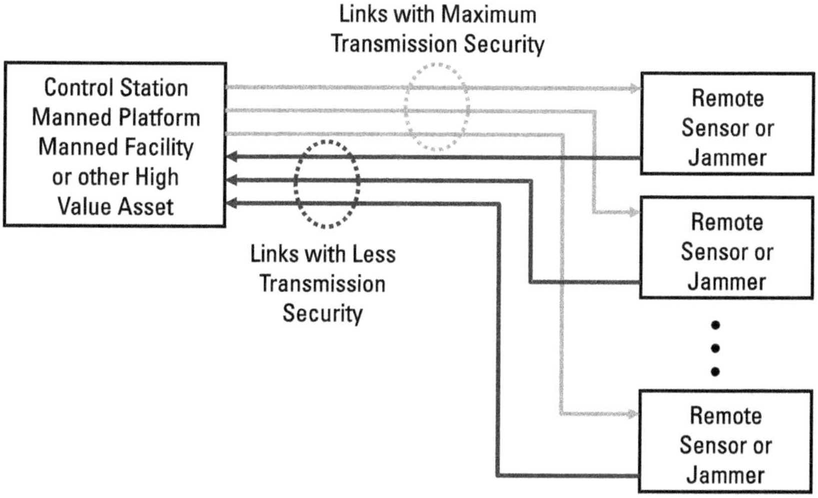

Figure 2.20 It is desirable to provide a higher level of transmission security on links from higher value assets.

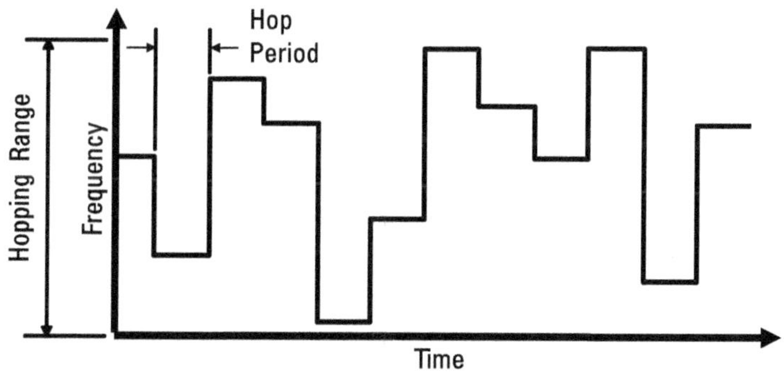

Figure 2.21 A frequency hopping signal moves its full transmit power to a new frequency many times during a message.

information signal is twice the bit rate while that of the spread signal is twice the chip rate. The power in the signal is distributed across this much wider spectrum. This creates a noise-like signal that literally has its energy spread across a wide frequency range in real time. Without ever receiving full power at a single frequency, it is much more difficult to determine that a signal is present. Detecting this signal requires either energy detection or very sophisticated processing to time-collapse the high rate bit stream chips to form a narrow frequency determinant. Thus, this technique is the favored approach to providing transmission security. As discussed next, the wider the signal is spread, the greater the transmission security.

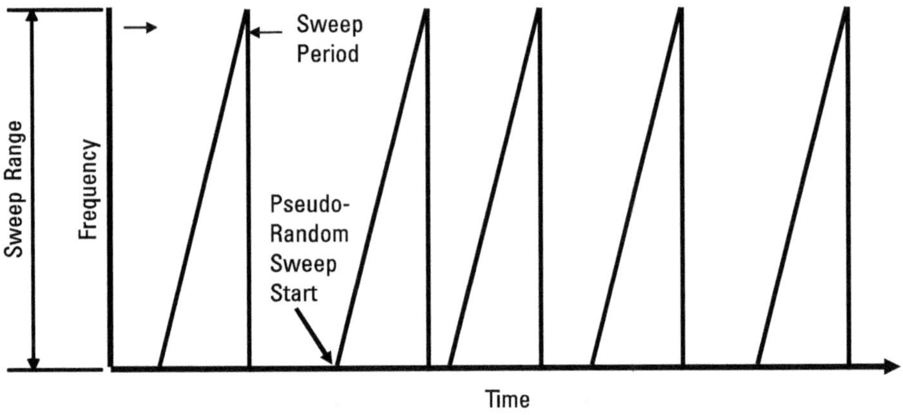

Figure 2.22 A chirped signal sweeps its full transmit power over a large frequency range very rapidly.

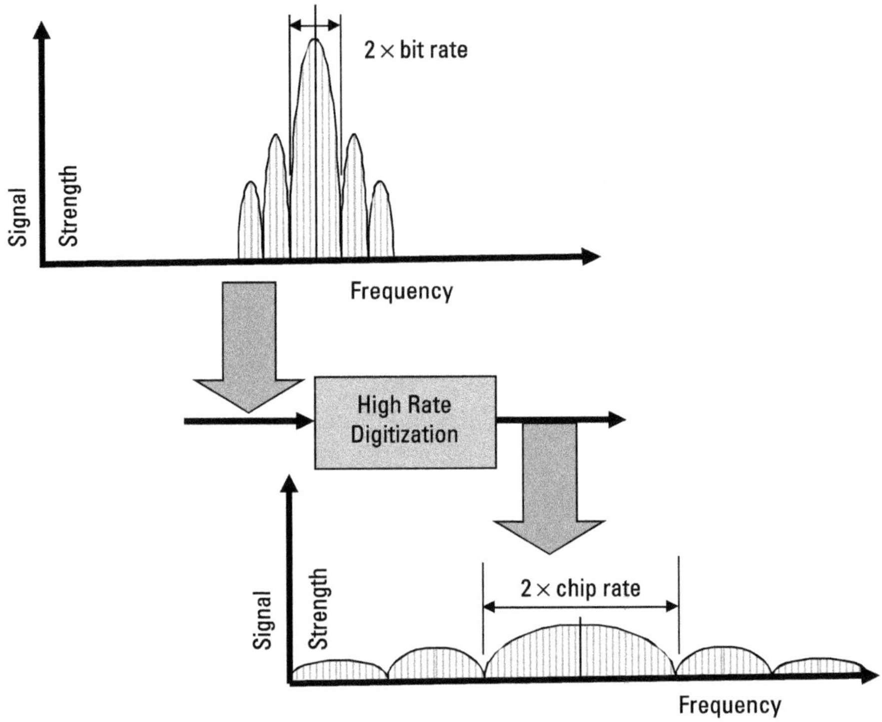

Figure 2.23 Direct sequence spread spectrum modulation spreads the signal over a wide frequency range, reducing its power at any single frequency.

It is important to realize that transmission security techniques do not provide dependable message security. Under normal circumstances, each of the spreading techniques used will make it difficult for an enemy to recover transmitted information. However, for each technique, there are conditions under which a sophisticated enemy can read the content of the message without despreading the signal. These circumstances involve short-range receivers or the use of highly sensitive receivers and sophisticated signal processing.

2.7.1 Transmission Security Versus Transmission Bandwidth

The SNR in a receiver is inversely proportional to the system bandwidth. This means that the ability of a receiver to detect a spread spectrum signal is degraded by the amount that the signal is spread. Without transmission security, a signal can be received in a bandwidth matched to the basic information modulation. However, if a signal is spread by (for example) a factor of 1,000, the receiver bandwidth must by 1,000 times as wide to capture the full signal power as shown in Figure 2.24. This causes a reduction in receiver sensitivity of 30 dB: {10 $\log_{10}$[bandwidth factor]}. This loss of receiver sensitivity has a fairly linear relationship to the accuracy with which the direction of arrival of a signal can be determined. We need to be a little careful with this generality, because there are processing gains associated with various emitter location approaches that depend on the specifics of signal modulations. However, the general rule remains true: the level of transmission security is a direct function of the factor by which the signal is spread.

2.7.2 Bandwidth Limitations

Now let us consider how much spreading can be applied to a signal. That depends on the bandwidth of the unspread signal. A narrowband transmitter, such as that in a command link, may be only a few kilohertz wide. For

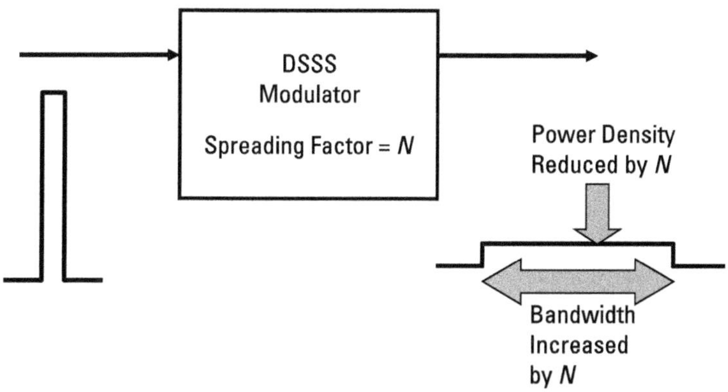

Figure 2.24 Spreading the spectrum of a signal reduces its detectability and the ability to geolocate the transmitter proportionally to the spreading factor.

example, the command signal might have 10,000 bits per second. Depending on the modulation used, the command link bandwidth might be 10 kHz. With a spreading factor of 1,000, the command link is still only 10 MHz wide. However, a real-time digital imagery data link might be 50 MHz wide. Even if video compression can be used, it will probably still be about 2 MHz wide. If you spread this by a factor of 1,000, the resulting signal would be 2 GHz wide.

Not only is the required transmitter power proportion to the link bandwidth, but amplifiers and antennas start to lose significant efficiency when they approach 10% bandwidth. The 10% bandwidth at 5 GHz is 500 MHz. Note that microwave links (e.g., at 5 GHz) usually require directional antennas to achieve good performance. Highly mobile tactical platforms connect much more easily with links using nondirectional antennas. This makes links in the UHF frequency range (perhaps 500 to 1,000 MHz) much more desirable. The 10% bandwidth is only 50 MHz for 500 MHz links. The point is that it is difficult to provide a high degree of transmission security to a high data rate link. The higher rate link will need to have a lower spreading ratio to fit within the practical link bandwidth.

2.8 Cyber Warfare Versus EW

At the time of this writing, there was a great deal of discussion in defense-related literature about cyber warfare. As in all new fields of interest, there is sometimes heated discussion about definitions, and there are those who lump cyber warfare and EW together in various ways. This discussion will eventually be resolved as all such discussions are. As the focus of this book is technical, we will emphasize the underlying principles and let others resolve the linguistic disagreements.

We have been discussing various aspects of the movement of digital information for military purposes. This background information is key to understanding and dealing with challenges and trade-offs in net-centric warfare as well as traditional command and control. In this section, an attempt is made to relate this information flow to its application in cyber warfare and EW.

2.8.1 Cyber Warfare

The term cyber is defined in many places on the Internet. The consensus is that cyber refers to information moved from computer to computer over the Internet, that is, within the network of computers comprising the Internet. Cyber warfare is defined (sometimes in much detail) as measures to use this Information Superhighway to gain a military advantage by gathering militarily significant information from an enemy or interfering with the enemy's ability to move information over the Internet or other networks or to process information within a computer.

2.8.2 Cyber Attacks

Again, from the literature, cyber warfare is conducted by use of malware, which is software whose purpose is to cause harm. This includes:

- *Viruses:* Software that can replicate itself and spread from one computer to another. Viruses can be used to load up computers with so much information that they have inadequate free memory to perform their intended functions. Viruses can also cause desired information to be deleted or can modify programs in highly undesirable ways.
- *Computer worms:* Software that takes advantage of security vulnerabilities to spread itself automatically to other computers through networks.
- *Trojan horses:* Software that seems harmless, but attacks a computer's data or functioning. This malware relates to the way that hostile code is introduced into a computer or network. A Trojan horse program is described as providing some valuable benefit, which it may well do. However, hidden in the downloaded software are other programs that have highly undesirable features.
- *Spyware:* Software that gathers and exports data from a computer for hostile purposes.

There are a number of other terms that are used to describe various techniques used to attack the ability of a computer to do its job, using the Internet to gain access to the victim's computer.

Hackers are a concern to everyone who uses the Internet, which is why we use complex, hard-to-remember passwords and spend money on firewalls. However, in cyber warfare, these attacks are designed and applied by professionals for significant military purposes. These professionals are very good at what they do, and every easy fix is soon overcome, requiring continuous, sophisticated defensive effort.

2.8.3 Parallels Between Cyber Warfare and EW

EW is described as having three major subfields and another closely related field:

- Electronic warefare support (ES), which involves hostile intercept of enemy transmissions.

- Electronic attack (EA), in which enemy electronic sensors (radars and communication receivers) are degraded either temporarily or permanently by the transmission of signals designed for that purpose.
- Electronic protection (EP), which is a set of measures designed to protect friendly sensors from enemy electronic attack actions.
- Decoys, which are not literally part of EW, but which are considered along with EW because they cause enemy missile and gun systems to acquire and track invalid targets.

The elements of cyber warfare are parallel to these EW-related subfields. As shown in Table 2.2, each of these fields has a parallel technique in cyber warfare:

- ES can be compared to spyware. Actually, spyware is also like signal intelligence (SIGINT). Both ES and SIGINT collect information which an enemy does not want collected. Note that the differences between these two fields are discussed in Chapter 10.
- EA denies an enemy information by transmitting jamming signals into enemy receivers. If the target receiver is a radar, the jamming can either cover the signal the radar receiver needs to receive (i.e., the reflected signal from a target) or can deceive the radar with waveforms that cause the processing subsystem(s) in the radar to determine that the target is at a false location:

Table 2.2
Comparison of EW and Cyber Warfare Functions

Operational Function	EW	Cyber Warfare
Collect information from enemy	EW support, which listens to enemy signals to determine enemy capabilities and operating mode	Spyware, which causes information to be exported to a hostile location
Electronically interfere with enemy's operational capability	Electronic attack, which either covers received information or causes processing to give inaccurate outputs	Viruses, which reduce available operating memory or modify programs to prevent proper processing outputs
Protect friendly capabilities from enemy's electronic interference	Electronic protection, which prevents enemy jamming from impacting operational capabilities	Passwords and firewalls, which prevent malware from penetrating a computer
Cause enemy systems to initiate undesired actions	Decoys, look like valid targets, which are acquired by missile or gun systems	Trojan horses, are hostile software accepted by enemy computers because they appear valid and beneficial

- Cover radar jamming is very similar to the way some types of computer viruses use up the available memory in a computer. This saturates the computing capability effectively covering the desired information.
- Deceptive radar jamming transmits signals that cause the computer processing to come to the wrong conclusions, just as viruses modify code in the target computer so that it will give incorrect or meaningless outcomes.
- Communication jamming covers the signals from which a target receiver is trying to extract information. Spoofing involves transmitting false signals that look like correct signals, but contain false information. These two EW functions parallel the effect of computer viruses that saturate or modify the code in target computers.

- EP comprises a set of measures in friendly sensors (radar receiver/processors or communication receivers) to reduce or eliminate the loss of information or function. This parallels the functions of password protection and firewall measures to protect computers against malware.
- Decoys are physical devices that return radar signals that appear to be valid reflections from significant targets. They parallel the function of Trojan horses because both fool enemy systems, causing them to initiate actions which are detrimental to system operation.

2.8.4 Difference Between Cyber Warfare and EW

The difference between cyber warfare and EW has to do with how the hostile function is introduced into an enemy's systems. As shown in Figure 2.25, cyber attack requires that the malware enter the system as software. That is to say, that the system is entered from the Internet, a computer network, a floppy disk, or a flash drive. As shown in Figure 2.26, EW enters the enemy systems' functionality electromagnetically. ES receives transmitted signals from hostile transmitting antennas and EA enters enemy receivers and processors to do its mischief through the enemy's receiving antennas.

It is true that modern threat systems are very software intensive, but if you look at, for example, the various components of the Russian S-300 surface to air system, you will note that every vehicle (command vehicles, radar vehicles, launchers, and so forth) all have communication antennas so that signals between computers can be moved to where they are needed in a dynamic engagement scenario. The tactical effectiveness and survivability of every ele-

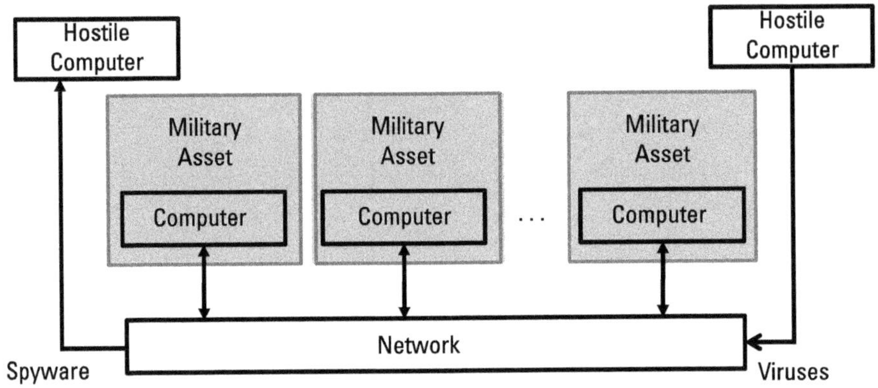

Figure 2.25 Cyber warfare involves attacks on military assets through networks, including the Internet.

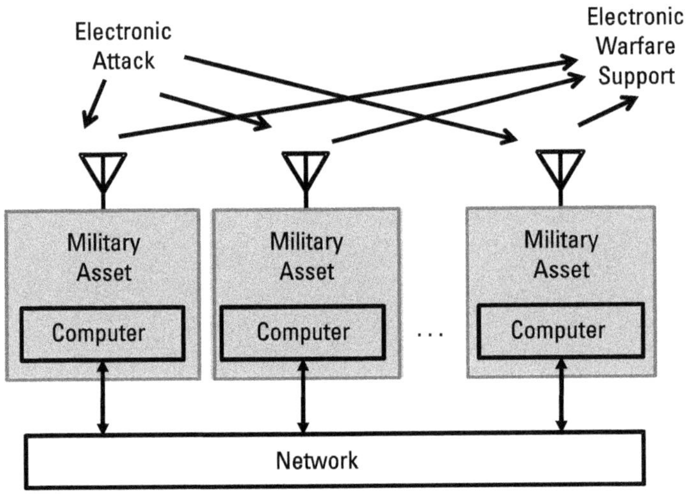

Figure 2.26 Electronic warfare involves attacks on military assets through electromagnetic propagation.

ment of the system depends on its mobility, which requires electromagnetic interconnectivity. Thus, they are vulnerable to EW attack.

2.9 Bandwidth Trade-Offs

Bandwidth is an important parametric trade-off for any communication network. In general, the greater the bandwidth, the faster information can be transported from one location to another. However, the greater the bandwidth, the greater received signal power required to provide adequate received signal fidelity.

In digital communication, received signal fidelity is measured in terms of the accuracy of the received signal bits. The bit error rate (BER) is the ratio of incorrectly received bits to total bits received. As discussed in detail in Chapter 5, digital data cannot be directly transmitted; it must be modulated onto a radio frequency carrier. For a typical modulation scheme, Figure 2.27 shows the received bit error rate as a function of E_b/N_0. As discussed in Section 2.3.4, E_b/N_0 is the predetection SNR (RFSNR) adjusted for the bit rate to bandwidth ratio. In typical transmitted digital links, the received BER varies between 10^{-3} and 10^{-7}. From the figure, you can see that, in this range, the BER increases about an order of magnitude for each decibel reduction in the RFSNR. This rate of change of BER with RFSNR is the same for all modulations used for digital data.

In cases where the BER must be less than this range, error correction techniques are used to correct bit errors.

2.9.1 Bit-Error Critical Cases

In Chapter 5, we will discuss video compression. With each of the techniques discussed, the presence of bit errors reduces the fidelity of the recovered imag-

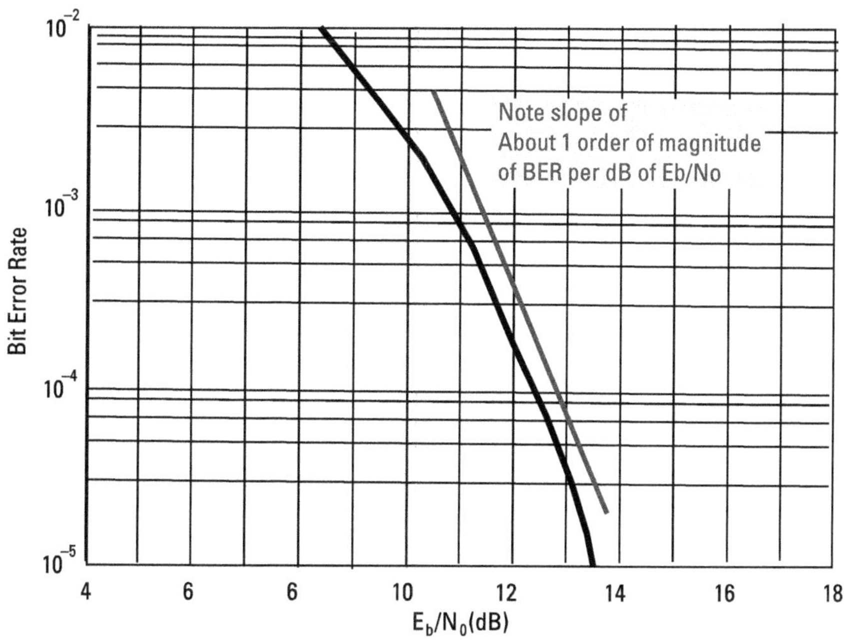

Figure 2.27 The bit error rate in a received signal is an inverse function of E_b/N_0.

ery. In some cases, the impact of even a single bit error can cause significant loss of data.

Other BER critical cases include encrypted signals in which bit errors can cause loss of synchronization and command links which typically have very low tolerance for errors.

2.10 Error Correction Approaches

As shown in Figure 2.28, errors can be corrected by rebroadcasting received signals back to the transmitter and checking bit for bit and then rebroadcasting if necessary. This requires a two-way link and adds latency to the system that varies with the momentary conditions (transmission distance, interference, jamming, and so forth). Errors can also be reduced by majority encoding in which redundant transmissions are compared and the version with the maximum number of agreements is output. Similar to this is sending multiple identical messages and eliminating erroneous data through powerful parity encoding. Both of these approaches add significant numbers of transmitted bits. The third approach shown in the figure is the use of error correction codes.

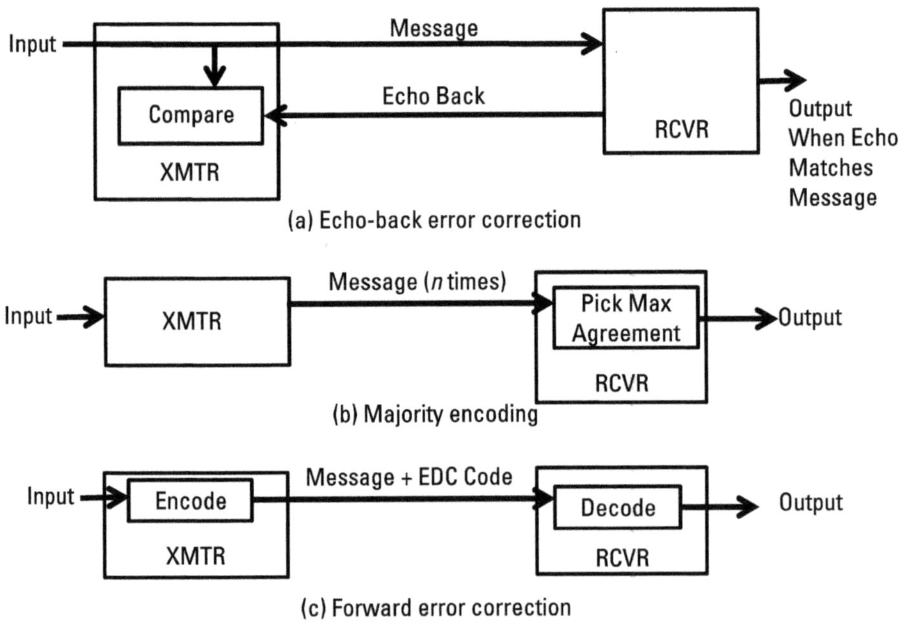

Figure 2.28 Bit errors can be corrected by several techniques.

2.10.1 Error Detection and Correction Codes

If an error detection and correction (EDC) code is used, received errors can be corrected (up to a limit set by the power of the code). The more EDC bits that are added to the data, the higher percentage of bit errors can be corrected.

Figure 2.29 demonstrates the operation of a simple hamming code encoder. If the first input bit is a one, the first 7-bit code is placed into a register. If a bit is a zero, all zeros are entered. When all bits are encoded, the register is summed and the sum is sent. Figure 2.30 shows the decoder. If a received bit

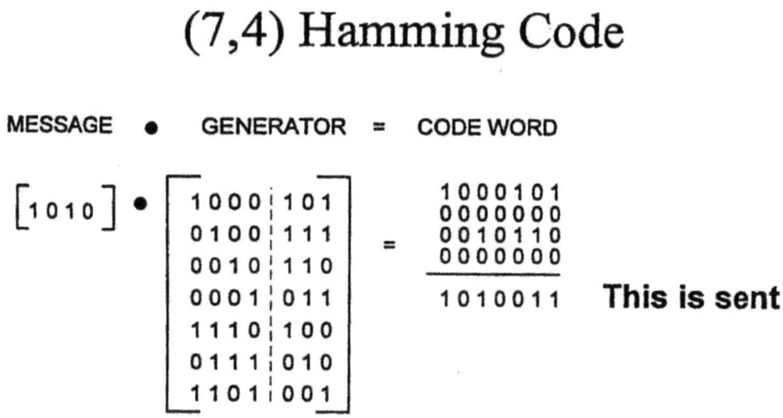

Figure 2.29 Hamming code encoder.

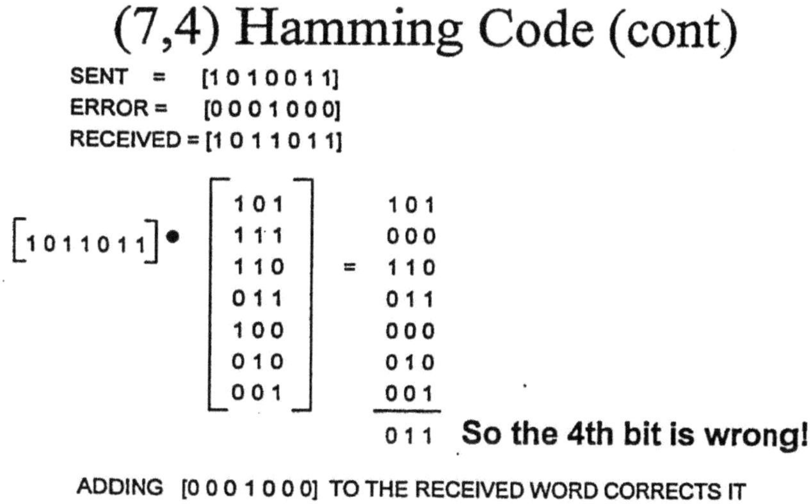

Figure 2.30 Hamming code decoder.

is one, the corresponding 3-bit code is entered into a register; if it is a zero, all zeros are entered. If all of the bits are correctly received, the register adds to zeros. In this example there is an error in the fourth bit of the received code, so the register sums to 011. This indicates that the fourth bit must be changed.

There are two classes of EDC codes: convolutional codes and block codes. Convolutional codes correct errors bit by bit, while block codes correct whole bytes (of, for example, 8 bits). Block codes do not care if one bit in a byte is bad or if all of them are bad; they correct the whole byte. In general, if the errors are evenly spread, convolutional codes are better. However, if there is some mechanism that causes groups of errors, the block codes are more efficient.

An important application for block codes is for frequency-hopping communication. Whenever the signal is hopped to a frequency occupied by another signal (highly likely in a dense tactical environment), all of the bits transmitted during that hop will be erroneous.

The power of a convolutional code is stated as (n, k). This means that n bits must be sent to protect k information bits. The power of a block code is stated as (n/k) meaning that n code symbols (bytes) must be sent to protect k information symbols.

2.10.2 Example of a Block Code

The Reed-Solomon (RS) code is a widely used block code. Examples of its application are Link 16 (for military interconnectivity) and television broadcast from satellites. The RS code can correct a number of bad bytes in its block equal to half the number of extra bytes in the block (above the number of data bytes included).

The version used in both of the above applications is the (31/15) RS code. It sends 31 bytes in each block including 15 information carrying bytes and 16 extra bytes for error correction. This means that it can correct up to 8 bad bytes of the 31.

Consider the use of this code with a frequency-hopped signal. Because the code can correct only 8 bad bytes, the data is interleaved to transmit no more than 8 of 31 bytes in a single hop. Figure 2.31 shows a simplified interleaving scheme; actually, the placement of the bytes in a modern communication

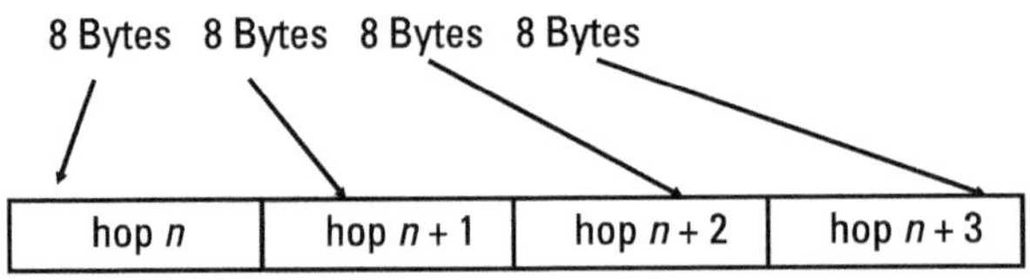

Figure 2.31 Interleaving places adjacent data into other parts of the signal stream to protect against systematic Interference or jamming.

system will be pseudo random. The resulting bit error rate will be effectively reduced to zero unless the RFSNR is low enough to cause a significant number of cases in which errors occur in multiple transmitted data blocks (for example, hops) carrying data from the same 31-byte code block.

2.10.3 Error Correction Versus Bandwidth

In any forward error correction approach (majority encoding, redundant data, or EDC coding), the bit rate is increased. If majority encoding is used, the data rate will be at least tripled. If redundant data with strong parity is used, the rate could be increased by a factor of five or six. With the above described (31/15) RS code, the data rate increases by 207%.

Receiver sensitivity varies inversely with the receiver bandwidth. As discussed in detail in Chapter 5, the typical bandwidth required to receive a digital signal is 0.88 times the bit rate. Therefore, doubling the bit rate with the same data throughput rate will decrease the sensitivity by 3 dB. Referring to Figure 2.1, this will typically increase the bit error rate by three orders of magnitude. This demonstrates what people in the digital communication business have said: error correction measures will probably hurt you more than they help you except where you have a very low tolerance to bit errors or there is significant interference or jamming.

2.11 EMS Warfare Practicalities

In this chapter, we have discussed a number of practical issues relating to the realm in which electromagnetic spectrum (EMS) warfare is fought and the nature of the physics involved. We have covered how information is moved from one point to another, and what an adversary can do to either prevent that movement or capture the information to support adverse outcomes.

2.11.1 Warfare Domains

There has been much discussion in the EW literature about terms, for example, whether the EMS is a domain or not. The discussion of what we call things goes on, but ignoring the battle of terms, there are some underlying truths on which we can agree.

EW has historically dealt with the electromagnetic spectrum (EMS) as it relates to kinetic threats:

- *Radars that locate targets guide missiles to those targets, and detonate warheads.* The purpose of EW has been to make the missiles unable to acquire or hit their targets. Thus, the limited goal of EW attack is to disable to receipt of the return signal from a radar target or to prevent missile uplink from delivering guidance information to the missile (Figure 2.32).

- *Enemy communications as they relate to the command and control of forces that can attack us kinetically.* The purpose of EW has been to prevent effective command and control by an enemy. Thus, the goal of EW attack is to prevent command and control signals from being properly received by the command headquarters or the remote military assets (Figure 2.33).

Computers and software are an integral part of almost every aspect of modern warfare, and cyber warfare attacks on those computers directly impacts kinetic attacks and the defenses against those attacks.

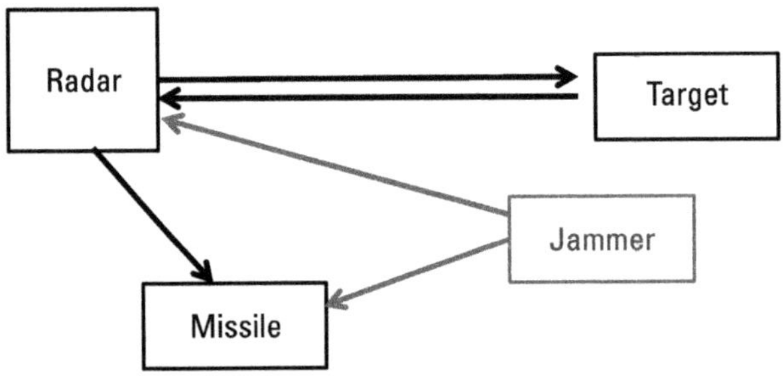

Figure 2.32 Classically, the jammer prevents the radar from acquiring or tracking its target or from guiding a missile to the target.

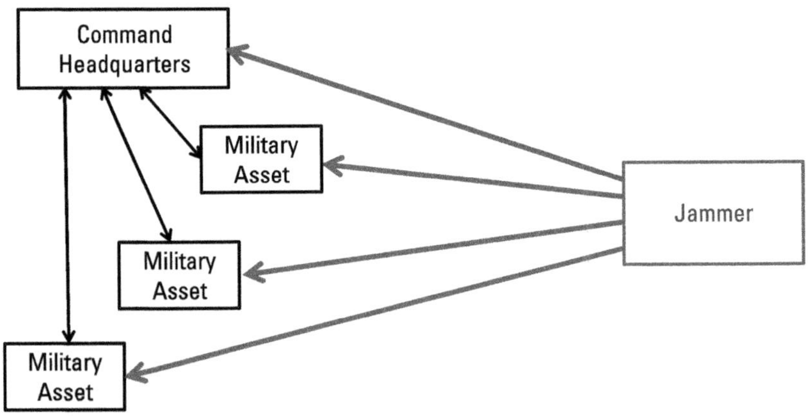

Figure 2.33 Classically, EW attack is to prevent an enemy from effectively commanding and controlling its military assets.

However, there is a new reality that has become a part of modern warfare. The EMS itself has now become a target of enemy actions. By denying us the use of the EMS, an enemy can inflict significant economic damage upon our society, without firing a single bullet or dropping a single bomb. Without the availability of the EMS, we cannot:

- Fly airliners or freight aircraft.
- Run our trains.
- Schedule the movement of freight by truck.
- Manufacture anything because we cannot get the materials to the factories.
- Get our goods to market.
- Power our homes and businesses.

The list goes on, and the dependence upon the EMS to make modern life work is increasing daily. An attack on our use of the EMS is strongly parallel to kinetic weapon attacks which have been a part of warfare throughout history.

Now consider the changes in modern warfare that incorporate significant uses of the EMS:

- Missile systems must now be characterized as "hide, shoot, and scoot" if they are to survive. This means all elements of the system must be interconnected through the EMS; interconnection by wire just will not work.
- Effective integrated air defense requires all elements to be mobile, hence interconnected in the EMS.
- Coordinated airborne attack, either kinetic or electronic, requires interconnected through the EMS.
- Naval operations cannot be effective without EMS interconnectivity.
- Without EMS interconnectivity, an army is just a bunch of people running around with guns, probably more dangerous to themselves than to the enemy.

Denying an enemy secure and dependable access to the EMS is a very effective attack on the enemy's whole military capability and can degrade their whole national economic activity.

Net-centric warfare is now an established buzzword for the way we will conduct future military operations. This approach maximizes the effectiveness of our active or passive EW operations. Without secure and dependable EMS access, there is no network and hence no net-centric warfare.

Cloud computing is well established in the commercial world and is growing in military importance. As shown in Figure 2.34, this approach allows us to move much software and data away from specific operating locations. The advantage is that distributed military hardware at operational locations can be made smaller, lighter, less power-consuming, cheaper, and less vulnerable to capture and exploitation by an enemy. However, this comes at the cost of increased dependence on the secure and dependable availability of the EMS.

The nature of EMS warfare is shown in Figure 2.35. In contrast to Figures 2.32 and 2.33, the actual target of EMS warfare is access to the EMS itself, not the reduction of the effectiveness of the associated kinetic weapons.

2.12 Steganography

Steganography is defined as hidden writing, and has been around for centuries. However, with the advent of digital communication, it has taken on a whole new life. If you look up steganography on your Web browser, you will get many evenings of entertainment, including detailed history, theory, countermeasures, and available software products to implement and detect it. As usual with this kind of subjects, we will focus on its utility in electronic and information warfare and particularly on its applicability to spectrum warfare.

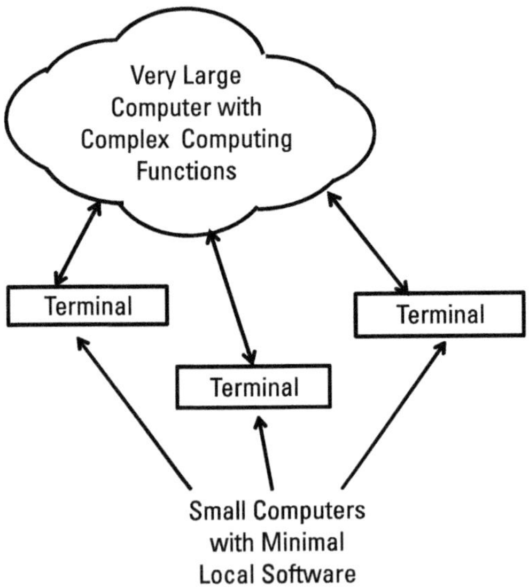

Figure 2.34 Cloud computing moves most software away from the point of use to a large scale central computing facility accessable by data links.

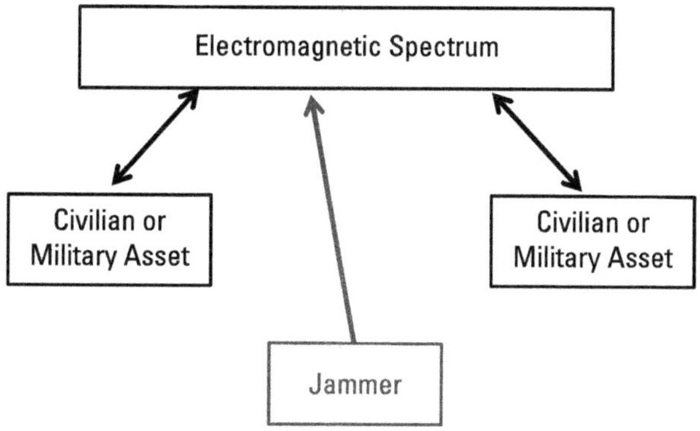

Figure 2.35 In EMS warfare, the direct objective is to deny an adversary the use of the electromagnetic spectrum.

2.12.1 Steganography Versus Encryption

This comparison is analogous to the difference between transmission security and message security in transmitted signal paths. When we use spread spectrum techniques, particularly high-level direct sequence spread spectrum (DSSS), the signal received by an enemy that does not have access to the pseudo-random spreading code is noise-like. That is, the signal appears to be only a slight increase in the noise level in the direction of the transmitter. Thus, without special equipment and techniques, an enemy will not even be able to detect that a transmission has taken place. However, encryption keeps an enemy from being able to recover the information sent. The spread spectrum modulation provides transmission security, which prevents an enemy from locating and attacking the transmitter. Encryption is also required because it keeps the enemy from learning our secrets after sophisticated means are used to detect the signal (see Figure 2.36).

Steganography deals directly with the information we are sending, either by hard copy or electronic means. It covers our secret messages with seemingly unrelated data, as shown in Figure 2.37, so an enemy will not even know

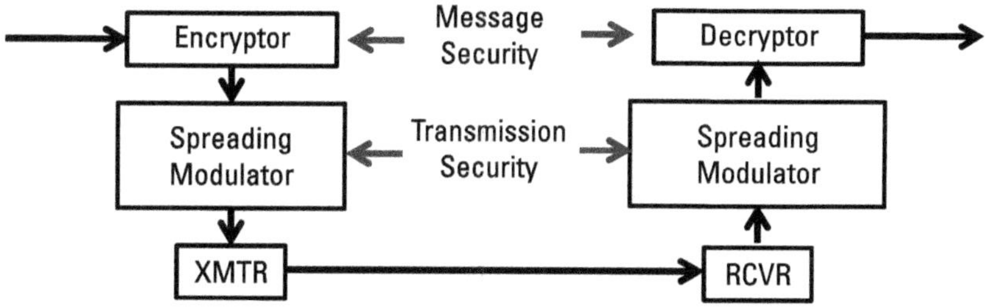

Figure 2.36 Spread spectrum communication provides transmission security, while encryption provides message security.

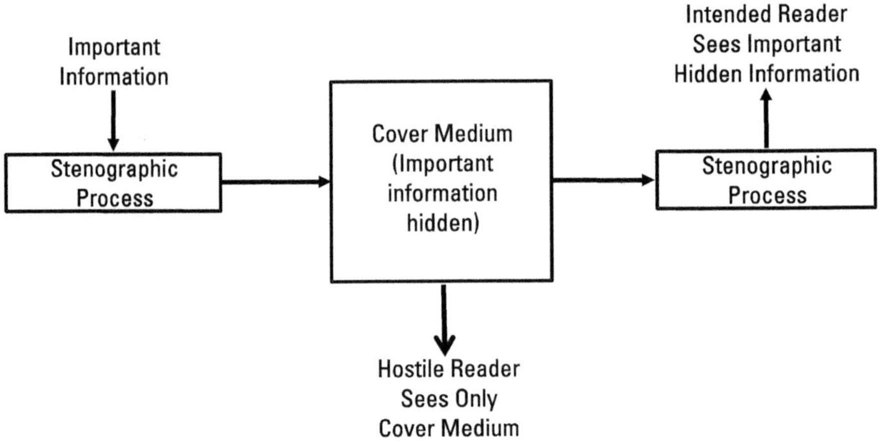

Figure 2.37 Steganography provides the equivalent of transmission security in hard copy or electronically delivered messages.

that we are conducting important (typically military) communication. This, in effect, provides transmission security. Encryption has the same function as mentioned earlier: protecting our information if the enemy discovers our hidden messages. However, an encrypted message displays random letters or bits, making it obvious that we are hiding something. This tells the enemy that we are communicating important information and may trigger an effort to analyze and ultimately recover our information. Steganography, if successful, will deny the enemy this operational advantage.

2.12.2 Early Stenographic Techniques

One early technique discussed in some articles was to shave the head of a messenger, tattoo a message on his bald head, and let his hair grow back. His head was then shaved to recover the message. Other techniques have included writing innocuous messages in which some pattern of letters scattered through the message contained the hidden information. There was also the use of microdots and invisible inks in seemingly innocent written communication. One particularly interesting approach (in a World War II spy movie) was to have a musician write a song in which the placement of a particular note (B flat in this case) carried the coded message.

2.12.3 Digital Techniques

Digital signals provide many opportunities to hide information in the format of the data. One very effective technique involves digitizing a color picture and making subtle changes to the transmitted data. Consider one image digitization technique. The picture is carried as pixels (tiny spots of color). Each pixel is digitized with a code that records the density of three basic colors (say, red, yellow, and blue). By combining the densities of these primary colors (analogous to mixing paint), a very large array of colors can be produced (see Figure 2.38). If each primary color density is measured in 256 levels, it can be captured in 8 bits. The transmitted full color data has 24 bits (8 bits for each primary color). The transmitted data rate is then 24 times the pixels per frame times the frame rate. If there are 640 by 480 pixels in a frame and there are 30 frames per second, this means that (without compression) the data rate would be about 2.2×10^8 bits per second sent ($640 \times 480 \times 24 \times 30$). Note that there are data compression techniques that reduce the transmitted bit rate, but they do not prevent the use of steganography.

Now that the image is digitized, we can reduce the number of bits for one primary color and use the extra bits to send our hidden message. For example, let us reduce the digitization of blue by 1 bit as shown in Figure 2.39 for every fifth pixel. This will very subtly change the color of every fifth pixel in the full image. The person looking at the received image would never detect that subtle change (without very specialized equipment). Using that one sacrificed bit of blue for hidden data will allow us to insert our covert messages at a 1.8-Mb rate ($640 \times 480 \times 6$), which allows a significant amount of hidden information to be passed. Online articles on steganography typically show a cover picture sent along with a greatly different hidden picture. One article has a detailed picture of a tabby cat on a rug hidden in a picture of trees against a cloudy sky.

There are similar approaches that can be employed in digital text transmission.

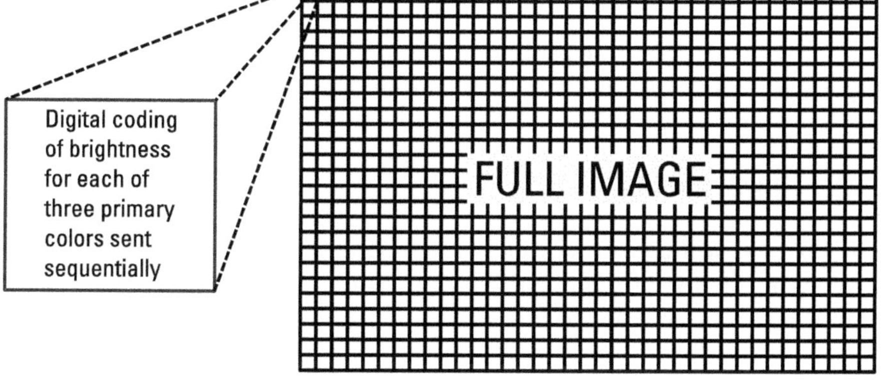

Figure 2.38 A digitized image is typically transmitted as pixels, each of which has coded brightness and color information.

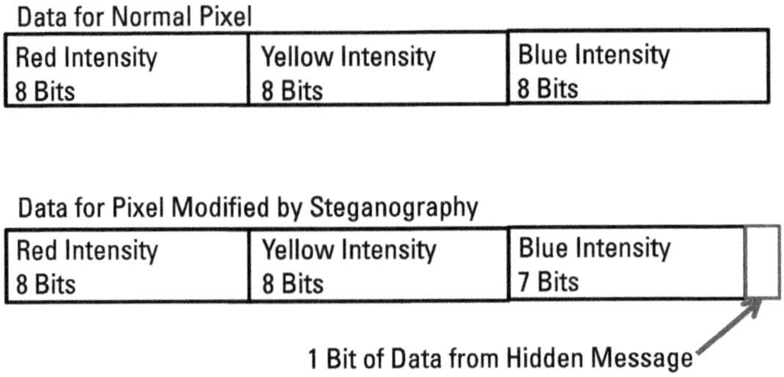

Figure 2.39 By use of a few bits in the digitized image, a second, hidden, image, or message can be carried in the transmitted signal.

2.12.4 How Does Steganography Relate to Spectrum Warfare?

First, it allows us to send important information from point A to point B without the enemy knowing that we are communicating. Another approach might be to imbed malware in seemingly innocent messages or graphics to initiate a cyber attack. Unless the steganography is detected, the targeted enemy will not know that a cyber attack is taking place.

2.12.5 How Is Steganography Detected?

This field is called steganalysis. With older techniques, like invisible ink, the approaches included careful inspection under magnification and the use of developing agents and/or ultraviolet light. In World War II prisoner of war camps, prisoners were required to send letters on special paper that was (secretly) designed to clearly show the presence of invisible inks. In digital communication, steganography can be detected by comparing an original of the cover art with the modified art (containing stenographic messages). Also, sophisticated statistical analysis can detect the presence of modified text or graphics. In every case, steganalysis is an expensive and time-consuming process.

2.13 Link Jamming

The links we will consider jamming are digital and the propagation mode is assumed to be line of sight. For your reference, the three major propagation modes important to EW operations are discussed in Chapter 6.

2.13.1 Communication Jamming

First, here are some basics (discussed in detail in Chapter 6):

- You jam the receiver, not the transmitter. Any jamming involves causing an undesired signal to enter a target receiver with enough power to keep the receiver from properly recovering the desired information from the signal it is trying to receive (see Figure 2.40).
- The decisive factor in jamming is the jamming-to-signal ratio (J/S). This is the ratio of jamming signal power to desired signal power at the point in the target receiver at which the information is recovered from the signal modulation.
- For digital signals, a J/S factor of 0 dB is enough, and a jamming duty cycle of 20% to 33% is usually adequate to stop all communication. The most dependable way to make a digital signal nonrecoverable is by creation of a high enough bit error rate (that is, the percentage of bits recovered incorrectly).
- It may be practical to disable the synchronization of the received digital signal to stop communication with lower J/S and/or duty cycle, but there are some very robust synchronization schemes that would make this very difficult to accomplish.
- There are some situations in which less J/S and much lower jamming duty factor are satisfactory to make communication over the target link ineffective. This depends on the nature of the information carried by the link.

The magnitude of J/S is given by the formula:

$$J/S = ERP_J - ERP_S - LOSS_J + LOSS_S + G_{RJ} - G_R$$

where ERP_J is the effective radiated power (ERP) of the jammer (dBm), ERP_S is the ERP of desired signal from transmitter (dBm), $LOSS_J$ is the propagation loss from jammer to target receiver (decibels), and $LOSS_S$ is the propagation loss from desired signal transmitter to target receiver (decibels). G_{RS} is the gain

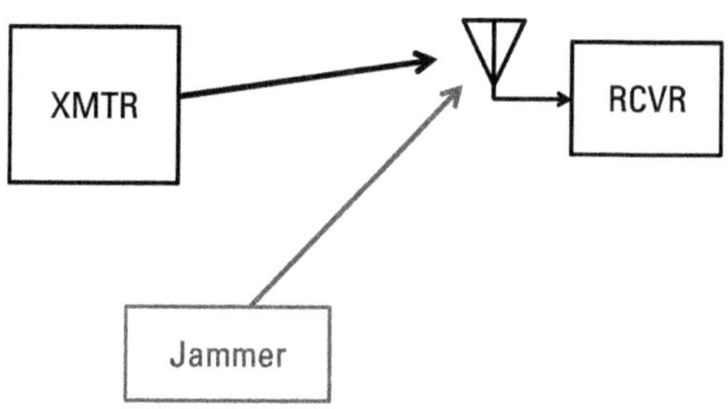

Figure 2.40 When jamming a data link, the jammer must transmit to the receiving location.

(decibles) of the receiving antenna toward the jammer, and G_R is its gain toward the desired signal transmitter.

2.13.2 Required J/S for Jamming Digital Signals

For any radio frequency modulation carrying digital information, there is a curve of received bit error rate (BER) versus energy per bit/noise per hertz (E_b/N_0). Note that E_b/N_0 is the predetection signal to noise ratio (RFSNR) adjusted for the bit rate to bandwidth ratio. As the RFSNR reduces, the bit error rate increases. Each type of modulation has its own curve with the general shape as shown in Chapter 5, but all of them approach 50% bit errors as the RFSNR becomes very low. Figure 2.41 is a variation of this curve with J/S on the horizontal axis, increasing to the left. Thus, as the J/S increases, the BER also increases until it approaches 50% errors. As shown in the figure, if J/S is 0 dB, most of the bit errors that can be caused have been caused because we are over the knee of the curve. Adding significantly more jamming power will cause very few additional errors.

2.13.3 Protections Against Link Jamming

There are several ways that links are protected against jamming; three important technique are:

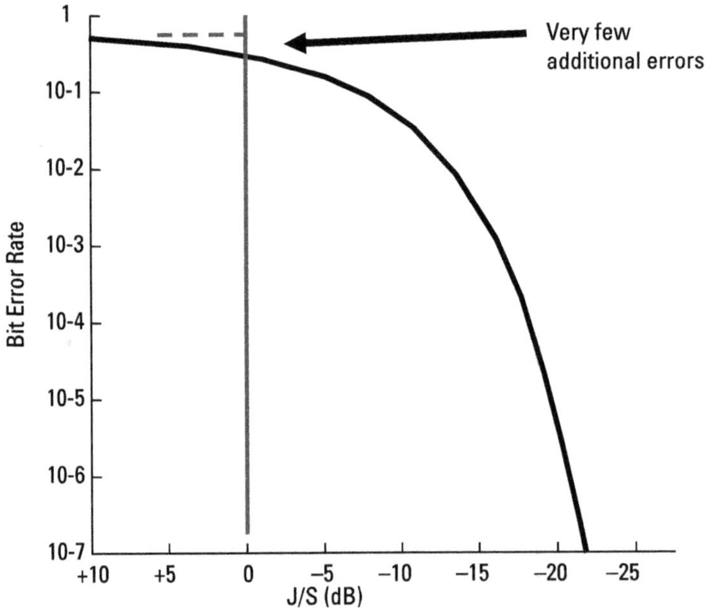

Figure 2.41 When a jammer achieves 0 dB J/S, it creates almost the maximum possible bit errors.

- *Spread spectrum modulations:* A special modulation can be added to a signal to cause its energy to be spread over a wide bandwidth. These low probability of intercept (LPI) techniques, including frequency hopping, chirp, and direct sequence spread spectrum, are discussed in more detail in Chapter 7. Among their other attributes, these techniques reduce the link's vulnerability to jamming. That is, they reduce the J/S in the target receiver. The receiver has special circuitry to remove the spreading modulation, thus causing a processing gain for signals from the desired signal transmitters. Each type of spreading modulation is driven by a pseudo random code which is also available to the receiver. Jamming signals, which presumably do not carry the spreading modulation, will not benefit from this processing gain.

The actual ways that the signals are spread in the transmitter and despread in the receiver are discussed in Chapter 7. Figure 2.42 generalizes the process by showing a generic block called a spreading demodulator. The point to be made here is that the dispreading process can be considered to create a processing gain that, in effect, increases the strength of desired signals while not increasing the strength of signals that have not been spread. In fact, the spreading demodulation process actually spreads any received signal that does not already contain the appropriate spreading modulation or is driven by the wrong code.

When we talked about a J/S of 0 dB above, we were talking about the effective J/S, that is, the J/S after considering that the jamming signal will not benefit from the processing gain caused by the spreading demodulator. Thus, the effective radiated power of the jamming signal must be increased by an amount equal to the processing gain to achieve the same effective J/S.

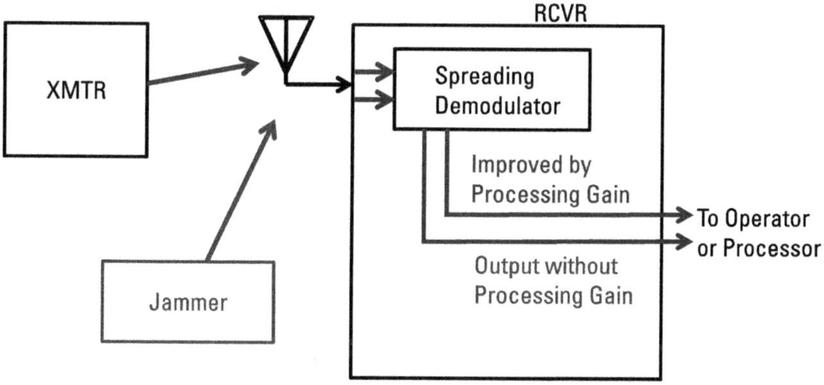

Figure 2.42 Spread spectrum signals are output from the target receiver with processing gain. Jamming signals are output at significantly lower level without the processing gain.

- *Antenna directivity:* In the equation for J/S above, there are two terms for receiver antenna gain. G_R is the receiving antenna gain in the direction of the desired signal transmitter and G_{RJ} is the receiving antenna gain the direction of the jammer. The use of directional antennas adds operational complexity to a network-centric system because the system components must know the locations of other elements of the system and track them. However, such antennas will significantly reduce the effective J/S achieved by a jammer. In calculations to predict J/S, it is common practice to assume that the target receiver's antenna is accurately oriented toward the desired receiver location. Because the jammer is at some other location, the target receiver antenna will show the jammer a side-lobe antenna gain except in the (very common) case that the target receiver has a nondirectional antenna. As you can see from the J/S equation above, the J/S is reduced by the difference between the receiving antenna gain to the desired signal (G_R) and to the jammer (G_{RJ}).

  Figure 2.43 shows a nulling antenna array. In such an array, the antennas have very wide beamwidth, covering a large angular segment, typically 360°. The processor creates phase shifts in the lines from the antennas in the array. These phase shifts can be set so that the output to the receiver sums the gain of all antennas to signals arriving from one direction, creating a narrow beam in the chosen direction. The phase shifts can also be adjusted to create nulls in one or more directions. If a null is directed toward a jammer, the effective jamming power is reduced by the depth of the null, reducing the J/S by that amount.
- *Error correcting codes:* Error detection and correction (EDC) codes add extra bits to transmitted digital signals, as discussed in Section 2.10. The

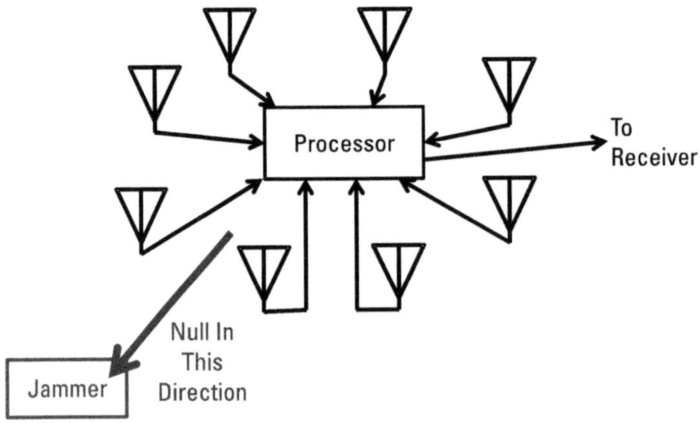

Figure 2.43 An array of antennas can create a null in the direction of a jammer.

receiver uses these extra bits to detect and correct bit errors up to some limit of BER, which is determined by the power of the code (basically, the percentage of additional bits). This means that the jammer must create more errors to have an adequate postcorrection BER to stop effective communication. Thus, more J/S is required. When an EDC is used, it is normal to rearrange blocks of bits before transmission to reduce the effectiveness of intermittent jamming. This can require high duty cycle jamming.

2.13.4 The Net Impact on Link Jamming

The effects of spread spectrum, antenna directivity, and/or error correction are to reduce jamming efficiency. For effective jamming of digital signals, we want to achieve 0 dB J/S, but this is the effective J/S after the impact of any antijamming techniques. This means that more jamming power needs to be delivered

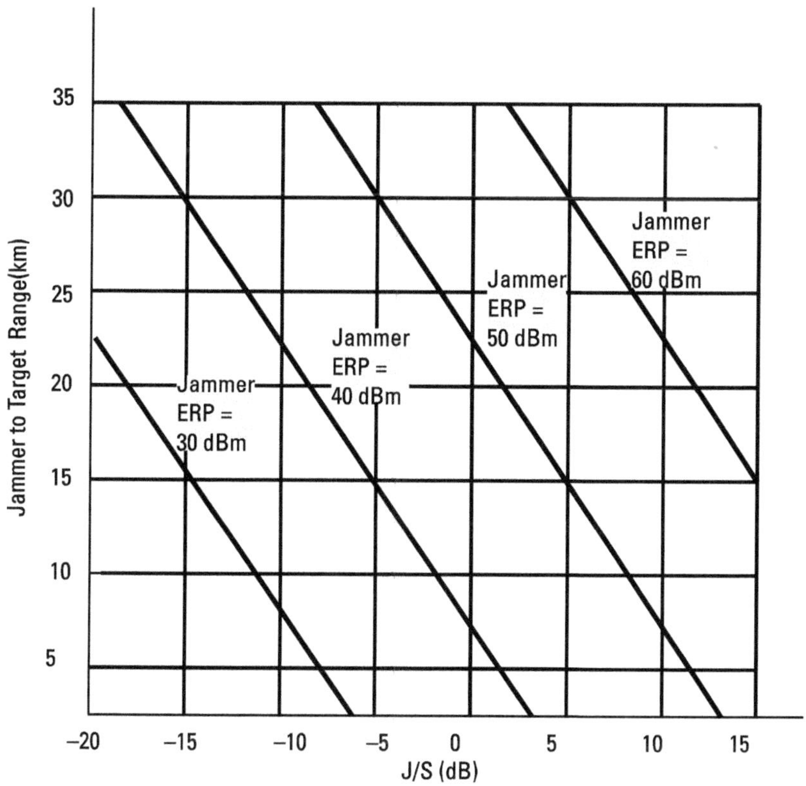

Figure 2.44 The J/S achieved at the target receiver input is a function of jammer ERP and jammer to receiver range.

to the target receiver. There are two basic ways to increase the received jamming power: increasing the jamming ERP and moving the jammer closer to the target receiver. Figure 2.44 shows the impact of both variables on J/S against a specific hostile link that is jammed. The target link has a desired transmitter ERP of 100W (+50 dBm) and operates over a range of 20 km. Each of the curves on the chart uses a different jammer ERP. To use this chart, start with the jammer to target range, move right to the curve for the jammer ERP, and then move down to the J/S achieved. For example, if the jammer to target range is 15 km and the jammer ERP is 40 dBm (10W), the J/S achieved would be –5 dB. One thing made obvious from this chart is the impact of stand-in jamming, in which the jammer is moved close to the target receiver.

3

Legacy Radars

This chapter discusses older threat systems primarily as a baseline against which the newest threats can be discussed in Chapter 4.

3.1 Threat Parameters

During the Cold War, a number of surface-to-air missiles were developed to attack high-flying aircraft at altitudes above the altitude at which anti-aircraft guns were effective. The most successful of these was the SA-2 Guideline missile, a command-guided weapon controlled to its target by the Fan Song radar. These weapons were organized into integrated air defense networks that comprised acquisition and tracking radars and multiple missile launchers along with ZSU-23 anti-aircraft guns guided by gun dish radars.

These weapons were used with great effectiveness during the Vietnam War and were joined by several other missile systems over the next few years in several regional conflicts.

There have also been air-to-air missiles and anti-ship missiles that are radar-guided.

In this book, we use the NATO designations for these weapons, which are also known by their Soviet designations. Some of these weapons are still in use and most have been upgraded in many ways.

It is important to clarify here that this discussion is not intended as a comprehensive threat briefing. There is much information about all of these weapons systems available on the Internet. The ranges of various weapons are described along with the general capabilities added with each cycle of upgrades.

Some parameters (e.g., operating frequency range, effective radiated power, and modulation parameters) are not described or are poorly described in open literature. In this chapter, we will mention these weapons only generally and will develop a set of typical parameters to support our discussions of the various electronic warfare (EW) techniques employed against them.

Some of the important parameters are found in open source literature (textbooks, technical magazine articles, and online articles), and some are not. In Chapter 4, we will move on to modern threats, which are even less well described in open literature. The open literature descriptions generally have a range of values for each threat parameter that they do describe and completely ignore other parameters. Where available from the literature, we will pick typical values that seem to make sense. Where a parametric value does not appear in open literature, we will calculate a typical value from the information that is available.

It is important to understand that all of the parameters of all of these weapons and their associated radars can be found in classified references. Because this is an unclassified book, that information is not available to us here. However, our set of typical parameters will allow us to discuss the EW techniques and drive to numerical solutions to problems. Those answers would be accurate only in reference to threat systems that happened to have our chosen typical values. Because no real-world system will have all of those parametric values, our answers will be wrong. However, they will be exactly wrong. Thus, when you take the calculations discussed to the real world, dealing with real threats, you can look up the real parameters of those real threat weapon systems and plug them into the equations discussed in this book to get the real answers.

When we address new systems in Chapter 4, there are even fewer parameters given in open literature, but the same approach will allow us to determine the changes in techniques and EW systems that are required as the threat parameters are upgraded.

The specific threat parameters that we will either capture or calculate for each threat type are:

- Lethal range;
- Operating frequency;
- Effective radiated power;
- Pulse width;
- Pulse repetition frequency;
- Antenna side-lobe isolation;
- Radar cross section of a vulnerable target.

3.1.1 Typical Legacy Surface-to-Air Missile

The SA-2 is an ideal candidate for a typical threat because it was widely employed and is still around, with various updates over the years. Table 3.1 shows the typical parametric values that we will use for a typical legacy surface-to-air missile. These are based on an analysis of the SA-2 from open literature. Because many values are not clearly stated in open source literature, the following rationale is given for the choice of each value in the table.

Lethal range for the SA-2 is most commonly stated in open literature as about 45 km, and *maximum altitude* is typically given as 20 km. However, there are shoot-down engagements described at greater altitudes.

Operating frequency is given as E-, F-, and G-bands for various SA-2 models. Where the operating bands are given, we will pick a frequency for our typical parameter value at a round number near the middle of that band. For our typical SA-2 operating frequency, we will use 3.5 GHz (in F-band).

Transmitter power is stated in open literature as 600 kW (for the E- and F-band versions), which converts to decibel form as

$$\text{Transmitter power in dBm} = 10\log_{10}(\text{power in milliwatts}) =$$
$$10\log_{10}(600{,}000{,}000) = 87.8 \text{ dBm}$$

For convenience in working problems in this chapter, we will round this to 88 dBm.

Antenna boresight gain for threat radars is not commonly found in open literature, but their beamwidth parameters are given. For the SA-2 Fan Song radar, the antenna beam angular dimensions of the two scanning fan beams is giv-

Table 3.1
Typical Legacy SAM Parameters

Parameter	Value
Lethal Range	45 km
Maximum altitude	20 km
Operating frequency	3.5 GHz
Transmitter power	88 dBm
Antenna boresight gain	32 dB
Antenna beamwidth	$+2° \times 10°$
Effective radiated power	+120 dBm
Side-lobe level	–21 dB
Pulse width	1 μs
Pulse repetition frequency	1,400 pps
Target radar cross section	1 m^2

en as 2° × 10°. The gain of a nonsymmetrical antenna beam can be calculated from the formula:

$$G = 29{,}000/(\theta_1 \times \theta_2)$$

where G is the boresight gain ratio and θ_1 and θ_2 are the 3-dB beamwidths in two orthogonal directions.

Calculating the antenna gain from information given for the SA-2 in open literature:

$$G = 29{,}000/(2° \times 10°) = 29{,}000/1{,}450$$

Converting this to decibels, 10 × 10 $\log_{10}$ (1,450) = 31.6 dB.

For convenience in working problems in this chapter, we will round this to 32 dB.

Effective radiated power is not easily found in open literature; however, the effective radiated power is defined as the product of the transmitter power and the antenna gain. For a radar, we use the antenna boresight gain. Thus, the effective radiated power of the SA-2 Fan Song radar is:

$$87.8\,\text{dBm} + 31.6\,\text{dB} = 119.4\,\text{dBM}$$

For convenience in working problems, we will use the rounded numbers given above:

$$88\,\text{dBm} + 32\,\text{dB} = 120\,\text{dBM}$$

Pulse width (PW) for the SA-2 is given in open literature as 0.4 to 1.2 μs. We will use 1 μs for our typical value.

Pulse repetition frequency (PRF) is given in open literature as 1,440 pulses per second in tracking mode. For convenience, we will take 1,400 pulses per second for our typical PRF value.

Antenna side-lobe level is not handily found in open literature, so we will use the midpoint of a table of antenna side-lobe levels from [1], in which the relative side-lobe level for ordinary antennas is listed as −13 to −30 dB. Side-lobe level here is defined as the average side-lobe level outside of the radar antenna's main beam, as compared to the peak boresight gain of the main beam. We will use 21 dB (close to the midpoint of his given range) as our typical value for the SA-2 antenna average side-lobe level.

The radar cross section of a vulnerable target varies widely for threat radars, but 1 m^2 appears often in example problems in tutorials and radar discussions. Therefore, we will use 1 m^2 as our typical value.

3.1.2 Typical Legacy Acquisition Radar

A typical legacy acquisition radar is the Soviet P-12 Spoon Rest. Table 3.2 shows the parameters of this radar from numbers found in open literature or values derived from open literature parameters.

The Spoon Rest model D range is given in open literature as 275 km and its operating frequency as 150 to 170 MHz, so we will use 160 MHz as the typical value. Its transmitter power is given as 160 to 260 kW, so we use 200 kW as a typical value. The antenna beamwidth is given as 6°, from which the antenna boresight gain can be calculated to be 29 dB from the formula:

$$G = 29{,}000/BW^2$$

where G is the antenna boresight gain and BW is the 3-dB beamwidth of the antenna.

The 200-kW transmitter power is 83 dBm, and because the ERP of a radar is normally assumed to be the product of its transmitter power and boresight gain, the ERP is 112 dBm.

Because the side-lobe levels are not readily found from open literature, we use the same value found in [1]. The pulse width and pulse repetition frequency are from open literature, and we assume the same 1-m^2 minimum target RCS.

3.1.3 Typical Anti-Aircraft Gun

The Soviet schilka, ZSU 23-4 automatic anti-aircraft gun (AAA) is taken as a typical legacy anti-aircraft gun. Table 3.3 shows the parameters of this weapon from open literature. The radar on the tracked platform is the gun dish, which

Table 3.2
Typical Legacy Acquisition Radar Parameters

Parameter	Value
Range	275 km
Maximum altitude	20 km
Operating frequency	160 MHz
Transmitter power	83 dBm
Antenna boresight gain	29 dB
Antenna beamwidth	6°
Effective radiated power	+112 dBm
Side-lobe level	–21 dB
Pulse width	6 μs
Pulse repetition frequency	360 pps
Target radar cross section	1 m^2

Table 3.3
Typical Legacy Anti-Aircraft Gun Parameters

Parameter	Value
Lethal range	2.5 km
Maximum altitude	1.5 km
Operating frequency	15 GHz
Transmitter power	70 dBm
Antenna boresight gain	41 dB
Antenna beamwidth	1.5°
Effective radiated power	111 dBm
Side-lobe level	–21 dB
Target radar cross section	1 m^2

has a 1-m diameter antenna and operates in the J-band. We pick 15 GHz for the typical AAA frequency as this is a round number in the middle of the J-band.

Since the transmitter power for the gun dish is not readily found in open literature, we use the 10-kW typical power of the German Wurzburg radar as typical for a short-range AAA radar. This is 70 dBm. The gain of a 1-m dish antenna at 15 GHz can be calculated from the formula:

$$G = -42.2 + 20\log(D) + 20\log(F)$$

where G is the antenna boresight gain in decibels, D is the diameter of the dish in meters, and F is the operating frequency in megahertz.

For a 1-m dish at 15 GHz, this calculates to be (rounded) 41 dB. Therefore, the ERP is (rounded) 111 dBm.

The antenna beamwidth of 1.5° is calculated from the formula:

$$20\log\theta = 86.8 - 20\log D - 20\log F$$

where θ is the 3-dB beamwidth in degrees, D is the antenna diameter in meters, and F is the operating frequency in megahertz.

For a 1-m dish at 15 GHz, the value of 20 log θ is 3.3.

The beamwidth is then found from:

$$\theta = \text{antilog}(20\log\theta/20) = \text{antilog}(3.3/20) = 1.5^\circ(\text{rounded})$$

The antenna side lobes and minimum target RCS are set at the same values used in Tables 3.1 and 3.2. The modulation parameters are not readily found in open literature.

3.2 EW Techniques

In this chapter and in Chapter 4, we will discuss the following EW activities and the associated calculations:

- Detection, intercept, and emitter location;
- Jamming for self-protection;
- Remote jamming to protect other assets;
- Chaff and decoys for asset protection;
- Antiradiation missiles.

In each case, we will develop the appropriate formulas and work example problems using the typical parametric values described in Section 3.1.

The specific answers that we will calculate for each threat type will include:

- Intercept range;
- Jamming-to-signal ratio (J/S);
- Burn-through range;
- Decoy simulation of radar cross section.

3.3 Radar Jamming

This section and the rest of this chapter is a review of radar jamming, which is covered in more detail (including formula derivations) in whole chapters in [2, 3]. The purpose here is to support the discussions of the EW impact of the new generations of threat radars presented in Chapter 4. Another convenient reference for more detailed tutorial coverage of radar jamming is a series in [4].

Radar jamming approaches are differentiated by geometry and by techniques. First, we will cover the geometric considerations: self-protection and remote jamming. This includes decibel formulas for the J/S and burn-through range associated with both types of jamming. In the following discussion, all jamming power is assumed to be within the radar receiver's bandwidth, and the radar is assumed to use a single antenna for transmit and receive. More complex cases will be considered later. As previously stated in Chapter 1, you will note that each of the decibel formulas in this section includes a numerical constant (for example, −103). This number combines conversion factors allowing values to be input in the most convenient units. The rather large resulting number is converted to decibel form. A very important consideration in the use of all decibel formulas is that *the input values must be entered in the specified units to get the correct answer.*

Another important point about these formulas is that all of these decibel formulas have inputs in various units: frequency in megahertz, power in dBm, and so forth, which seem to be added with no respect for the differences of units. Although troubling to some, these units can be combined because there are unit conversions hidden in the numerical constants. It is common practice to take these hidden conversions on faith, but they are dealt with in any rigorous derivation of the decibel formulas presented (without derivation) in this book.

3.3.1 Jamming-to-Signal Ratio

First, consider the power a radar receiver receives from the skin return from a target. As shown in Figure 3.1, the transmitted power is focused toward the target by the radar's antenna. The effective radiated power (in decibel form) is the transmitter power increased by the main beam boresight gain. Because a typical radar uses a directional antenna to transmit and receive signals, the propagation mode is line of sight (see Chapter 6). The skin return power in the radar receiver is called S and is given (in dBm) by the formula:

$$S = -103 + ERP_R - 40\log R - 20\log F + 10\log\sigma + G$$

where ERP_R is the radar effective radiated power toward the target in dBm, R is the range from the radar to the target in kilometers, F is the radar's transmitting frequency in megahertz, σ is the radar cross section of the target in square meters, and G is the main beam boresight gain of the radar antenna in decibels.

The power received by the radar from the jammer is called J and is given (in dBm) by the formula:

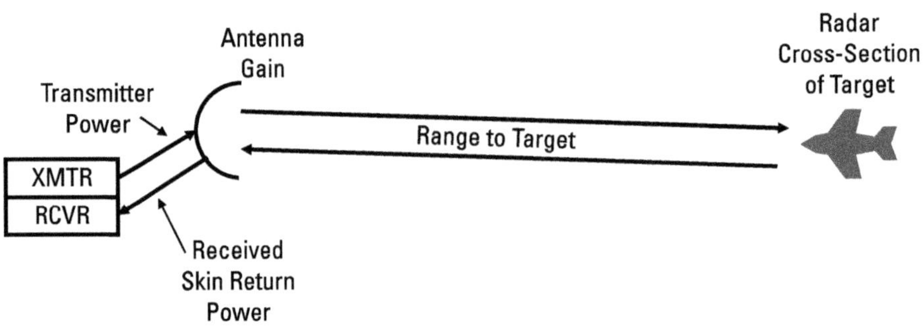

Figure 3.1 Radar skin return power is calculated from the radar transmitter power and antenna gain, the range to the target, and the target radar cross section.

$$J = -32 + ERP_J - 20\log R_J - 20\log F + G_{RJ}$$

where ERP_J is the jammer effective radiated power toward the radar in dBm, R_J is the range from the jammer to the radar in kilometers, F is the jammer's transmitting frequency in megahertz, and G_{RJ} is the gain of the radar's antenna (in decibels) in the direction toward the jammer.

3.3.2 Self-Protection Jamming

As shown in Figure 3.2, a self-protection jammer is located on the target being detected or tracked by a radar. This means that the distance from the jammer to the radar is R and the gain of the radar antenna toward the jammer and the target are the same (we will call this gain G). By subtracting the expression for S from the expression for J and simplifying, we get the following formula for the J/S produced by a self-protection jammer:

$$J/S = 71 + ERP_J - ERP_R + 20\log R - 10\log\sigma$$

where 71 is a constant, ERP_J is the effective radiated power of the jammer in dBm, ERP_R is the effective radiated power of the radar in dBm, R is the range from the radar to the target in kilometers, and σ is the radar cross section of the target in square meters.

Let us consider a specific self-protection jamming situation as shown in Figure 3.3 using the parameters listed in Table 3.1. A threat radar is tracking a target aircraft with 1-m^2 radar cross section at 10 km. The ERP of the jammer,

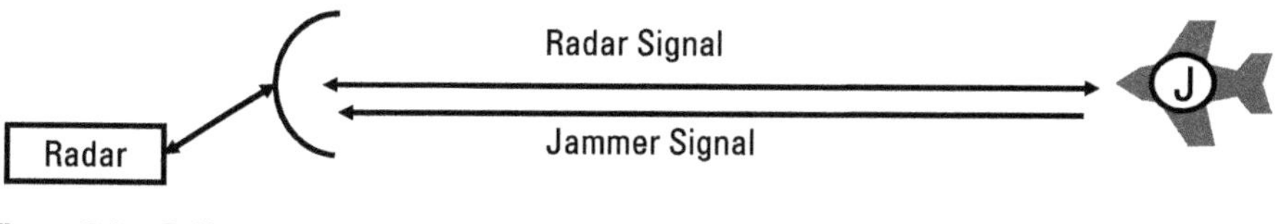

Figure 3.2 Self protection jamming protects a target by use of an on-board jammer.

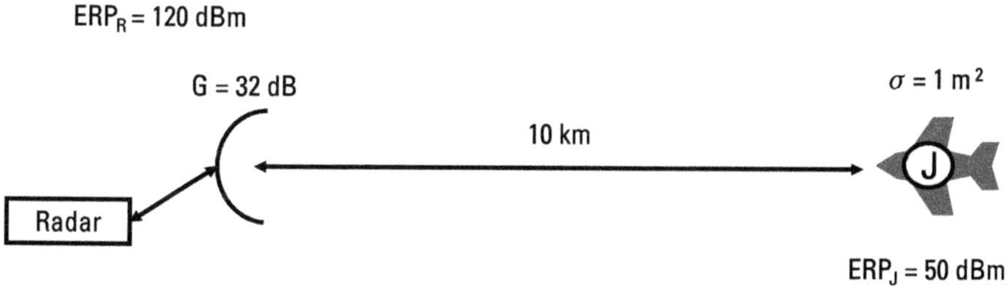

Figure 3.3 Self protection jamming problem.

located on the target aircraft, is 100W or +50 dBm. The radar ERP is +120 dBm. The radar antenna boresight gain is 32 dB, and the boresight of the antenna is pointed directly at the target.

Plugging these values into the *J*/*S* formula above gives:

$$J/S(\text{in decibles}) = 71 + 50\,\text{dBm} - 120\,\text{dBm} + 20\log(10) - 10\log(1)$$
$$= 71 + 50 - 120 + 20 - 0 = 21\,\text{dB}$$

where 71 is a constant, ERP_J is the effective radiated power of the jammer in dBm, ERP_R is the effective radiated power of the radar in dBm, R is the range from the radar to the target in kilometers, and σ is the radar cross section of the target in square meters.

3.3.3 Remote Jamming

In remote jamming, the jammer is not located at the target. The classical case of remote jamming is stand-off jamming as shown in Figure 3.4. The jammer (in a special jamming aircraft) is beyond the lethal range of the weapon controlled by a tracking radar. The jammer protects the target aircraft which is within that lethal range. Note that the stand-off jammer typically protects multiple targets from acquisition by multiple radars. This means that the jammer cannot be in the main beam of all of the radars; hence, it is assumed to be broadcasting into the side lobes of all hostile radars.

All types of remote jammers will produce J/S according to the following formula:

$$J/S = 71 + ERP_J - ERP_R + 40\log R_T - 20\log R_J + G_S - G_M - 10\log\sigma$$

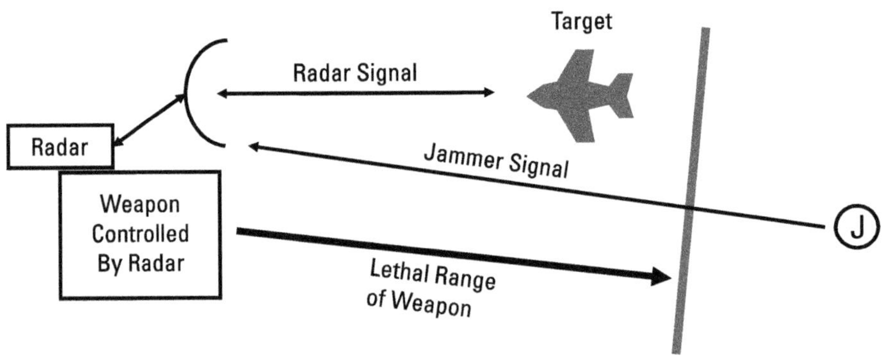

Figure 3.4 Standoff jamming protects a target within the lethal range of a radar controlled weapon using a jammer which is located beyond the lethal range.

where 71 is a constant, ERP_J is the effective radiated power of the jammer in dBm, ERP_R is the effective radiated power of the radar in dBm, R_T is the range from the radar to the target in kilometers, R_J is the range from the jammer to the radar in kilometers, G_S is the radar side-lobe gain (redefined from G_{RJ} above) in decibels, G_M is the radar main beam boresight gain in decibels, and σ is the radar cross section of the target in square meters.

Consider a radar trying to track a target aircraft that is 5 km from the radar, which has the boresight of its antenna on the target as shown in Figure 3.5. The jammer (on a stand-off jamming aircraft) is located in a side lobe of the radar antenna just a little beyond the maximum lethal range of the weapon system controlled by the radar.

The ERP of the jammer is much larger than that of the self-protection jammer. If its transmitter power is 1 kW and its antenna gain is 20 dB, its ERP is 80 dBm. The radar antenna boresight gain is 32 dB and its side-lobe isolation is 21 dB (both values are from Table 3.1). Thus, the side-lobe gain is 11 dB. The range to the stand-off jammer is 46 km (just barely beyond the 45-km lethal range from Table 3.1). The target aircraft RCS is 1 m^2.

Plugging these values into the remote jamming formula above gives:

$$\begin{aligned}J/S &= 71+80\,\text{dBm}-120\,\text{dBm}+40\log(5)-20\log(46)\\&\quad+11\,\text{dB}-32\,\text{dB}-10\log(1)\\&=71+80-120+28-33.3+11-32-0=4.7\,\text{dB}\end{aligned}$$

Figure 3.6 shows another case of remote jamming. This is stand-in jamming, in which the jammer is placed closer to the hostile radar than the target aircraft it is protecting. This jammer is also assumed to be broadcasting into the side lobes of the hostile radar.

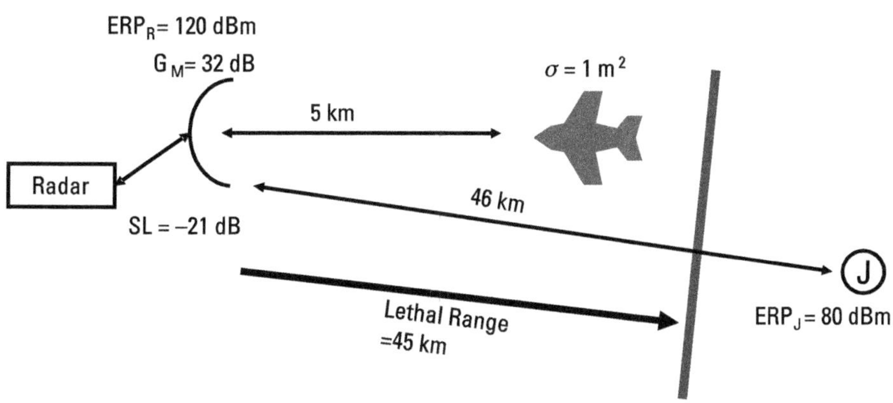

Figure 3.5 Standoff jamming problem.

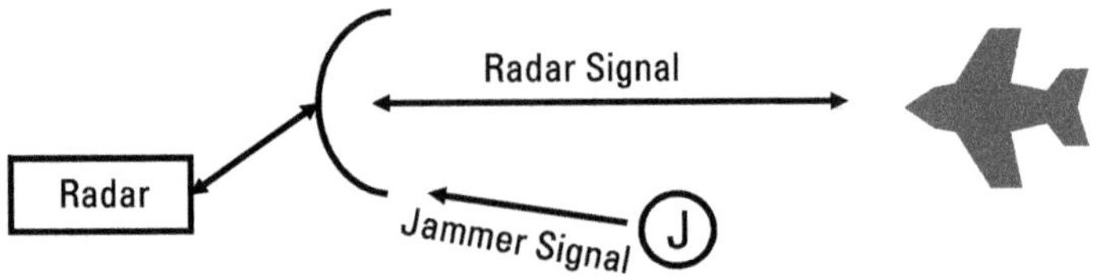

Figure 3.6 Stand in jamming protects a target using a jammer which is located closer to the radar.

Consider the situation shown in Figure 3.7 in which a 1-m^2 RCS aircraft is 10 km from a radar at the boresight of its antenna. A small, emplaced jammer is 500m from the radar in a side lobe with gain 21 dB below the boresight gain. The ERP of the jammer is 1W (30 dBm).

Plugging these values into the remote jamming formula above gives:

$$
\begin{aligned}
J/S &= 71 + 30\,\text{dBm} - 120\,\text{dBm} + 40\log(10) - 20\log(0.5) \\
&\quad + 11\,\text{dB} - 32\,\text{dB} - 10\log(1) \\
&= 71 + 30 - 120 + 40 - (-6) + 11 - 32 - 0 = 6\,\text{dB}
\end{aligned}
$$

3.3.4 Burn-Through Range

In both of the above equations, J/S is a positive function of range from the radar to the target. Thus, as the target approaches the radar, the J/S is reduced. When the J/S is small enough, the jammed radar can reacquire the target. It is common practice to determine some J/S value at which reacquisition might occur and define the range from the target at which this J/S occurs as the burn-through range.

This is illustrated in Figure 3.8 for self-protection jamming. Note that the radar skin return power increases as the fourth power of reducing range while the received jammer power increases only as the square of reducing range.

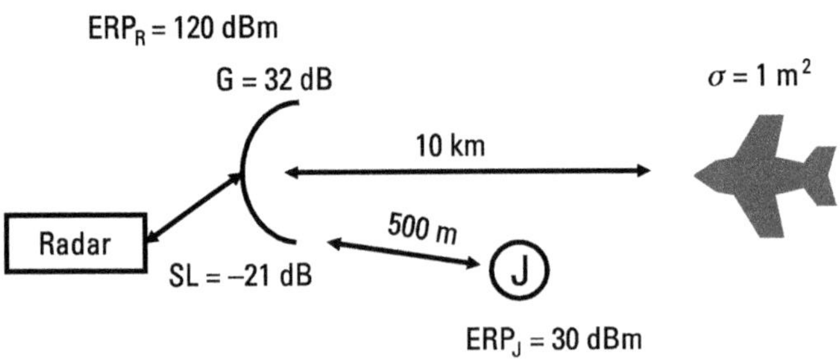

Figure 3.7 Stand in jamming problem.

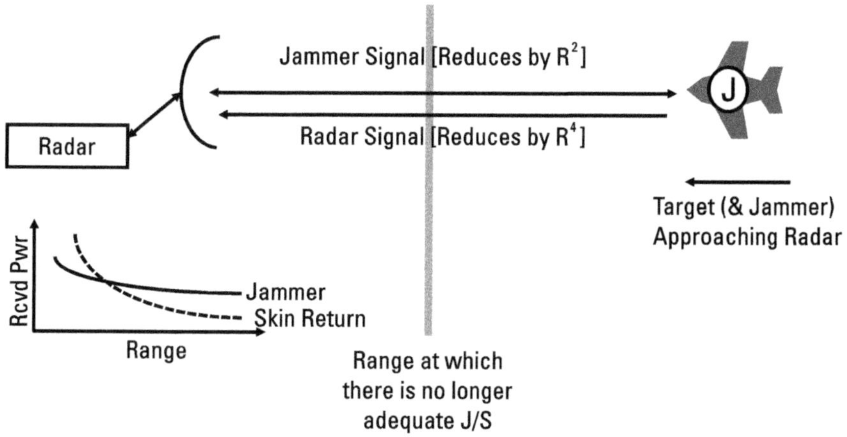

Figure 3.8 Self protection burn through occurs when the target is close enough to the radar that the radar can reacquire the target.

The equation for self-protection burn-through range is derived from the self-protection J/S formula as follows:

$$20\log R_{BT} = -71 + ERP_R - ERP_J + 10\log\sigma + J/S \text{ Rqd}$$

where R_{BT} is the burn-through range in kilometers, ERP_J is the effective radiated power of the jammer in dBm, ERP_R is the effective radiated power of the radar in dBm, σ is the target RCS, and J/S Rqd is the J/S value at which jammer reacquisition may take place.

The burn-through range in kilometers is found from the value of 20 log R_{BT} as:

$$R_{BT} = \text{antilog}\left[\left(20\log R_{BT}\right)/20\right]$$

Consider the self-protection jamming situation shown in Figure 3.3 with the target aircraft flying toward the radar. In Figure 3.9, the target has reached the range at which the J/S is reduced to the point at which the radar can reacquire the target in the presence of jamming. Note that the burn-through J/S depends on the type of jamming employed, and 0-dB J/S is often appropriate. We have arbitrarily set the burn-through J/S value at 2 dB for this example.

The jammer ERP is 50 dBm, the radar ERP is 120 dBm, σ is 1 m^2, and the required J/S is 2 dB. Plugging these numbers in to the self-protection burn-through equation above:

$$\begin{aligned}20\log R_{BT} &= -71 + 120\,\text{dBm} - 50\,\text{dBm} + 10\log(1) + 2\,\text{dB}\\ &= -71 + 120 - 50 + 0 + 2 = 1\end{aligned}$$

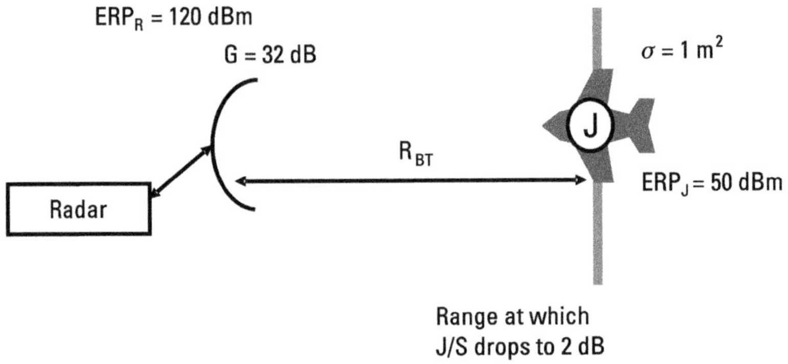

Figure 3.9 Self protection burn through problem.

Solving for R_{BT},

$$R_{BT} = \text{antilog}\left[(1)/20\right] = 0.056\,\text{km} = 56\,\text{m}$$

Figure 3.10 illustrates burn-through for any type of remote jamming. Note that it is common practice to assume that the stand-off or stand-in jammer does not move while the target approaches the radar. Thus, the received jammer power remains constant while the received skin return power increases by the fourth power of reducing range. Thus, the burn-through range refers only to the range from the radar to the target.

The formula for any kind of remote jamming burn-through is derived from the remote jamming J/S formula as:

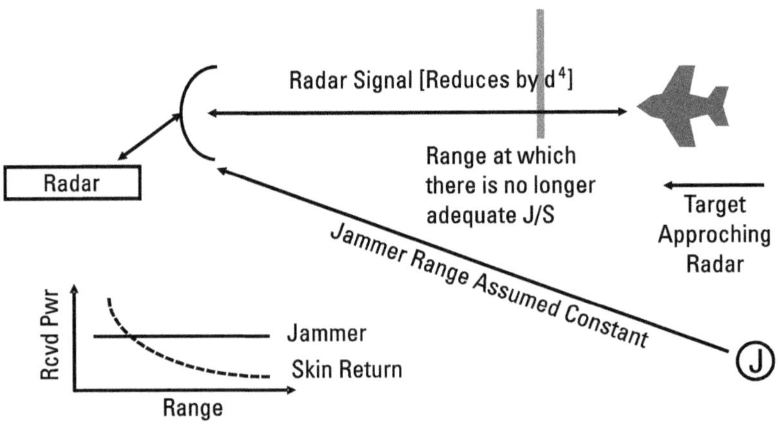

Figure 3.10 Remote jammer burn through occurs when the target is close enough to the radar that the radar can reacquire the target.

$$40\log R_{BT} = -71 + ERP_R - ERP_J + 20\log R_J + G_M - G_S + 10\log\sigma + J/S\,\text{Rqd}$$

The burn-through range in kilometers is found from the value of 40 log R_{BT} as:

$$R_{BT} = \text{antilog}\left[\left(40\log R_{BT}\right)/40\right]$$

Consider the stand-off protection jamming situation shown in Figure 3.5 with the target aircraft flying toward the radar and the stand-off jamming aircraft flying a small pattern in a fixed location in the radar side lobe. In Figure 3.11, the target has reached the range at which the J/S is reduced to the point at which the radar can reacquire the target in the presence of jamming. As in the self-protection example, we have arbitrarily set the burn-through J/S value at 2 dB.

The jammer ERP is 80 dBm, the radar ERP is 120 dBm, σ is 1 m^2, and the required J/S is 2 dB. Plugging these numbers in to the self-protection burn-through equation above:

$$\begin{aligned}40\log R_{BT} &= -71 + 120\,\text{dBm} - 80\,\text{dBm} + 20\log(46)\\&\quad + 32\,\text{dB} - 11\,\text{dB} + 10\log(1) + 2\,\text{dB}\\&= -71 + 120 - 80 + 33.3 + 32 - 11 + 0 + 2 = 25.3\end{aligned}$$

Solving for R_{BT},

$$R_{BT} = \text{antilog}\left[(25.3)/40\right] = 4.2\,\text{km}$$

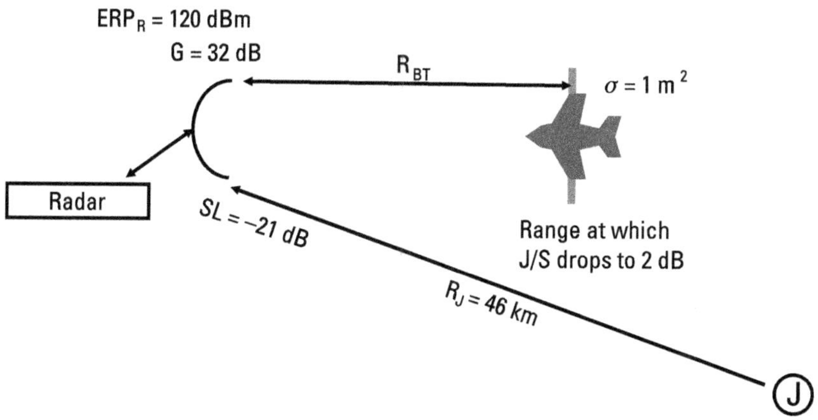

Figure 3.11 Remote jammer burn through problem.

3.4 Radar-Jamming Techniques

Radar-jamming techniques can be divided into cover and deceptive jamming. The jamming effectiveness of both types of techniques is stated in terms of the J/S as discussed above.

3.4.1 Cover Jamming

The object of cover jamming is to reduce the quality of the signal in the radar's receiver enough that the radar cannot acquire or track its target. It can be used in either self-protection or remote-jamming geometry. Cover jamming usually has a noise waveform, but sometimes other waveforms are used to overcome electronic protection (EP) features of the radar. These EP techniques will be covered in Chapter 4.

The equations for J/S and burn-through presented in Section 3.3 assumed that all of the jammer's power was within the bandwidth of the radar receiver. If a jammer uses noise that is wider in frequency than the effective bandwidth of the radar receiver, only the part that is within the radar's receiver bandwidth is effective. Jamming efficiency is the total jammer effective radiated power (ERP) divided by the effective jammer ERP. This is equal to the radar receiver bandwidth divided by the jamming bandwidth. For example, if the radar receiver bandwidth is 1 MHz and the jamming signal bandwidth is 20 MHz, the jamming efficiency is 5%.

3.4.2 Barrage Jamming

Barrage jamming is generated by a wideband jammer that broadcasts noise over a whole band of frequencies that is expected to contain one or more threat radars. This technique was frequently used in early jammers and is still an appropriate approach for many jamming situations. The great advantage of barrage jamming is that it does not require real-time information about radar operating frequencies. Look-through (i.e., interruption of jamming to look for threat radar signals) is not necessary. The problem is that barrage jamming typically has very low jamming efficiency. Most of the jamming power is wasted because the effective J/S is reduced by the efficiency factor, and the burn-through range is correspondingly increased.

3.4.3 Spot Jamming

When the bandwidth of the jamming signal is reduced to a little more than the target radar bandwidth and the jammer is tuned to the radar broadcast frequency, this is called spot jamming. As shown in Figure 3.12, spot jamming wastes little of its jamming power, so the jamming efficiency is increased significantly.

The spot width is enough to cover the uncertainty in target signal and set-on frequencies. (We will cover coherent jamming in Chapter 4.) Efficiency is still the radar bandwidth divided by the jamming bandwidth, but the ratio is more favorable. Schleher [1] defined spot jamming as jamming over a bandwidth less than five times the radar's bandwidth.

3.4.4 Swept Spot Jamming

If a narrowband jammer is swept across all of the frequency range that is expected to contain threat signals, as shown in Figure 3.13, it is called a swept spot jammer. The swept spot jammer, like the barrage jammer, does not require look-through and will jam any signal within the sweeping range. While the jammer is within a target radar's bandwidth, it will provide the same jamming efficiency as a set-on spot jammer. However, the jamming duty cycle will be reduced by the ratio of the spot bandwidth to the sweeping range. This can still

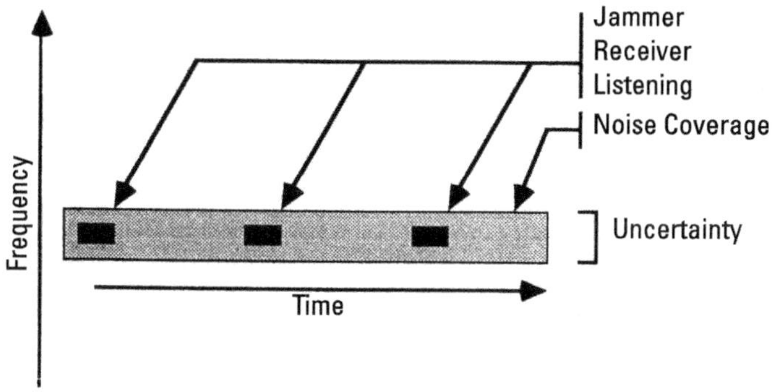

Figure 3.12 Spot jamming concentrates noise around the radar's operating frequency.

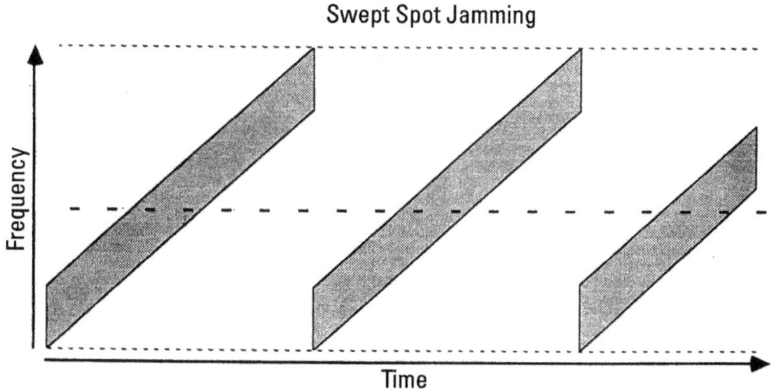

Figure 3.13 Swept spot jamming moves a narrow jamming band across the whole band in which the radar might operate.

provide adequate jamming performance against some radars in some situations. The spot bandwidth and sweeping range must be optimized for the situation.

3.4.5 Deceptive Jamming

A deceptive jammer makes a radar think it is receiving a valid skin return from a target, but the information that it derives from the received signal causes the radar to lose track on the target in range or angle. Because the deceptive jammer must key to the target signal at the target to submicrosecond accuracy, deceptive jamming is generally limited to self-protection applications. It is possible to do some deceptive techniques from a remote jammer, but it is very seldom practical. Thus, deceptive techniques will be discussed here as self-protection jamming. We will first discuss techniques that deceive the radar in range, then those that deceive it in frequency, and then those that deceive the radar in angle.

3.4.6 Range Deception Techniques

We will consider three range techniques: range gate pull-off (RGPO), range gate pull-in (RGPI), and cover pulses.

3.4.6.1 RGPO

An RGPO jammer receives each radar pulse and returns it to the radar with increased power. However, after the first pulse, it delays subsequent pulses by an increasing amount. The rate of change of delay from pulse to pulse is exponential or logarithmic. Because the radar determines the distance to a target from the round-trip propagation time of its pulses, the target seems to be moving away from the radar.

Figure 3.14 shows the early and late gates in the radar's processor. These are two time gates that are typically about the width of a pulse when the radar is tracking (longer during acquisition). The radar tracks range by balancing the energy from returned pulses in these two time increments. By delaying a stronger return, the jammer causes the energy in the late gate to dominate over the early gate, causing the radar to lose range track on the target.

The radar's resolution cell is the spatial volume in which the radar cannot resolve multiple targets. The center of this cell in range is the range at which the round-trip propagation time places a transmitted signal at the junction of the early and late gates. Thus, the radar assumes that the target is at the center of the cell. As shown in Figure 3.15 (in two dimensions), an RGPO jammer causes the radar to move its resolution cell out in range. Once the true target is out of the resolution cell, the radar has lost range tracking.

When the RGPO reaches its maximum delay, it snaps back to zero delay and repeats the process (many times). The radar will then have to reacquire its

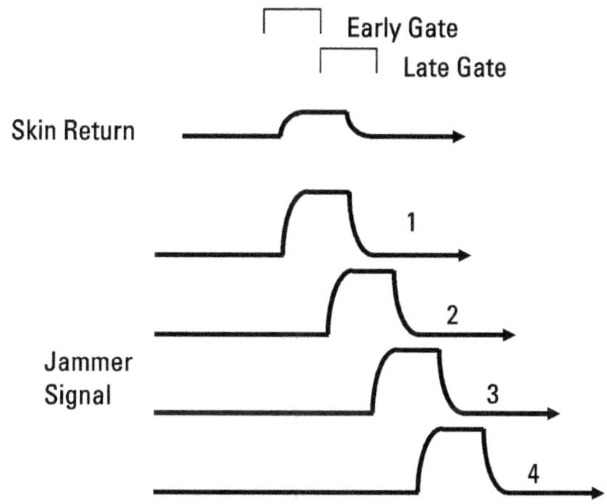

Figure 3.14 Range gate pull off involves sequential delay of the return pulse, which loads up the radar's late gate.

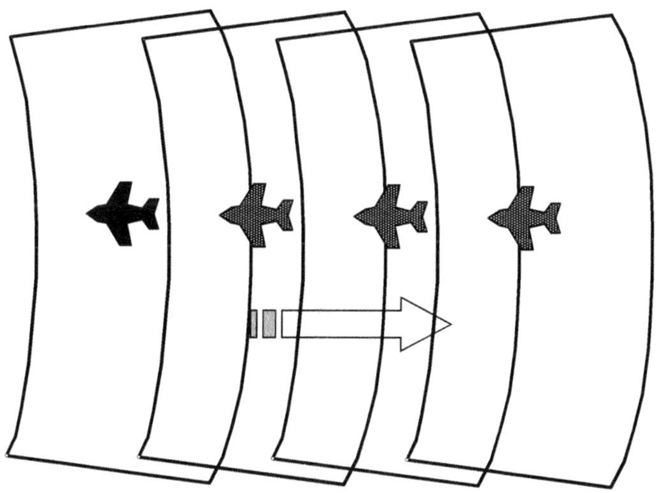

Figure 3.15 Loading up the late gate causes the radar's resolution cell to move out, making the radar think the target has moved farther away.

target in range, which takes several milliseconds, by which time the range track will have been pulled off again.

3.4.6.2 RGPI

Range gate pull-in (RGPI) is also sometimes called inbound range gate pull-off. It is used against radars that track in range using only the energy in the leading edges of its pulses. Thus, the early and late gates balance the leading edge energy. Because there is latency in the process of generating a deceptive jamming pulse, an RGPO jammer is unlikely to capture the tracking gates during the leading edge energy burst, so it will not deceive the radar. The RGPI jammer tracks the radar pulse repetition timing and generates a stronger return pulse that anticipates the next pulse by an exponentially or logarithmically increasing amount as shown in Figure 3.16. This loads up the early gate and makes the radar think that the target is approaching.

Note that RGPI jammers work fine when the radar has a constant pulse repetition frequency (PRF) or when it has a low-level staggered PRF. However, a random PRF cannot be tracked, so RGPI will not work against this type of signal.

3.4.6.3 Cover Pulses

While not technically deceptive jamming, cover pulses are intimate with the timing of pulses at the target, so they are discussed here. If the jammer has a pulse train tracker, it can output a long pulse centered on the radar's skin return pulse. This denies the radar range information and thus prevents range tracking.

3.4.7 Angle Deceptive Jamming

When a radar's range track is broken, several milliseconds may be required to re-establish tracking, after which the range track must be broken again. However, if the angle track is broken, the radar must typically return to a search mode to locate the target in angle, which can take seconds. Older radars required movement of the antenna beam to track targets in angle. Consider the received power versus time diagram for a conically scanned radar shown on the top line of Figure 3.17. The antenna movement describes a cone. When the antenna is pointed closer to the target, the received signal is stronger and when it is

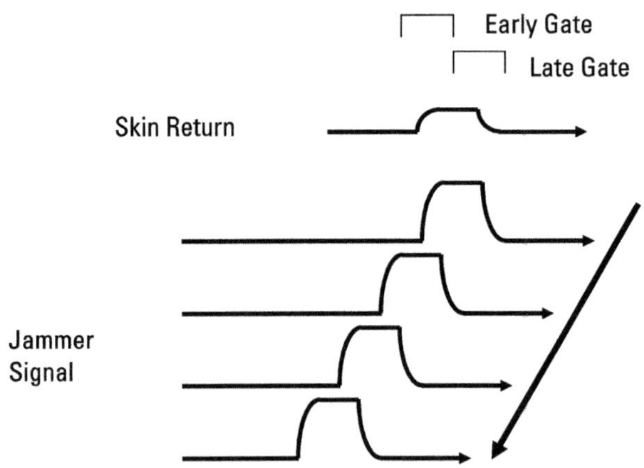

Figure 3.16 Range gate pull in involves sequentially increased anticipation of the return pulse, which loads up the radar's early gate.

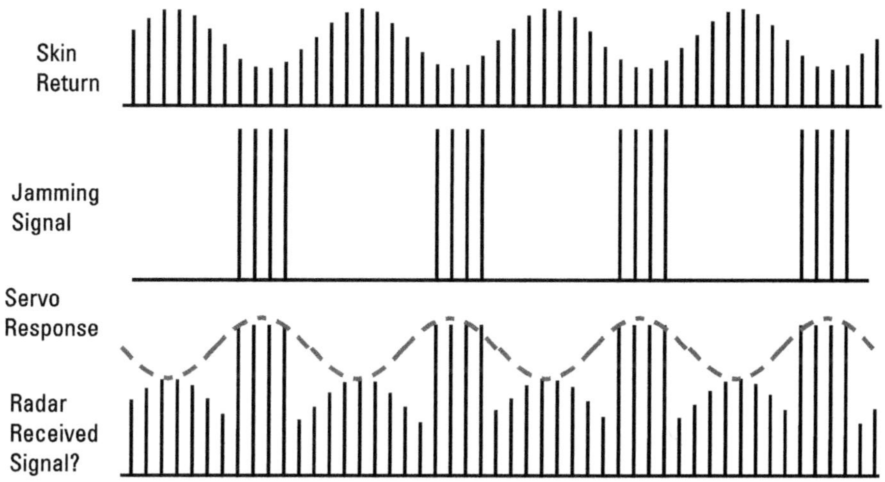

Figure 3.17 Inverse gain jamming causes a radar to correct its angle guidance in the wrong direction.

pointed away from the target the signal is weaker. The radar moves the center of its scanning pattern in the direction of the maximum return power to center the target in the scan. Both the radar receiver and a radar warning receiver on the target see this same power versus time plot. If a jammer located on the target transmits a burst of strong pulses (synchronized with the radar's pulses) at the weakest signal strength time (see the second line of Figure 3.17), the radar will see a power versus time plot as shown on the third line of the figure. Because the radar develops guidance signals from this information, the processing will see the power data in its (narrow) servo response bandwidth as shown by the dash line. Hence, the radar will direct its scan axis away from the target, breaking the angle track. This is called inverse gain jamming.

If the radar has a nonscanning illuminator, but scans its receiving antenna, the jammer on the target will be unable to know the phase of the sinusoidal power variation with time. Thus, the jammer is unable to time its pulse bursts to the minimum received power times. However, if the jammer times its bursts slightly faster or slower than the known scanning rate of the radar antenna, the jamming can still break up the angle tracking by the radar. This will still allow effective jamming, although not as effective as though the bursts were optimally timed.

Figure 3.18 shows angle jamming of a track-while-scan (TWS) radar. On the first line, the skin return from the TWS radar shows a burst of pulses as the beam passes the target. The radar will use angle gates to determine the angular location of the target. It will move the angle gates to equalize the power in the (in this case) right and left gates. The intersection between these two gates represents the angle to the target. If a jammer on the target generates a series of

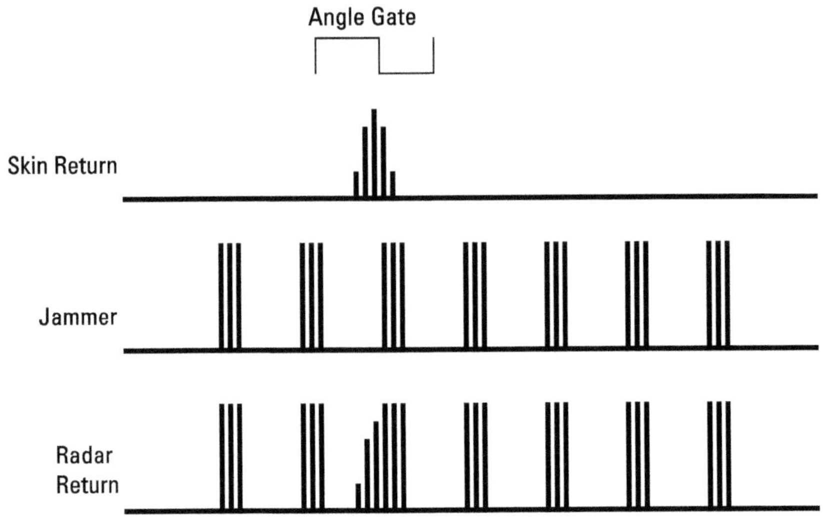

Figure 3.18 Inverse gain jamming causes a track while scan radar to move away from its target in angle.

synchronized pulse bursts as shown on the second line of Figure 3.18, the radar will see the combined power versus time curve shown on the third line. This will load up one side of the angle gate, causing the radar to move away from the angle of the target.

3.4.7.1 AGC Jamming

Because of the huge dynamic range over which a radar must operate, it must have automatic gain control (AGC). AGC is implemented by measuring the received power level at some point in a circuit and adjusting a gain or loss at an earlier stage of the circuitry to equalize the signal strength at the measurement point. To be effective, the AGC circuit must have a fast attack/slow decay characteristic. The first line of Figure 3.19 shows a sinusoidal power versus time curve as would be generated in a conically scanned radar return. If a strong narrowband jamming signal is added to the skin return, the high-level pulses will capture the AGC, so that the sinusoidal signal from the conically scanning antenna will be significantly reduced as shown on the second line of the figure. The sinusoidal signal will actually be reduced much more than shown in the figure, making it impossible for the radar to track targets in angle.

3.4.7.2 Other Angle Jamming Examples

There are several other examples of angle jamming; for example, inverse gain can be used against a lobing radar. However, the above angle jamming descriptions show how angle jamming works and will support our later discussions. One important point is that the above examples were for radars that must move

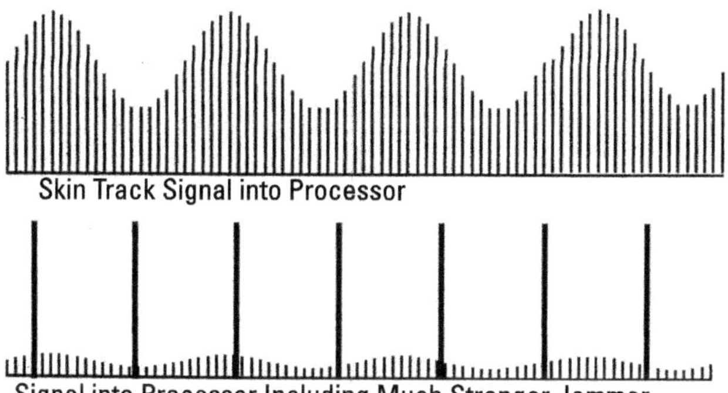

Figure 3.19 AGC jamming generates strong, narrow pulses at about the target signal modulation rate to capture the radar's AGC.

their antennas and receive multiple skin return pulses to support angle tracking. There is an important class of radars called monopulse radars that get complete angle information from each received skin return pulse. These types of radars and jamming techniques effective against these types of radars are covered in Section 3.4.9.

3.4.8 Frequency Gate Pull Off

It is often important to deceive a radar in frequency. The received frequency of a skin return signal is determined by the transmitted frequency and the rate of change of range between the radar and the target. The first line of Figure 3.20 shows the signal strength versus frequency for skin returns from a Doppler radar. Note that internal noise in the radar shows at the lower frequency range of the return. There are also multiple ground returns. If this is an airborne radar, the largest and highest frequency (i.e., highest velocity) ground return would be from the ground the aircraft is passing over. Lesser returns are from terrain features being passed. These returns are at lower Doppler frequencies because of the offset angle of the terrain feature from the flight path of the aircraft. Finally, we see the target return which is at the frequency related to the closing velocity between the radar and the target. The radar will place a velocity gate around the target return frequency to allow the target to be tracked. If the jammer places a signal in the velocity gate and then sweeps the jamming signal away from the target return frequency, the radar will be caused to lose velocity track on the target. This technique is called velocity gate pull-off.

Note that some radars can discriminate against range gate pull-off jamming by correlating the rate of change of range (caused by the range gate pull-off) to the Doppler shift of the skin return. In this case, it may be necessary to perform both range and velocity gate pull-off.

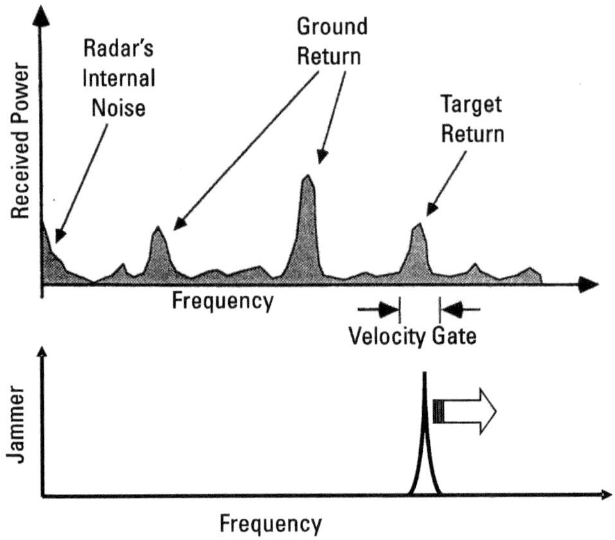

Figure 3.20 Frequency gate pull off places a jamming signal in the radar velocity gate, captures the gate and moves it off of the target return.

3.4.9 Jamming Monopulse Radars

In Section 3.4.7, we discussed the angle deception of radars, which must determine the angular position of a target from multiple pulse returns. Now we consider monopulse radars, which get angular information from every pulse return. Monopulse radars determine target angle by comparing signals in multiple receiving sensors. Figure 3.21 shows only two sensors; however, actual monopulse radars have three or four sensors to allow two-dimensional angle tracking. The sensor outputs are combined in sum and difference channels. The sum channel establishes the level of the returned signal and the difference channel provides angle tracking information. Note that the difference response is typically linear

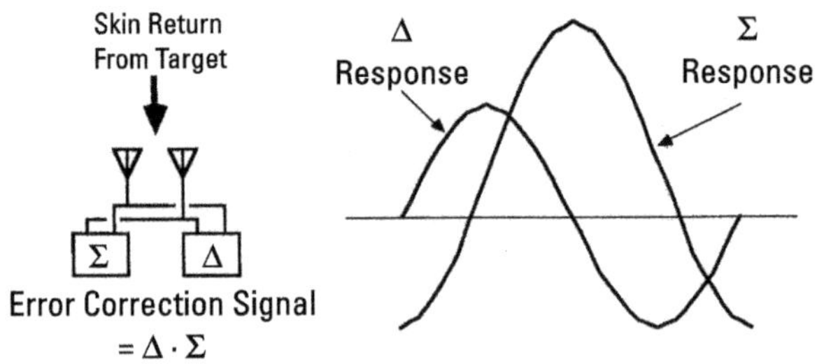

Figure 3.21 Monopulse radars derive angle information from each pulse by use of multiple sensors.

across the 3-dB width of the sum response. The guidance input is the difference response minus the sum response.

Jamming techniques shown so far in Section 3.4 actually improve the angle tracking effectiveness of monopulse radars by increasing the signal strength received from the target location. However, there are several techniques that do work against monopulse radars. These include the following:

- Formation jamming;
- Formation jamming with range denial;
- Blinking;
- Terrain bounce;
- Cross-polarization;
- Cross eye.

3.4.10 Formation Jamming

If two aircraft fly formation inside the radar's resolution cell as shown in Figure 3.22, the radar will be unable to resolve them, seeing in effect a single target between the two real targets. The difficulty with this technique is that is can be very challenging to keep both aircraft within the resolution cell.

The width (i.e., cross-range) dimension of the resolution cell is:

$$W = 2R\sin\left(BW/2\right)$$

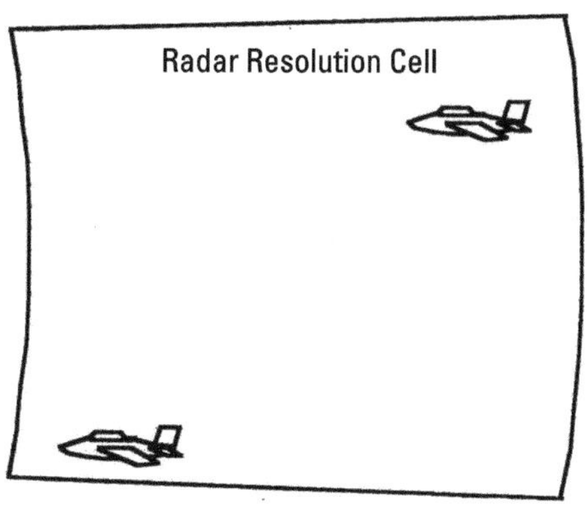

Figure 3.22 Formation jamming involves flying two aircraft within the radar's resolution cell. The radar will "see" only one target half way between the two real targets.

where W is the width of the cell in meters, R is the range from the radar to the target in meters, and BW is the 3-dB beamwidth of the radar antenna.

The depth (i.e., down-range) dimension of the cell is:

$$D = c\left(PW/2\right)$$

where D is the depth of the cell in meters, PW is the radar pulse width in seconds, and c is the speed of light (3×10^8 meters per second).

For example, if the target is 20 km from the radar, the radar pulse width is 1 μs and the radar antenna beamwidth is 2°, the resolution cell is 698m wide and 150m deep. Figure 3.23 compares the dimensions of the resolution cell for this radar at various radar to target ranges.

3.4.11 Formation Jamming with Range Denial

Self-protection jamming, because it is emitted from the radar's target, enhances the monopulse radar's angle tracking. However, it can deny the radar range information. If both aircraft jam with approximately the same power as shown in Figure 3.24, the radar will be unable to resolve the two targets in range, so they will be required to station keep only within the cross-range dimension of the resolution cell to prevent the radar from resolving its two targets. At long ranges, the resolution cell is much wider than its depth, so this technique can simplify station keeping.

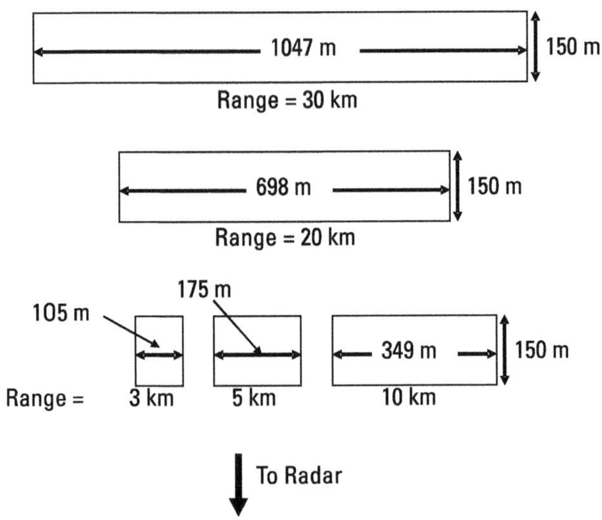

Figure 3.23 The shape of the radar's resolution cell varies significantly with the radar to target range. This is for 1 μ sec PW and 2° BW.

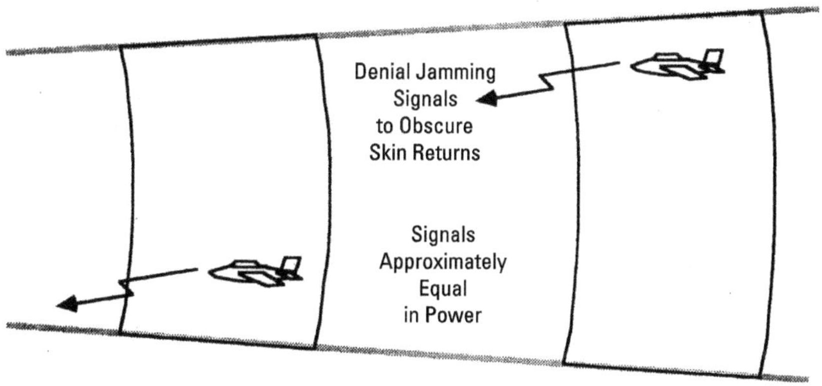

Figure 3.24 If each aircraft jams equally to deny the radar range information, the two aircraft must only hold formation within the cross range dimension of the radar resolution cell.

3.4.12 Blinking

If two aircraft in the radar's resolution cell alternate their jamming at a moderate rate (0.5 to 10 Hz) as in Figure 3.25, an attacking missile will be guided alternately to one or the other. As the missile approaches the two aircraft, it will be retargeted with an increasingly large angular offset. Because the missile's angular guidance is limited in loop bandwidth, it will be unable to follow one of the target changes and will fly off to one side.

3.4.13 Terrain Bounce

If an aircraft or missile rebroadcasts a radar's signal with significant gain from an antenna pointed down toward the water or land over which it is flying (as shown in Figure 3.26), the monopulse tracker will be caused to track below the protected platform. This will make the weapon miss the target.

3.4.14 Cross-Polarization Jamming

If a parabolic radar antenna reflector has significant forward geometry, it will have small lobes (called Condon lobes) that are cross-polarized to the main antenna feed. In general, the greater the curvature of the antenna, the larger the Condon lobes will be. As shown in Figure 3.27, these lobes can become dominant if the radar is illuminated by a very strong jamming signal cross-polarized to the primary radar signal.

Figure 3.28 shows the operation of a cross-polarization jammer. It receives the radar signal in two antennas that are orthogonally polarized. In this figure, one is vertically polarized and the other is horizontally polarized. The signal received by the vertically polarized antenna is rebroadcast with horizontal

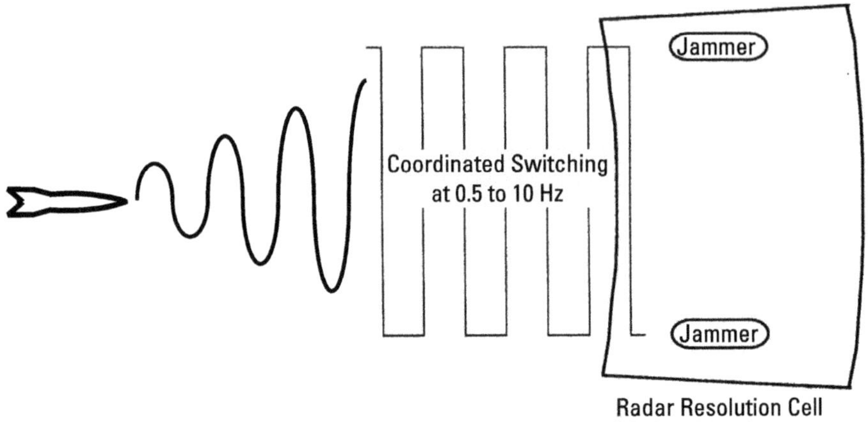

Figure 3.25 Blinking jamming involves sequencing jammers on two aircraft to force the tracking radar to switch between targets until the missile guidance is over stressed.

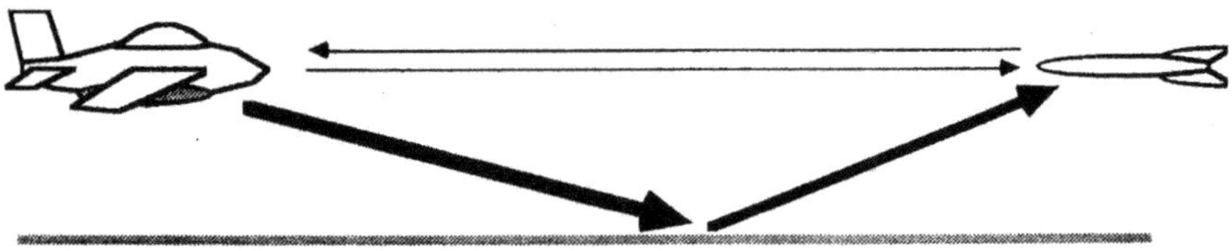

Figure 3.26 Terrain bounce jamming reflects a strong return signal from the earth or water causing the radar to track below the target.

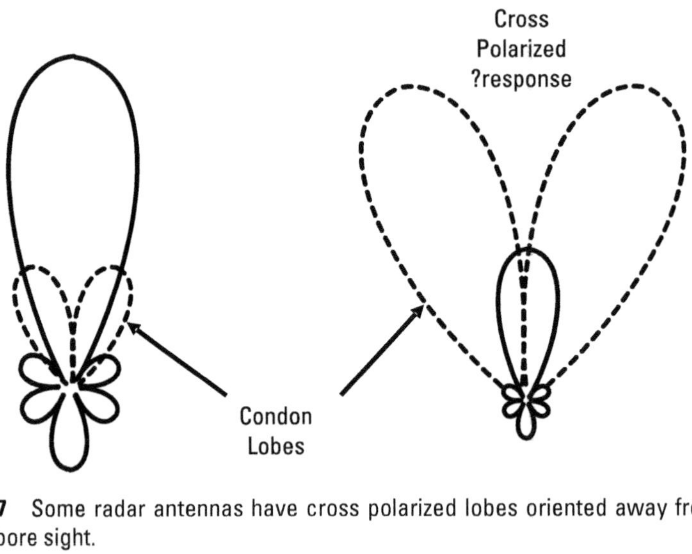

Figure 3.27 Some radar antennas have cross polarized lobes oriented away from the co-polarized bore sight.

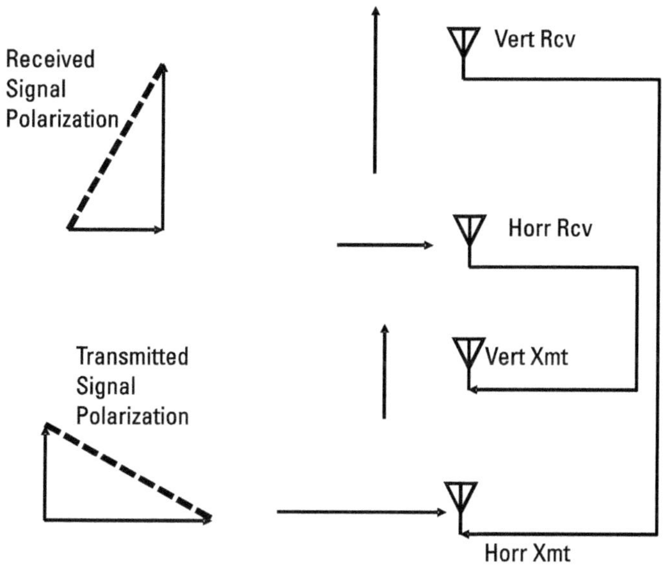

Figure 3.28 Cross-pol jamming generates a strong cross polarized return signal which causes the radar to track the target in one of its Condon lobes.

polarization and the signal received by the horizontally polarized antenna is rebroadcast with vertical polarization. This causes the jammer to produce a signal that is cross-polarized to the received signal regardless of the received signal polarization. The jamming signal thus produced is amplified by a large enough factor to produce a J/S of 20 to 40 dB.

When the strong cross-polarized signal reaches the radar, it will capture one of the Condon lobes. The radar will then move its antenna so that the captured Condon lobe is aimed at the target. This causes the radar to lose its track on the target.

In general, this type of jamming is not effective against radars that have flat plate phased array antennas, as they do not have the forward geometry to produce Condon lobes. However, if the phased array has significant beam-shaping from variable illumination, it may have Condon lobes.

If the radar antenna is protected by a polarization filter, it will be immune to cross-polarization jamming.

3.4.15 Cross-Eye Jamming

The configuration of a cross-eye jammer is shown in Figure 3.29. The signal received by an antenna at point A is amplified 20 to 40 dB and rebroadcast from an antenna at point B. Likewise, signals received by an antenna at point B are amplified and rebroadcast from an antenna at point A, but there is a 180° phase shift in this circuit. For the jammer to be effective, these two signal paths

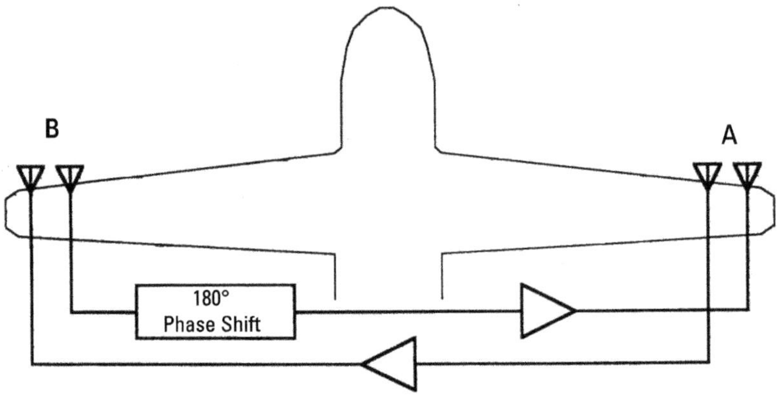

Figure 3.29 Cross eye jamming broadcasts the radar signal received at location A from location B and simultaneously broadcasts the signal received at location B from location A with a 180° phase shift.

must be exactly the same length. Because points A and B must have significant spacing for the jamming to be effective, the cables are long. It is extremely difficult to maintain adequate balance in these two sets of cables over variations of temperature and frequency. The two cable paths must maintain the 180° relationship within an electrical degree or two for effective jamming. This is a differential electrical length of the order of a tenth of a millimeter.

To mitigate this problem, the system can be configured as shown in Figure 3.30. Nanosecond switches allow a single cable to be used from a single antenna at each of the two locations, and it is easy to maintain phase matching within the (quite small) box. The switches alternate the signal path between the phase-shifted and nonshifted branches many times during reception of each radar pulse. Because the radar receiver must be optimized to receive the radar's pulse, it will average the square waves shown below the pulse in the figure. Thus, the signals from the two jammer antennas will be seen by the radar as two simultaneous pulses that are 180° apart in phase.

The path from the radar to antenna A to antenna B and back is exactly the same length as the path from the radar to antenna B to antenna A and back. This does not require that the A-B baseline be perpendicular to the path from jammer to radar. Thus, the radar will receive two signals 180° out of phase. As shown in Figure 3.21, this will cause a null at the radar's sensors. The result will be that the sum response will be below the difference response, which will change the sign of the "difference – sum equation." This will cause the radar to correct its tracking angle away from the target rather than toward the target.

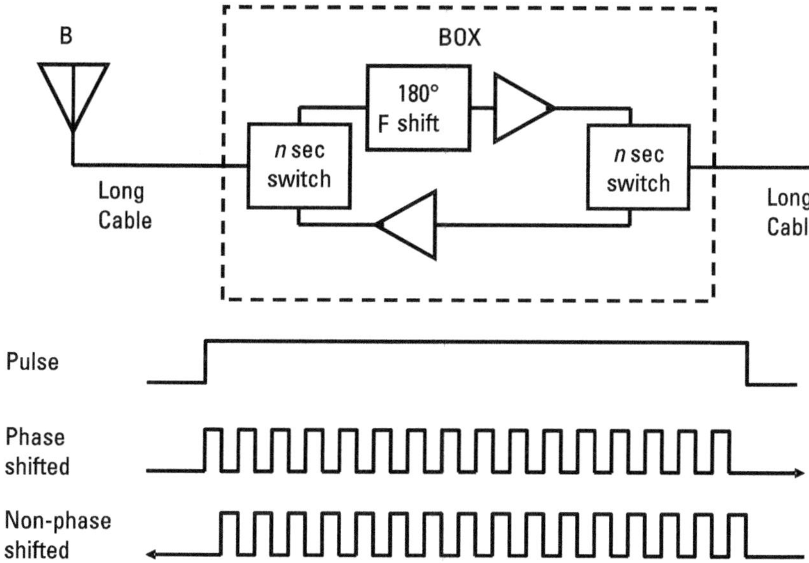

Figure 3.30 Nanosecond switches allow single cables from each of the antennas to time share signals in both directions, eliminating critical cable length matching.

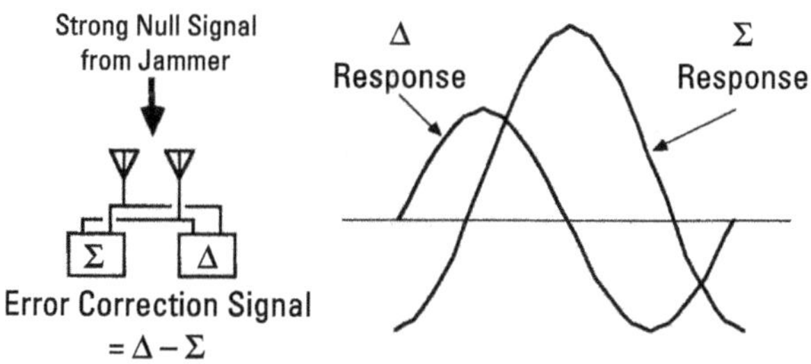

Figure 3.31 The null from the cross-eye jammer makes the sum response less than the difference response, reversing the direction of the mono-pulse tracking response.

When a video camera has been co-bore-sighted with a monopulse radar, it shows the target moving out of the picture at a high rate of speed when cross-eye jamming is applied. This is the result of the monopulse radar being forced rapidly away from its intended target.

The effect of the cross-eye jammer is often described in literature as a distortion of the wavefront of the skin return signal as shown in Figure 3.32.

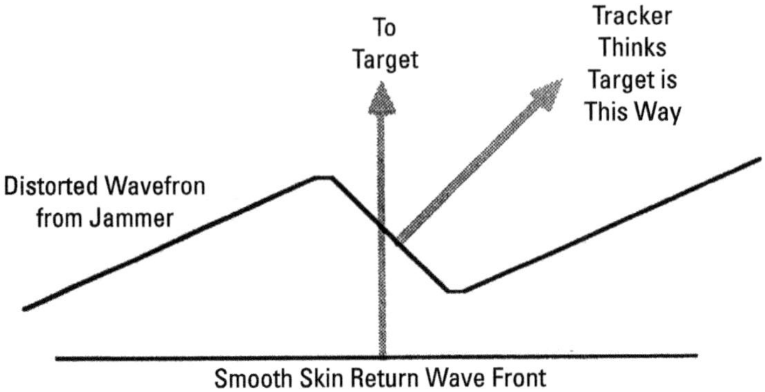

Figure 3.32 Since the phase-shifted and non-phase-shifted signal arrive at the mono-pulse tracking sensors at the same time, they cause a null which forces the tracker away from the target.

References

[1] Schleher, D. C., *Electronic Warfare in the Information Age*, Norwood, MA: Artech House, 1999.

[2] Adamy, D., *EW 101: A First Course in Electronic Warfare*, Norwood, MA: Artech House, 2001.

[3] Adamy, D., *EW 102: A Second Course in Electronic Warfare*, Norwood, MA: Artech House, 2004.

[4] Adamy, D., "EW 101," *Journal of Electronic Defense*, May 1996-April 1997.

4

Next Generation Threat Radars

4.1 Threat Radar Improvements

There has been a great deal of activity during the last decade in the development of new threats. The new threats have been designed to overcome the countermeasures that have been successful against legacy weapons over the years. These new developments include more capable weapons and radars. As stated in Chapter 3, this is not intended as a threat briefing. Classified sources have such information and it is changing constantly. However, this being an unclassified book, that information is not available to us here. Our approach in this chapter is to discuss the technical assets of new threats in general terms. We will cover each class of changes in threats and threat radars in generic terms, and its electronic warfare (EW) impact. The EW part of the discussion will focus on:

- What is no longer practical for EW systems and tactics?
- What new EW tactics are required?
- What new EW system capabilities are required?

Rather than dealing with the classified issues, we will cover the generic changes. If a specific parameter is changed, what is required of EW systems? This chapter has tables and graphs showing the impact of various levels of change in threat parameters. Then, when you are approaching a specific real-world problem, you can look up the parameters of a specific new-generation threat in classified sources and determine the specifications of new EW equipment and tactics required to counter the new real-world threat.

That said, there are important features of these new weapons and radars that are available from open literature, and those features mean that the way we have been conducting EW is no longer adequate.

Clearly stated, there is significant change in the threats that must be countered by EW. We cannot conduct EW operations in the way we have been for the last several decades. It is clear from open literature that:

- Missiles have significantly increased range. This impacts stand-off jamming.
- Threat radars have significant electronic protection (EP) capabilities. This requires new equipment and new tactics.
- New weapons have improved hide, shoot, and scoot capabilities. This reduces reaction times.
- New threat radars have increased effective radiated power (ERP). This improves their jamming-to-signal ratio (J/S) and burn-through range.
- There are significant changes in radar processing. This requires increased complexity in EW processing tasks.
- Many new threats include active electronically steered arrays. This increases EW processing complexity and also impacts the required jamming power.

Another new development is significant improvement in the sensors and guidance in heat-seeking missiles. This requires significant changes in flares and infrared (IR) jammers. Note that this issue and the necessary changes in EW systems and tactics in the infrared (IR) spectrum are covered in Chapter 9.

Radio frequency (RF) spectrum EW tactics are changed in several ways:

- Stand-off jamming has significant challenges.
- Self-protection jamming is impacted by home-on-jam weapons.
- Decoys and other off-board assets have an increasing role.
- ES is impacted by LPI radars.

In this chapter, we will cover electronic protection, the genealogy of weapon and radar updates, new missile capabilities (from open literature), and new threat radar parameters (again from open literature). Then we will look at each anticipated threat upgrade feature and show (with tables and graphs) how a range of parametric values impacts various EW activities.

The flow of this chapter is:

- Electronic protection;
- Surface-to-air missile (SAM) upgrades;
- Acquisition radar upgrades;
- AAA upgrades;
- Required new EW techniques.

4.2 Radar Electronic Protection Techniques

Although electronic protection (EP) is one of the subfields of EW, it is unlike ES or EA in that it does not typically involve specific EW hardware. It is, rather, a number of features of sensor systems that are designed to reduce the effectiveness of enemy jamming. Thus, we say that EP does not protect your platform, but rather protects your sensors. We discussed EP techniques to protect communication systems in Chapter 6. In this chapter, we will cover radar EP.

Table 4.1 lists the principle radar EP techniques and the EA techniques against which they provide protection.

As we discuss each of these techniques, it will be necessary to get into related subjects, such as the way the radar processes data. You will also see that what we are calling EP techniques are sometimes incorporated in radars for other reasons and provide antijamming protection as an additional benefit. As we go through these techniques, you will note that the amount of antijam protection depends on the details of the implementation and that some techniques attack more than one type of jamming.

Table 4.1
Electronic Protection Techniques

Technique	Protect Against
Ultralow side lobes	Radar detection and side-lobe jamming
Side-lobe cancellation	Side-lobe noise jamming
Side-lobe blanking	Side-lobe pulse jamming
Anti-cross-polarization	Cross-polarization jamming
Pulse compression	Decoys and noncoherent jamming
Monopulse radar	Many deceptive jamming techniques
Pulse Doppler radar	Chaff and noncoherent jamming
Leading-edge tracking	Range gate pull-off
Dicke-Fix	AGC jamming
Burn-through modes	All types of jamming
Frequency agility	All types of jamming
PRF jitter	Range gate pull-in and cover pulses
Home-on-jam modes	All types of jamming

4.2.1 Useful Resources

Some useful references are a textbook recommended for those who want to go into the math behind electronic protection techniques [1] and another book that is very helpful in understanding radar operation [2].

4.2.2 Ultralow Side Lobes

Figure 4.1 shows the gain pattern of a typical radar antenna. Note that the angular variation of gain is shown in two views. The top view is a polar plot of gain versus angle. If you go to an antenna manufacturer's Web site and look up the gain pattern for a specific antenna, you will see a family of curves like this. The curves are generated by placing the antenna in an anechoic chamber and rotating it on a turntable. There is a carefully calibrated transmitting antenna in a conical section of the chamber and all of the chamber's surfaces are covered with radio absorptive material. Thus, the antenna on the turntable only receives direct waves from the transmitter. All reflections from the antenna and elsewhere are absorbed at the chamber walls. If the antenna under test is rotated 360° in the horizontal plane, the resulting received power level is proportional to the antenna gain toward the transmitting antenna. The displayed curve of relative received power is then the horizontal antenna pattern. The antenna can then be reoriented 90° on the turntable and rotated to determine the vertical antenna pattern. The Web site may have a whole family of curves over a range of frequencies in various planes around the antenna.

The lower curve in the figure shows the angle from the boresight on the abscissa and the gain on the ordinate. On this curve, the boresight gain and the relative level of the first side lobe are defined. The boresight and side-lobe gains

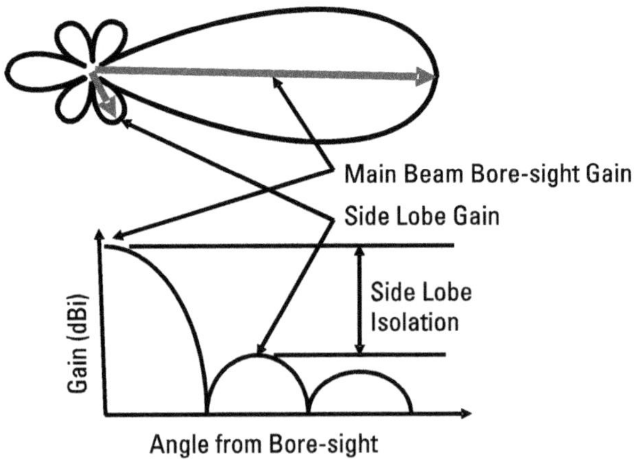

Figure 4.1 Antenna side lobes allow radar detection and jamming from any direction.

are properly stated in dBi (decibels relative to isotropic) and the relative side-lobe level is properly stated in decibels.

The gain pattern is normally defined relative to the main beam boresight gain. The boresight is defined as the direction the antenna is intended to point. This is almost always the direction to which the antenna has its maximum gain, either for transmission or reception.

This gain pattern is a sine(x)/x pattern near the boresight. There is a null at the edge of the main beam and there are side lobes in all other directions. Beyond the first one or two side lobes, side lobes are determined by reflections from structure. There is often a large back lobe. The nulls between the lobes are much narrower than the side lobes, so if we consider the average side-lobe level, we have a reasonable estimate of the radar antenna transmit or receive gain that will be encountered in an EW interaction away from the radar's main lobe.

There is no crisp definition of ultralow side lobes. This merely means that the antenna side lobes are much lower than might be expected from a normal antenna. Schleher [1] has given a range of values that are reasonable, even though some specific antennas may vary from this:

- "Ordinary" side lobes as 13 to 30 dB below the peak main beam (or boresight) gain with average side-lobe peak gain as 0 to –5 dBi;
- "Low" side lobes as 30 to 40 dB below the boresight gain with peak gain of –5 to –20 dBi;
- "Ultralow" side lobes as more than 40 dB below the boresight with less than –20 dBi gain.

4.2.3 EW Impact of Reduced Side-Lobe Level

To detect the presence of a radar that has not yet acquired a target, the receiver (e.g., a radar warning receiver) must have adequate sensitivity (including its antenna gain) to receive the radar side-lobe signal. The receiver sensitivity in this case demands enough received signal power to determine the direction of arrival of the signal, and support analysis of signal parameters to determine the radar type and operating mode. As shown in Figure 4.2, the radar ERP applicable to the side-lobe intercept problem is the transmitter output (sometimes called the tube power) increased by the average side-lobe gain. The signal from the radar diminishes as the square of distance from the radar. Therefore, a reduction of side-lobe gain by10 dB (i.e., 10-dB less effective radiated power in the side lobe direction) reduces the detection range by a factor of the square root of 10 (i.e., 3.16) for any fixed receiver sensitivity level; 20-dB side-lobe isolation would decrease the detection range by a factor of 10. Note that Chapter 5 includes a complete discussion of radio propagation model.

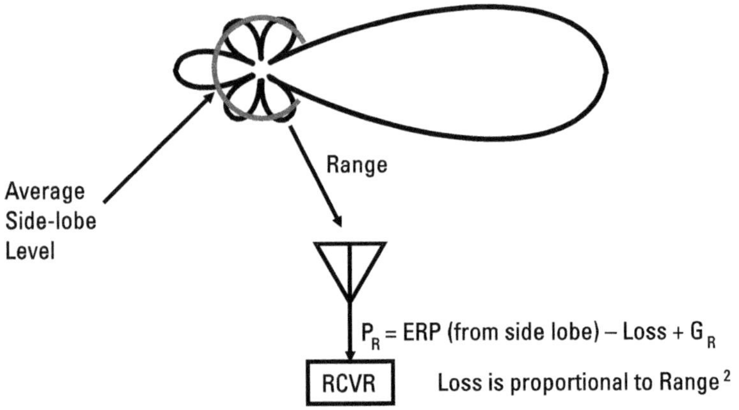

Figure 4.2 Signals received by an intercept receiver away from the antenna main lobe direction are reduced by the radar's average side lobe isolation.

As discussed Section 3.3.3, stand-off jamming is normally performed into a radar's side lobes, because a single jammer, for example, an EA6B aircraft pod will typically jam a number of radars. As shown in Figure 4.3, stand-off jamming-to-signal ratio (J/S) is a function of the relative effective radiated power (ERP) of the jammer and radar, the ratio of the fourth power of range to the target (R_T) to the square of the distance from the jammer to the radar (R_J), and the ratio of the average side-lobe gain (G_S) to boresight gain (G_M) of the radar antenna. Thus, if everything else remains the same, a reduction of side-lobe gain of 10 dB will reduce the range (to the jammer) at which a particular J/S

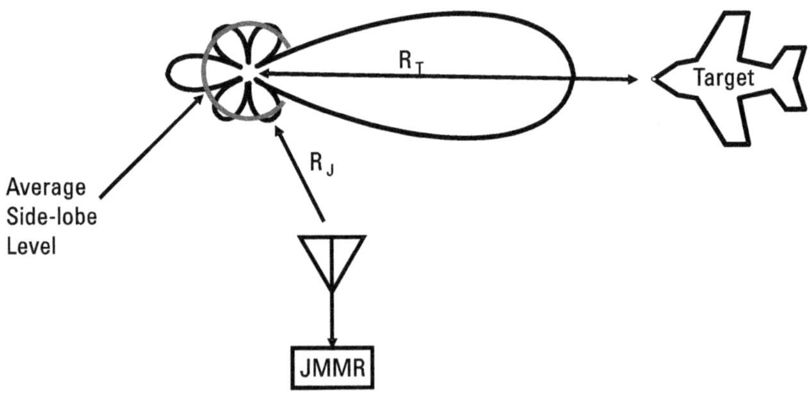

$$\text{J/S is proportional to } \frac{\text{Jammer ERP}}{\text{Radar ERP}} \quad \frac{R_T^4}{R_J^2} \quad \frac{G_S}{G_M}$$

Figure 4.3 The J/S achieved by a side lobe jammer is reduced by the side-lobe isolation of the radar's antenna.

can be achieved by a factor of 3.16. A 20-dB side-lobe isolation would decrease the stand-off jamming range by a factor of 10.

4.2.4 Side-Lobe Cancellation

As shown in Figure 4.4, a side-lobe canceller (SLC) requires an auxiliary antenna which receives signals from the direction of the main radar antenna's important side lobes. These are the side lobes close to the main beam. The auxiliary antenna has greater gain in the side-lobe direction than the side lobes of the main antenna beam. Thus, the radar can determine that the signal arrives from the side-lobe direction and can discriminate against it.

This technique is also called coherent side-lobe cancellation (CSLC) because the (jamming) signal is reduced in the input to the radar's receiver by coherently cancelling it. As shown in Figure 4.5, the jamming signal from the auxiliary antenna is used to generate a copy which is shifted by 180 electrical degrees. The process of making a phase shifted copy of a signal requires some sort of a phase locked loop circuit, and to have high-quality phase control (i.e., very close to 180°), this must have a narrow loop bandwidth. Note that a wide loop bandwidth allows fast response, but a high-quality lock requires a narrow loop and hence has a slower response. The narrow loop requires a continuous signal, for example, a noise-modulated CW signal such as used in a stand-off noise jammer. It is important to understand that the closer the phase-shifted

Figure 4.4 A coherent side lobe canceller removes CW signals which are stronger in the side lobes than in the main beam of the radar antenna.

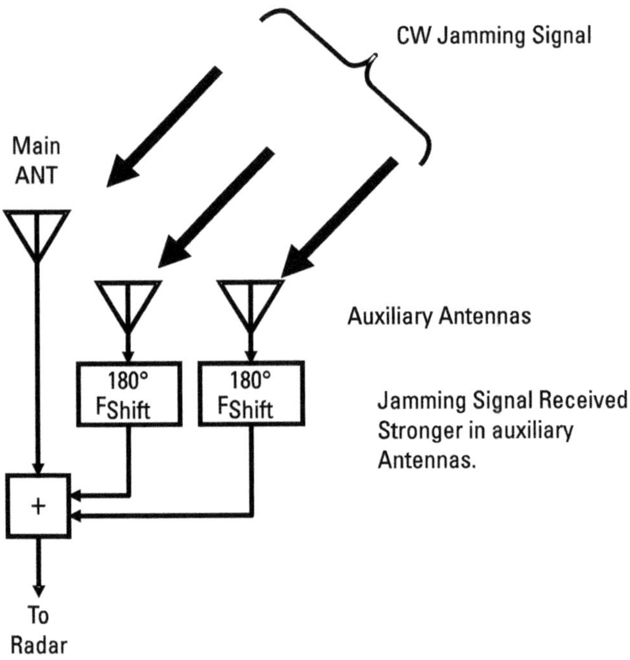

Figure 4.5 Inputs from auxiliary antennas are added to the output of the main antenna 180° out of phase.

signal is to exactly 180° out of phase with the jamming signal, the greater the reduction of the jamming signal into the radar receiver will be.

Each jamming signal that is cancelled requires a separate antenna and phase-shift circuit. Because there are two auxiliary antennas in Figure 4.5, this radar would be able to cancel two CW side-lobe jammers.

It is interesting to note that the Fourier transform of a pulse signal (i.e., the pulse signal viewed in the frequency domain) has a large number of distinct spectral lines as shown in Figure 4.6. The top part of the figure shows a pulse signal in the time domain (as it would be viewed on an oscilloscope) and the bottom part of the figure shows the same signal in the frequency domain (as it would be viewed on a spectrum analyzer). Note that the main lobe of the frequency response is 1/PW wide, where PW is the pulse width in the time domain. Also note that the spectral lines are separated by the pulse repetition frequency (PRF). PRF = 1/PRI where PRI is the pulse repetition interval in the time response. Thus, a single pulse signal broadcast into the side lobes of a radar protected by a side-lobe canceller can capture several coherent side-lobe cancellation circuits, making the CSLC ineffective against noise jamming. That is why it is sometimes appropriate to add pulsed signals to side-lobe jamming noise.

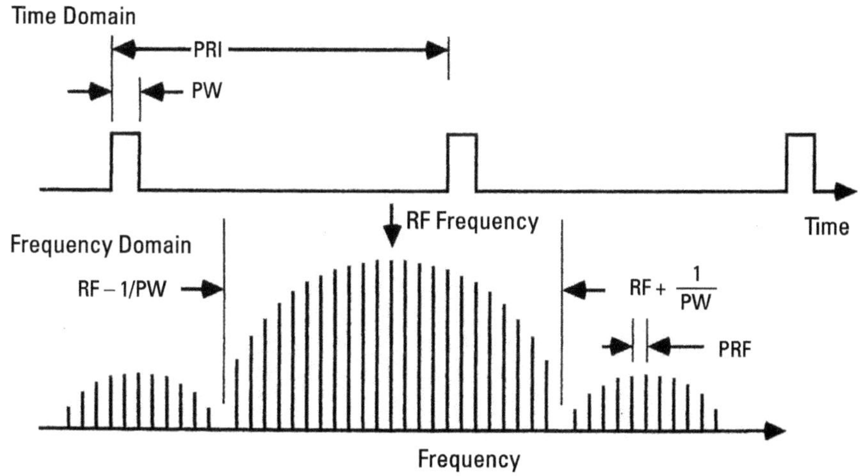

Figure 4.6 A pulse signal has many spectral lines when viewed In the frequency domain.

4.2.5 Side-Lobe Blanking

The side-lobe blanker (SLB) (see Figure 4.7) is similar to the side-lobe canceller in that it uses an auxiliary antenna that covers the angular area of major side lobes as shown in Figure 4.8. The difference is that it is intended to diminish the effect of side-lobe pulse jamming. If a pulsed signal is received in the auxiliary antenna at a higher level than it is received by the main radar antenna, the radar knows it is a side-lobe jamming signal, rather than a skin return from the

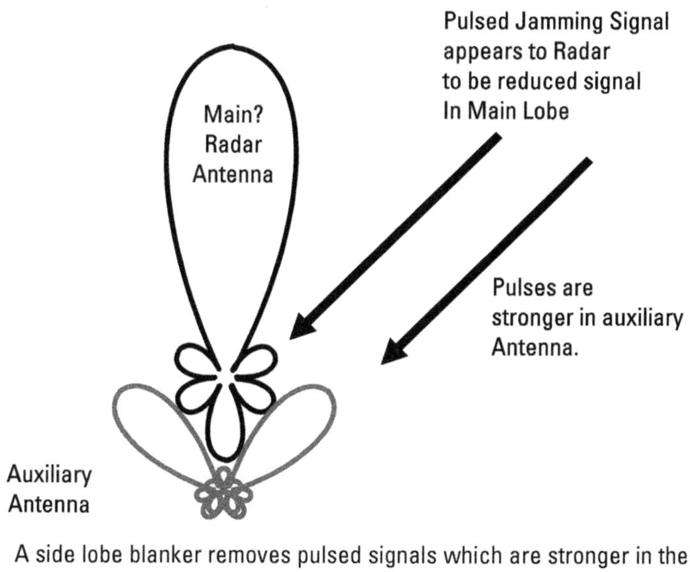

Figure 4.7 A side lobe blanker removes pulsed signals which are stronger in the side lobes than they are in the main beam.

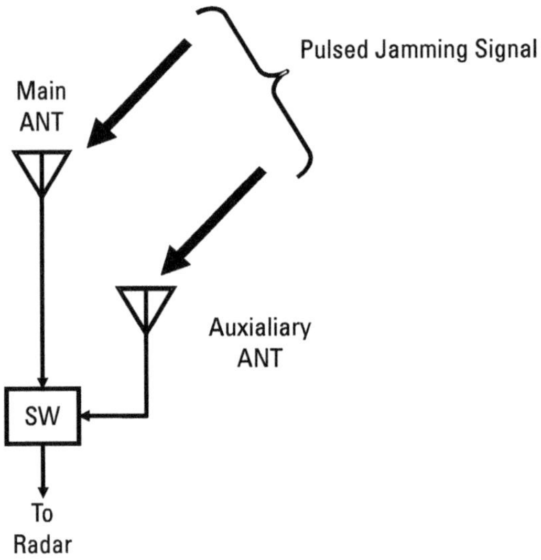

Figure 4.8 The output of the main radar antenna is blanked during a pulse which is stronger in the auxiliary antenna.

radar's transmitted signal. The radar then blanks the input to its receiver during the jamming pulse with the circuit shown in the figure.

This type of EP is also useful in any type of pulse signal receiver, for example, some control links and some types of identification friend foe (IFF) systems receive pulses. These systems can be jammed with false pulses, which would be removed by the SLB.

The problem that this technique gives the radar is that it cannot receive its own return signal during the time that any pulse is present in its side lobes. Thus, a jammer can disable the radar (or data link or IFF) by use of cover pulses, which blank the radar just when it needs to be looking for a return pulse. Because a side-lobe jammer (e.g., a stand-off jammer) is not at the target, it does not know the timing of enemy pulses to microsecond accuracy. Therefore, it cannot place pulses directly over enemy skin return pulses. This will require that side-lobe cover pulses be long enough to include this time uncertainty.

4.2.6 Monopulse Radar

Monopulse radars get direction of arrival information from every skin return pulse. Because this makes some kinds of deceptive jamming ineffective, it can be considered an EP technique. Note that the operation of monopulse radars is covered in Chapter 3.

Jamming techniques such as range gate pull-off or cover pulses provide range deception, but because they generate strong pulses from the direction of the target, they enhance angle tracking by monopulse radars. Angle deception techniques like inverse gain jamming, which generate strong pulses to fool radar tracking algorithms, likewise enhance monopulse angle tracking. These jamming techniques are discussed in Chapter 3.

In general, angle deception is more powerful than range deception. A radar can typically reacquire in range in milliseconds, while a significant pull-off in angle will require a return to the radar's acquisition mode. This may cause an angle reacquisition time of seconds.

A chaff cloud or a decoy, because it creates an actual, trackable object, works well against monopulse radars.

Monopulse radars point their antennas toward targets by adjusting in angle to balance the power received by multiple antenna feeds as in Figure 4.9. Effective angle jamming forces the radar to move its antenna in an improper direction in response to jamming signals, which distort the balance of the antenna feeds. For example, cross-polarization jamming causes a radar to point one of its cross-polarized Condon lobes at the target.

4.2.7 Cross-Polarization Jamming

Cross-polarization jamming was covered in Section 3.4.14, but to better understand cross-polarized Condon lobes, try this. Hold a pencil in your hand oriented 45° to the right and move your hand toward a wall at a 45° angle until the pencil touches the wall. Then move your hand in the direction that the pencil would move if it were reflected from the wall. You will notice that the pencil is now oriented 45° left in the direction of travel. The forward geometry of the wall and the diagonal angle of the pencil have caused the angle of the pencil relative to the forward motion of your hand to change 90°.

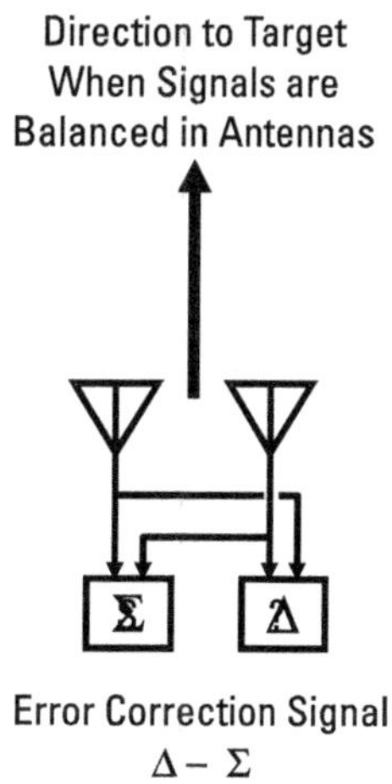

Figure 4.9 A monopulse radar has multiple antenna feeds and generates antenna pointing corrections from the difference of the two received signals normalized to the sum.

Now consider the vertically polarized signal arriving in the upper right portion of the parabolic dish reflector in Figure 4.10. The forward geometry of the dish causes a (weak) horizontally polarized reflection toward the antenna feed because this part of the dish is about 45° to the signal polarization. This effect causes each Condon lobe.

In his excellent but very technical (and now out of print) set of three books on applied ECM, Leroy Van Brunt provided detailed discussions of cross-polarization jamming [3]. He pointed out that cross-polarization jamming can be used with either on-frequency or noise jamming and is effective against both acquisition and tracking radars, including the two beam SA-2 track-while-scan radars in which the beams are cross-polarized to each other.

In addition to the two path repeater type cross-polarization jammer described in Chapter 3, there are jammers that sense the polarization of arriving radar signals and create a cross-polarized response with a signal generator as shown in Figure 4.11.

If a two-channel repeater cross-polarization jammer cannot achieve adequate antenna isolation, Van Brunt pointed out that time gating can be used to isolate the two cross-polarized signals from each other. The timing that he suggested in his text predates the availability of modern, extremely fast switches like those presented in the discussion of cross-eye jamming in Section 3.4.15. The time gated cross-polarization technique should work even better with today's technology.

4.2.8 Anti-Cross-Polarization

Radars that include features to reduce their sensitivity to cross-polarized signals or to reduce their Condon lobes are said to have anti-cross-polarization EP. As

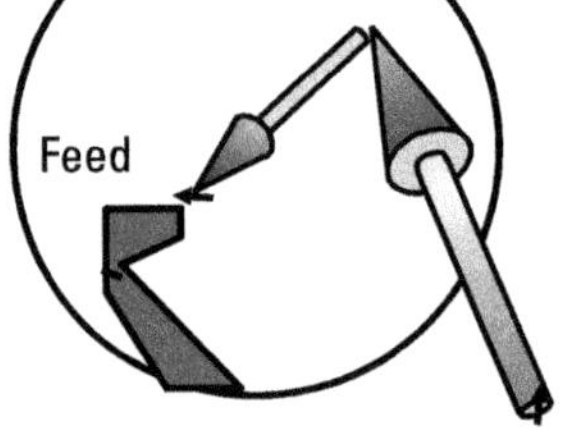

Figure 4.10 The forward geometry at the edges of a parabolic dish reflector cause off axis signals to change polarization by 90° when reflected into the antenna feed.

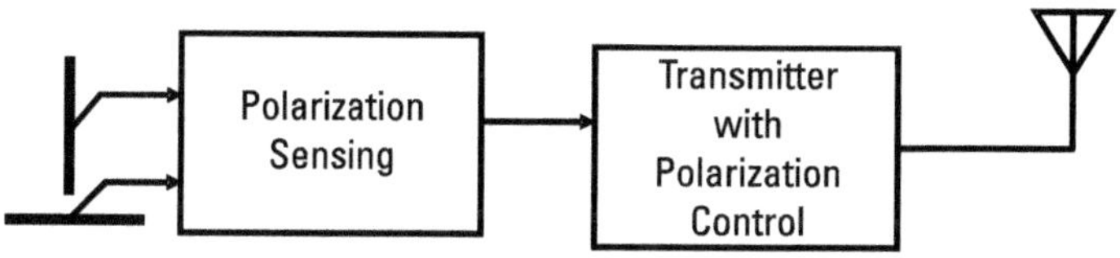

Figure 4.11 One technique for creating a cross polarized jamming signal involves sensing the polarization and generating a return signal with the proper polarization.

shown in Figure 4.12, a radar with cross-polarization isolation has very small Condon lobes. A radar antenna reflector that is a small part of a large parabolic surface will have its feed far from the reflector relative to the reflector diameter and the reflector will have little forward geometry (hence, low Condon lobes). If the reflector is a larger part of a smaller parabolic surface, its feed will be relatively close to the reflector and the reflector will have more forward geometry and hence higher Condon lobes. If the radar antenna is a flat phased array, it will typically have almost nonexistent Condon lobes because it has no forward geometry to create the cross-polarized response. However, if there is differential gain in its array antenna elements for beam-shaping, it can have Condon lobes. The antenna geometry impact on Condon lobes is illustrated in Figure 4.13.

Another way to implement anti-cross-polarization EP is with a polarization filter across the throat or feed of the antenna or across the phased array.

4.2.8.1 Polarization Canceller

This related EP technique is also described in Van Brunt's series [3]. It involves use of two orthogonally polarized auxiliary antennas, and can be very effective against a single circularly or diagonally polarized jammer. Its circuitry discriminates against the component of the jamming signal that is not copolarized

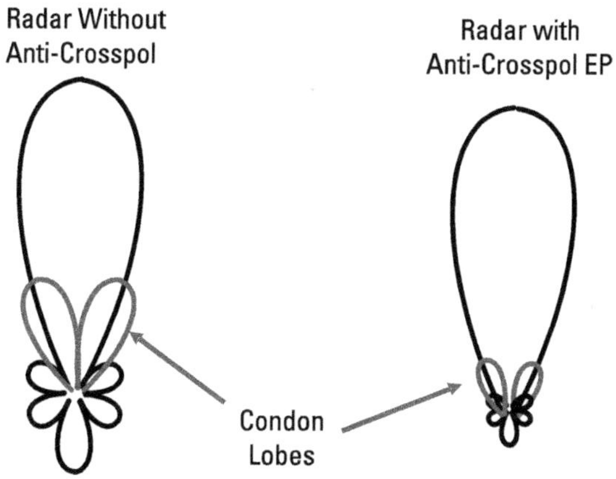

Figure 4.12 A radar with anti-crosspol EP has greatly reduced Condon lobes.

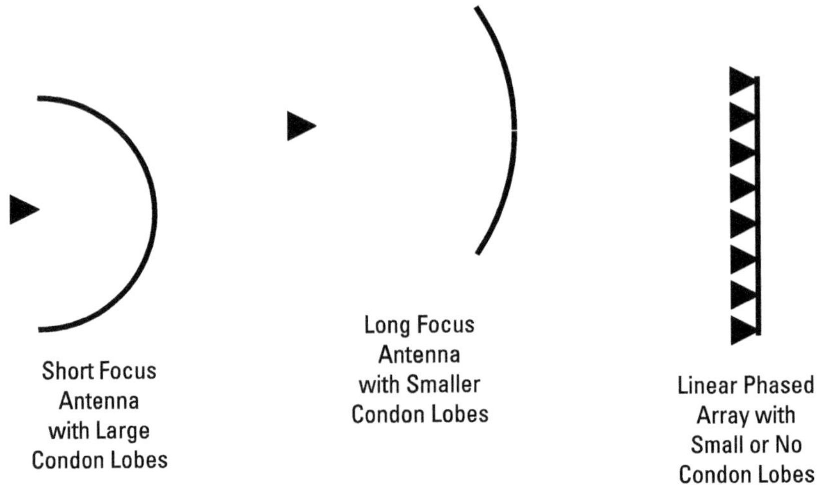

Figure 4.13 The geometry of a radar's antenna impacts the strength of its Condon lobes.

with the radar but passes the radar's skin return signal. Van Brunt noted that dual cross-polarized jamming channels as described above will defeat this EP technique.

4.2.9 Chirped Radar

The purpose of pulse compression is to reduce the range resolution distance for radars, but it also has the effect of reducing the effectiveness of jammers unless they mimic the pulse compression techniques of the target radars.

One of the types of pulse compression, linear frequency modulation on pulse (LFMOP) is also called chirp. A chirped radar has a linear frequency modulation across each pulse. It is called chirped because it sounds like a bird's chirp when received by some receivers. Figure 4.14 shows the block diagram of a chirped radar. These are normally thought of as long-range acquisition radars, with long pulses to provide the necessary signal energy. However, LFMOP can also be used in shorter-range tracking radars. Note that the return pulse into the radar receiver is passed through a compressive filter. The filter has a delay that varies with frequency. The filter slope matches the FM on the pulse (i.e., the frequency variation versus time curve is the same as the delay versus frequency curve). This has the effect of delaying each part of the pulse to the end of the pulse. Thus, after processing, the long pulse is collapsed into a much shorter pulse.

A radar's resolution cell is the region in which the radar cannot distinguish multiple targets. Figure 4.15 shows the resolution cell in two dimensions; actually it is a three-dimensional volume rather like a huge wash tub.

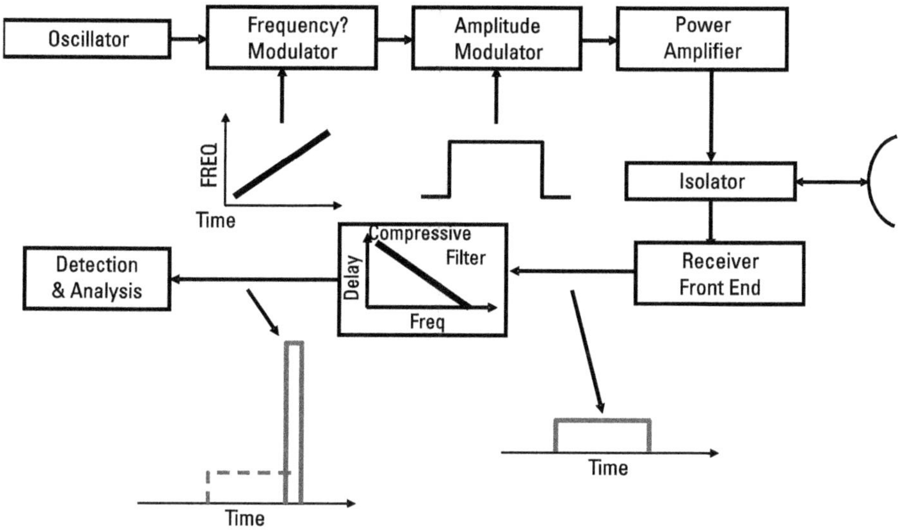

Figure 4.14 A chirped pulse has a linear frequency modulation on its pulse, which allows the received pulse to be shortened in receiver processing.

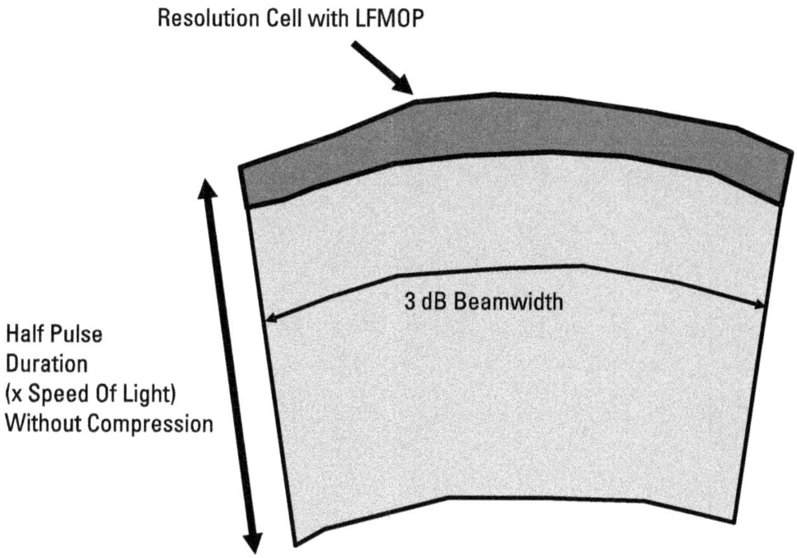

Figure 4.15 The radar's resolution cell is determined by the antenna beamwidth and the pulse duration with LFMOP, the effective pulse duration is significantly reduced.

As shown in the figure, the cross-range dimension of the cell is determined by the 3-dB beamwidth of the radar's antenna. The range resolution limitation is determined by the radar's pulse duration (one-sixth of a meter per nanosecond of pulse duration). A long pulse, while it has more energy, causes poor range resolution. The darker band at the top of the resolution cell in Figure 4.15 shows the reduced range uncertainty caused by LFMOP. Because the effective pulse is shorter after passing through the compressive filter, the range resolution is improved.

The amount of range compression is the ratio of the frequency modulation range to the inverse of the pulse width. Thus, a 10-μs pulse with 2 MHz of frequency modulation range would have its range resolution improved by a factor of 20.

The impact on jamming is shown in Figure 4.16. The black pulse is the radar signal with LFMOP; it is compressed by the compressive filter as shown at the right of the figure. The gray pulse is a jamming pulse without LFMOP. As shown in gray at the right of the figure, its energy does not build up at the end of the pulse. The radar processing is focused only on the time period that the compressed pulse is present, so the energy of the noncompressed jamming pulse is significantly below that of the compressed pulse. This has the effect of reducing whatever J/S that would otherwise be created. The J/S reduction is equal to the pulse compression factor. In the example above, this would be a 13-dB reduction of J/S.

If a jammer places the appropriate LFMOP on its jamming signal, this EP feature of the radar will be countered. A matching LFMOP can be created by a jammer using direct digital synthesis (DDS) or a digital RF memory (DRFM). Both of these technologies will be discussed in Chapter 8.

4.2.10 Barker Code

The block diagram of a radar with Barker code pulse compression is shown in Figure 4.17. A binary phase shift keyed (BPSK) modulation is placed on each

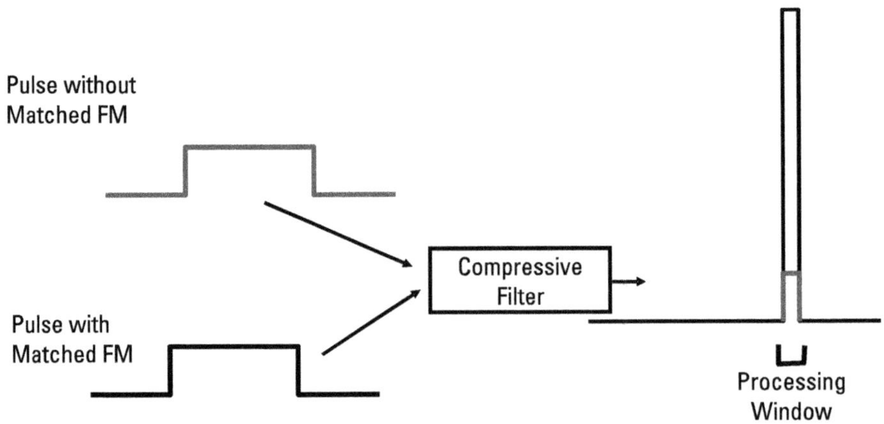

Figure 4.16 Unless jamming has the correct frequency slope, the effective J/S is reduced by the compression factor.

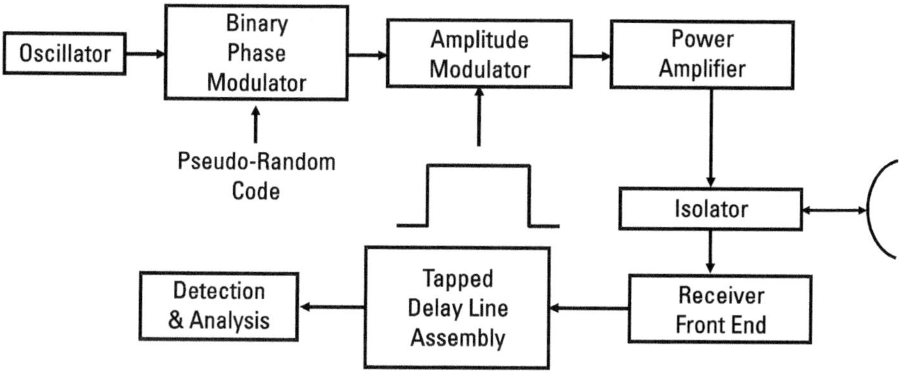

Figure 4.17 A binary frequency shift keyed code is modulated onto each pulse; a tapped delay line in the receiver reduces the effective pulse width, improving range resolution.

of a radar's pulses, and pulse compression is achieved by passing the returned pulses through a tapped delay line. The top of Figure 4.18 shows an example maximal length code with 7 bits. Radars typically use much longer codes. This code is 1110010, where the 0 bits are shifted 180° relative to the signal phase during the 1 bits. As the pulse passes through the tapped delay line, the sum of

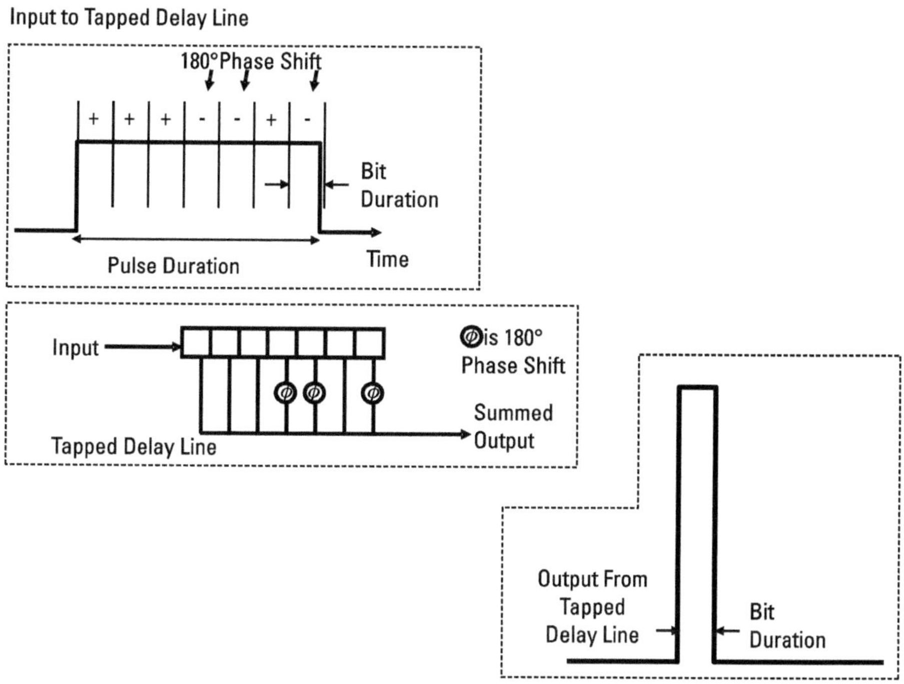

Figure 4.18 The coded pulse produces a large output from the delay line when all of its bits align to the taps.

the signals on all of the taps add to 0 or −1, except when the pulse exactly fills the shift register. Note that the fourth, fifth, and seventh taps have 180° phase shifts, so an exactly aligned pulse will cause all of the taps to add constructively. This causes a large output for the time of one bit duration. Therefore, the pulse duration after the tapped delay line is effectively 1 bit long. This compresses the pulse (and improves the range resolution) by the number of bits of the code placed on each pulse.

For example, if there were 31 bits in the code during each pulse, the range resolution would be improved by a factor of 31.

Now consider Figure 4.19. The black pulse is the radar signal with the proper binary code to match the tapped delay line; it is compressed by the delay line as shown in black at the right of the figure. The gray pulse is a jamming pulse without a code. As shown in gray at the right of the figure, its energy is not collapsed into the one bit duration output. Like LFMOP, digital code compression reduces the J/S that would otherwise have been achieved. The J/S reduction factor is the same as the compression factor. In the 31-bit code example above, this would cause 15-dB reduction in the effective J/S.

If a jammer places the appropriate binary code on its jamming signal (by use of a DRFM), this EP feature of the radar will be countered.

4.2.11 Range Gate Pull-Off

Recall from Chapter 3 that range gate pull-off (RGPO) deceptive jamming generates a false return pulse which is increasingly delayed (with each subsequent pulse) to convince the radar that the target is turning away from the radar, thus causing the radar to lose range track. RGPO does this by loading up the radar's

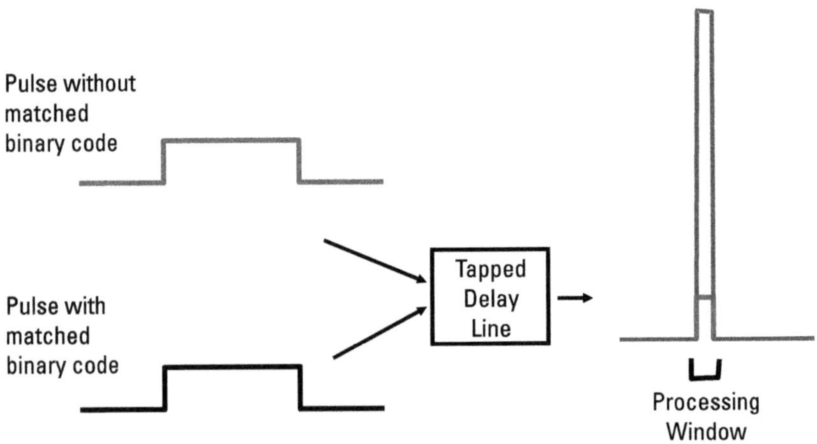

Figure 4.19 Unless jamming has the correct binary code, the effective J/S is reduced by the compression factor.

late gate with the larger energy of the jamming pulse. An EP technique used to defeat RGPO is leading-edge tracking. As shown in Figure 4.20, the radar tracks the target's range from the energy in the leading edge of the skin return. Assuming that there is some throughput latency in the RGPO jammer, the leading edge of the jamming pulse starts later than the leading edge of the true skin return. Schleher [1] has placed the value of about 50 ns on the maximum jamming process latency that would allow the RGPO jammer to capture the range tracking. Assuming more jammer latency than this, the radar processing will not see the jamming pulse and thus continues to track the true target range from the true skin return pulse.

The jamming technique used to overcome leading edge tracking is range gate pull-in (RGPI), also called inbound range gate pull-off. As shown in Figure 4.21, the jammer generates a false pulse that moves ahead in time, anticipating each pulse by an increasing amount. The false pulses move back through the true skin return pulse, capturing the radar's range tracking (even if the radar is tracking leading edges), and thus convinces the radar that the target is turning toward the radar. This causes the radar to lose range track. To perform RGPI, the jammer must have a PRI tracker that allows it to anticipate when the next pulse will occur. The radar EP effective against the RGPI technique is the use

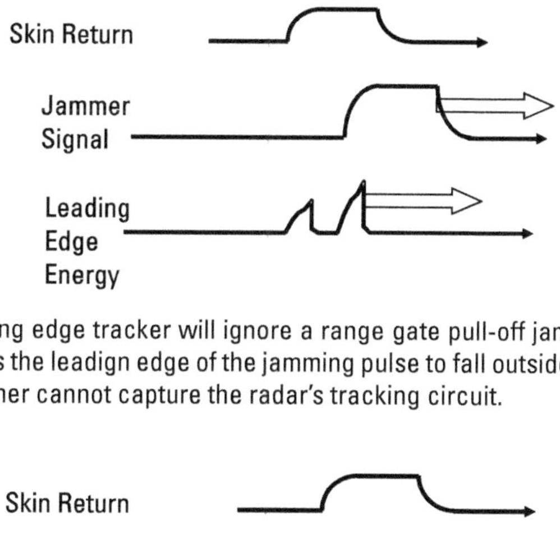

Figure 4.20 A leading edge tracker will ignore a range gate pull-off jamming signal latency in the jammer causes the leadign edge of the jamming pulse to fall outside of the leading edge late gate so the jammer cannot capture the radar's tracking circuit.

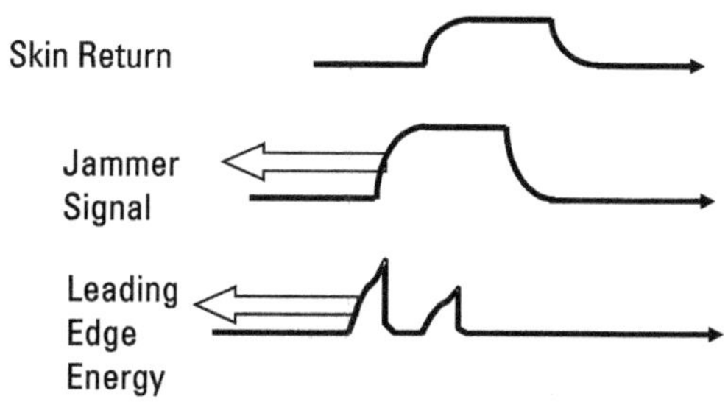

Figure 4.21 Range gate pull-in jamming generates a pulse that moves ahead of the actual skin return pulse, thereby capturing the leading edge tracking circuit.

of jittered pulses. With jittered pulses, the pulse to pulse spacing is a random function, so the jammer cannot anticipate the timing of the next pulse and therefore cannot generate a false pulse that anticipates the pulse in a smoothly increasing way.

4.2.12 AGC Jamming

In Chapter 3, we discussed automatic gain control (AGC) jamming in which a strong, narrow jamming pulse is generated at about the scanning rate of the target radar. The narrow jamming pulse captures the radar's AGC, causing the radar to turn down its gain to the degree that it cannot see the amplitude variations in the skin return from the radar antenna scan (see Figure 4.22). Thus, the radar cannot perform its angle tracking function. Because the jamming pulse has a low duty cycle, this technique allows effective jamming with minimal jammer energy. The EP against this AGC jamming technique is the Dicke fix as shown in Figure 4.23.

The Dicke fix involves a wideband channel with a limiter followed by a narrow channel with bandwidth matched to the radar's pulse. Because the narrow jamming pulse has wide bandwidth, it is clipped in the wideband channel. The radar's necessary AGC function is performed in the narrow channel and can thus not be captured by the previously limited narrow pulses.

4.2.13 Noise-Jamming Quality

The effectiveness of noise jamming is strongly impacted by the quality of the noise. Ideally jamming noise should be white Gaussian. Thus, the distortion

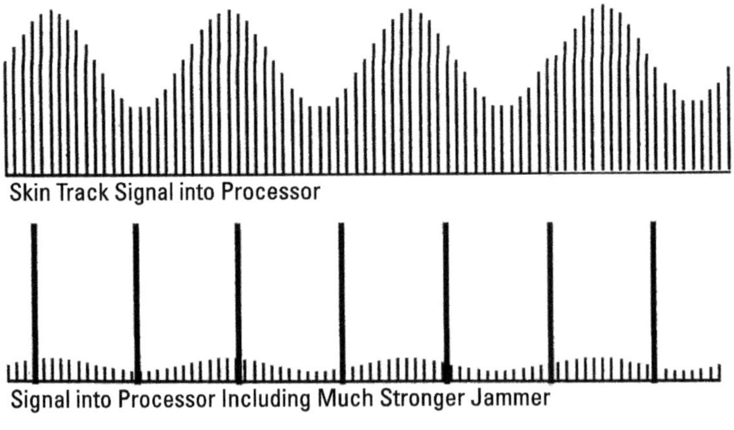

Figure 4.22 By transmitting strong, narrow pulses at about the radar antenna scanning rate, the AGC jammer captures the radar's AGC, reducing the amplitude variations from antenna scan to an unusable level.

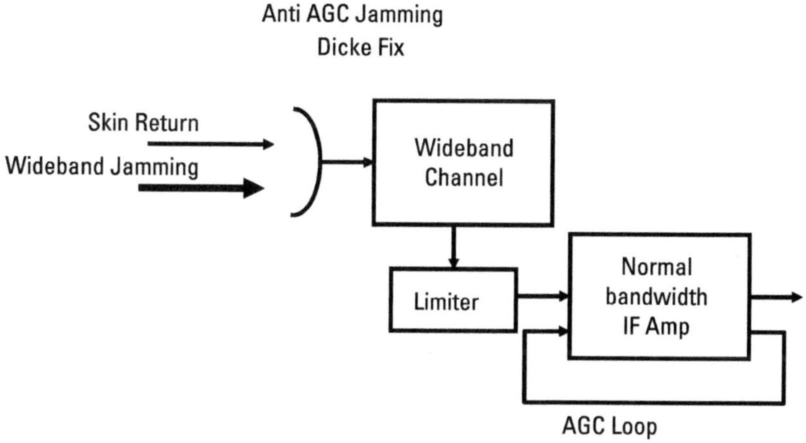

Figure 4.23 The Dicke fix feature in a radar limits the output of a wideband channel to reduce wideband signals before input to a narrow channel to protect the AGC function from strong wideband jamming.

from clipping in a saturated jammer amplifier can reduce the J/S in the target radar receiver by many decibels. One very efficient way to generate high-quality jamming noise is shown in Figure 4.24. A CW signal is frequency modulated by a Gaussian signal across a frequency band much wider than the radar receiver's bandwidth. Each time the jamming signal passes through the radar receiver's band, an impulse is generated. This series of randomly timed impulses causes high quality white Gaussian noise in the receiver.

Impulses, by their nature, are very wideband. Thus, the limiting in the wideband channel of the Dicke fix reduces the J/S in the narrowband channel. This is an effective EP against this noise-jamming technique.

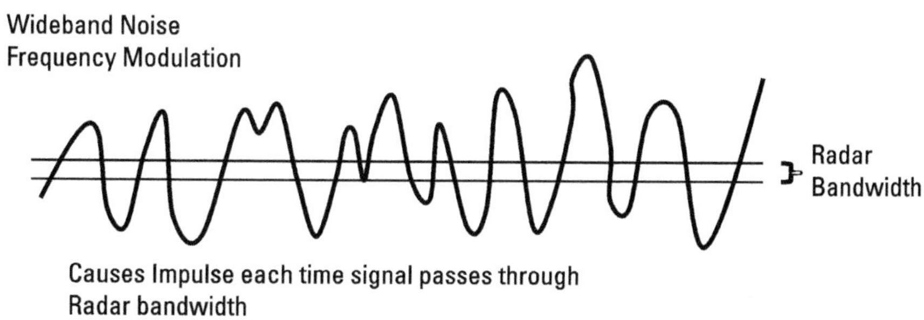

Figure 4.24 Wideband FM noise modulation causes ideal noise jamming in a radar's receiver by creating an impulse each time it passes through the radar bandwidth. The Dicke fix reduces the effectiveness of this jamming.

4.2.14 Electronic Protection Features of Pulse Doppler Radars

A pulse Doppler (PD) radar has inherent Electronic Protection (EP) features, including:

- It expects its return in a narrow frequency range, so it can discriminate against noncoherent jamming.
- It can see spurious outputs from jammers.
- It can see frequency spreading from chaff.
- It can see separating targets.
- It can correlate range rate and Doppler shift.

4.2.15 Configuration of Pulse Doppler Radar

Pulse Doppler (PD) radars are coherent, because each pulse is a sample of the same RF signal as shown in Figure 4.25. Thus, both the time of arrival and Doppler shift of received signals can be measured. The time of arrival allows determination of range to the target and the Doppler shift is caused by the radial velocity of the target relative to the radar. As will be discussed later, there are some significant ambiguity issues which must be overcome by PD radar processing.

The processor in a PD radar forms a matrix of range versus velocity as shown in Figure 4.26. The range cells show the time of arrival of received pulses relative to the transmitted pulse, and each cell is one range resolution deep. The time resolution (or the depth of a range cell) is half of the pulse width. This gives the PD radar a range resolution of:

$$\text{Range cell depth} = \left(\text{pulse width}/2\right) \times \text{Speed of light}$$

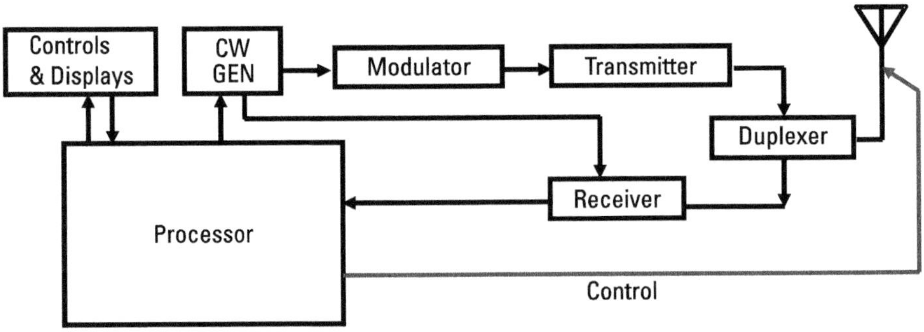

Figure 4.25 A pulse-Doppler radar is coherent and uses complex processing to deal with ambiguities.

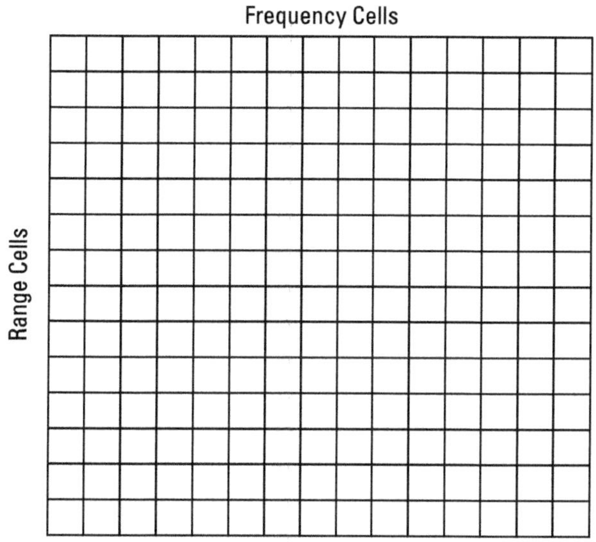

Figure 4.26 Pulse-Doppler radar processing allows generation of a range vs. return frequency matrix.

These range cells are contiguous during the whole time between pulses.

The velocity cells are fed by a bank of channelized filters or channelization by fast Fourier transform processing. The width of the velocity (i.e., Doppler frequency) channels is the bandwidth of each filter. The inverse of the filter bandwidth is the coherent processing interval (CPI), which is the time over which the radar processes the signal. Note that in a search radar, the CPI can be as long as the time the radar's antenna is illuminating the target. Thus, the frequency channels can be very narrow. For example, if the radar beam illuminates the target for 20 ms, the filters could be 50 Hz wide.

The number of pulses that are integrated by the radar determines its processing gain (above the noise level). The processing gain is:

$$\text{Processing Gain}(\text{in dB})\text{is } 10\log(CPI \times PRF) or 10\log(\text{PRF/filter BW})$$

4.2.16 Separating Targets

Consider the use of RGPO deceptive jamming (discussed in Chapter 3). Figure 4.27 shows the true return pulse and the false pulse generated by the jammer. In a conventional radar, the processor has an early and a late gate (rather than the contiguous range cells of the PD radar). The jamming pulse captures the range tracking of the radar because it has positive J/S. By delaying each subsequent jamming pulse, the jammer loads up the energy in the late gate, making the

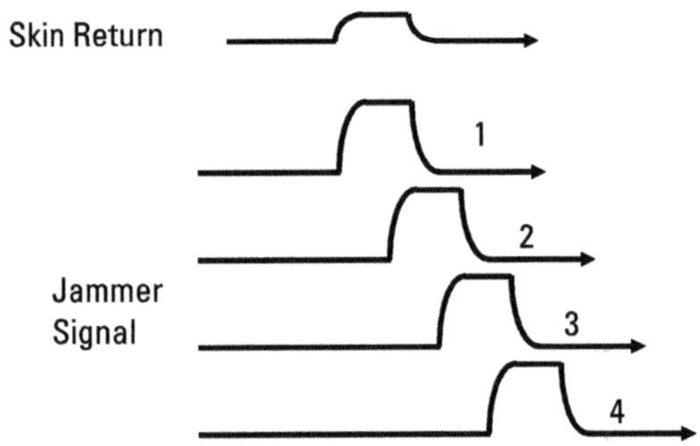

Figure 4.27 Range gate pull off involves sequential delay of the return pulse, which loads up the radar's late gate.

radar think the target is moving away. However, a PD radar can see both return pulses (i.e., separating targets). Each of the pulses is placed in the time versus velocity matrix as shown in Figure 4.28.

The true target return signal will move through a series of range cells with increasing range value. This increasing range indicates a radial velocity. The target return pulses will fall into the velocity cell corresponding to the Doppler shift caused by the true target range rate. However, the jammer pulses are increasing in apparent range because the jammer is delaying the returns. The Doppler frequency cell that holds each jamming pulse will be determined by the actual radial velocity of the jammer. Thus, the jammer pulses will fall into

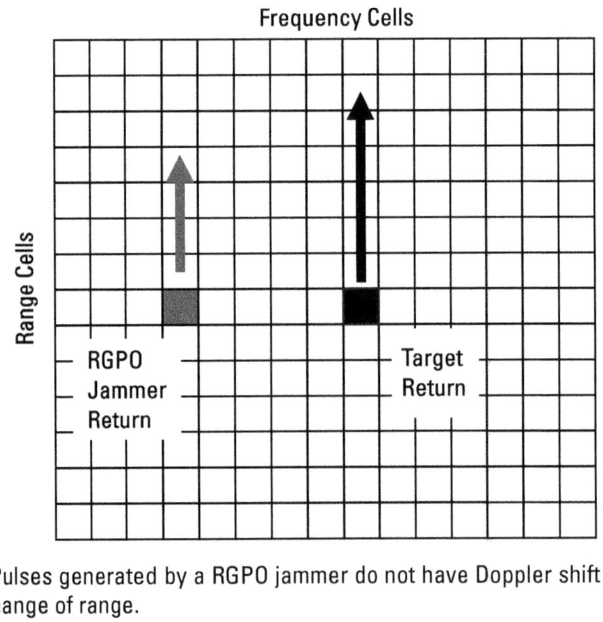

Figure 4.28 Pulses generated by a RGPO jammer do not have Doppler shift consistent with their rate of change of range.

velocity cells that do not correspond to the range rate that can be calculated from the changing range indicated in the range cells. This allows the PD radar to select the pulses for which the change-rate of range corresponds to the observed Doppler frequency. Hence, it will continue to track the target, defeating RGPO jamming.

The above discussion is simplified. Understand that in a dynamic engagement, the target range will most likely be changing, but the time history of the range cells occupied will indicate a radial velocity which agrees with the velocity value indicated by the Doppler filter that contains the return signals. For the jamming signal, the calculated and indicated range rates will be different.

Note that this also allows the PD radar to discriminate against RGPI jamming.

To overcome this advantage of the PD radar, the jammer must also apply velocity gate pull-off (VGPO) as explained in Chapter 3. The frequency offset must be coordinated with the rate of range gate pull off to fool the PD radar.

4.2.17 Coherent Jamming

As shown in Figure 4.29, the coherent return from a target will fall within a single Doppler cell. A wideband jamming signal (for example, barrage or noncoherent spot noise) will occupy multiple frequency cells, so the radar can discriminate in favor of the coherent target return. This means that a jammer, if it is to deceive a PD radar, must generate a coherent jamming signal.

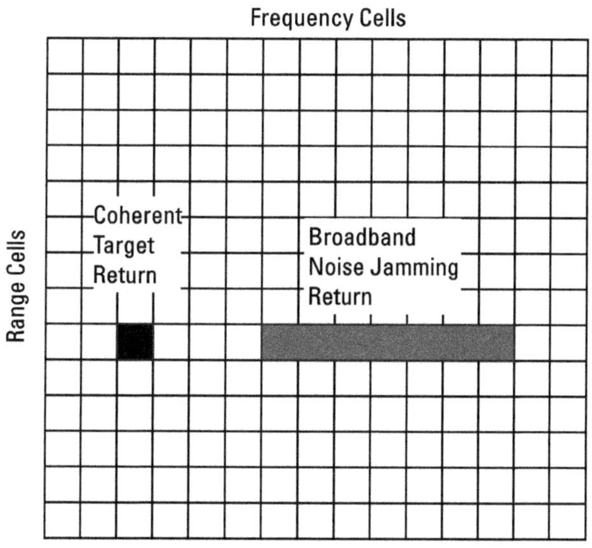

Figure 4.29 Coherent PD radars observe the target return in a single frequency cell, while broadband noise jamming occupies many frequency cells.

Note that the scintillation caused by a chaff cloud also spreads the radar signal. The PD radar can detect this frequency spreading and thus discriminate against the chaff return.

4.2.18 Ambiguities in PD Radars

The maximum unambiguous range of a radar is the distance for which a transmitted pulse can make a round trip at the speed of light before the next pulse is transmitted (see Figure 4.30).

$$R_U = (\mathrm{PRI}/2) \times c$$

where R_U is the unambiguous range in meters, PRI is pulse repetition interval in seconds, and c is speed of light (3×10^8 m/s).

For example, if the PRI is 100 μs, the unambiguous range is 15 km. The higher the pulse repetition frequency (PRF), the shorter the PRI and hence the shorter the unambiguous range. If the PRF is quite high, there will be many range ambiguities.

The Doppler shifted frequency of the return signal falls into a Doppler filter in the PD radar's processor.

The maximum Doppler frequency shift is:

$$\Delta F = (v_R/c) \times 2F$$

where ΔF is the Doppler shift in kilohertz, v_R is the rate of change of range in meters per second, and F is the radar operating frequency in kilohertz.

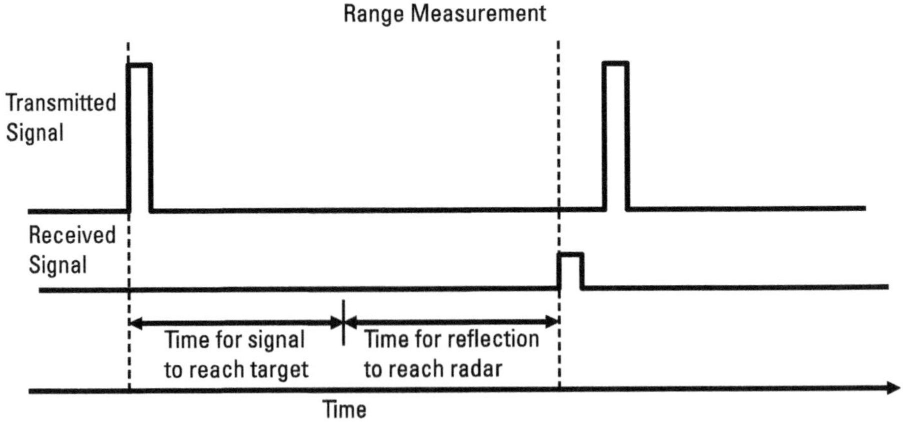

Figure 4.30 The maximum unambiguous range is the range at which the radar pulse can make a round trip to the target at the speed of light before another pulse is transmitted.

For example, if a 10-GHz radar were designed to handle an engagement with a maximum range rate of 500 m/s (a little over mach 1.5):

$$\Delta F = \left(500\,\text{m/s}/3\times10^{8}\,\text{m/s}\right)\times2\times10^{7}\,\text{kHz} = 33.3\ \text{kHz}$$

The spectrum of a pulsed signal has spectral lines spaced at frequency increments equal to the PRF as shown in Figure 4.31. If the PRF is low, for example, 1,000 pps, the spectral lines are only 1 kHz apart. If the PRF is high, for example, 300 kpps, the spectral lines are 300 kHz apart. Each of these lines will also be Doppler shifted and will cause frequency responses in the processing matrix (i.e., frequency ambiguities) if they are less than the maximum Doppler frequency shift for the design engagement. The lower the PRF, the greater the frequency ambiguity. A PRF of 1,000 pps will have many ambiguous responses less than 33.3 kHz, while a PRF of 300 kpps will be totally unambiguous within the frequency range of the processing matrix.

As shown in Figure 4.32, the range is ambiguous if the PRI is less than the round-trip time to the maximum target range of the processing matrix and the frequency is ambiguous if the PRF is less than the maximum Doppler shift in the matrix (i.e., the frequency of the highest Doppler filter).

4.2.19 Low, High, and Medium PRF PD Radar

There are three types of pulse Doppler radars, differentiated by PRF. These are illustrated in Figure 4.33.

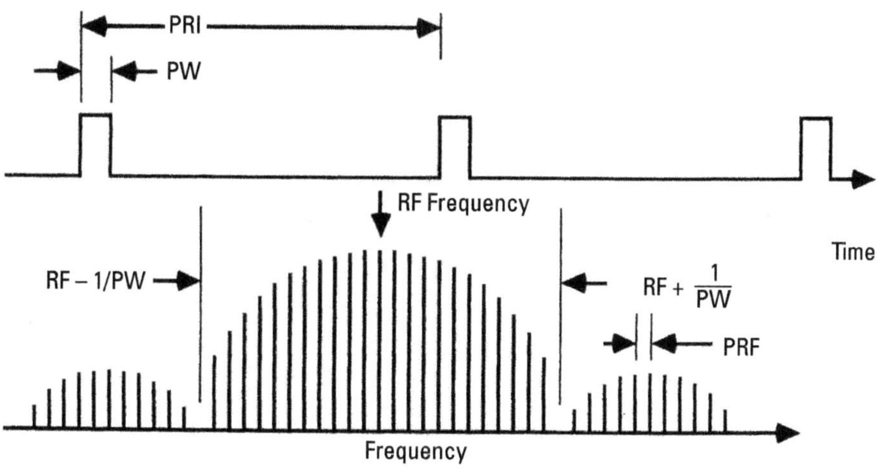

Figure 4.31 In the frequency domain, a pulse signal has spectral lines separated by a frequency equal to the PRF.

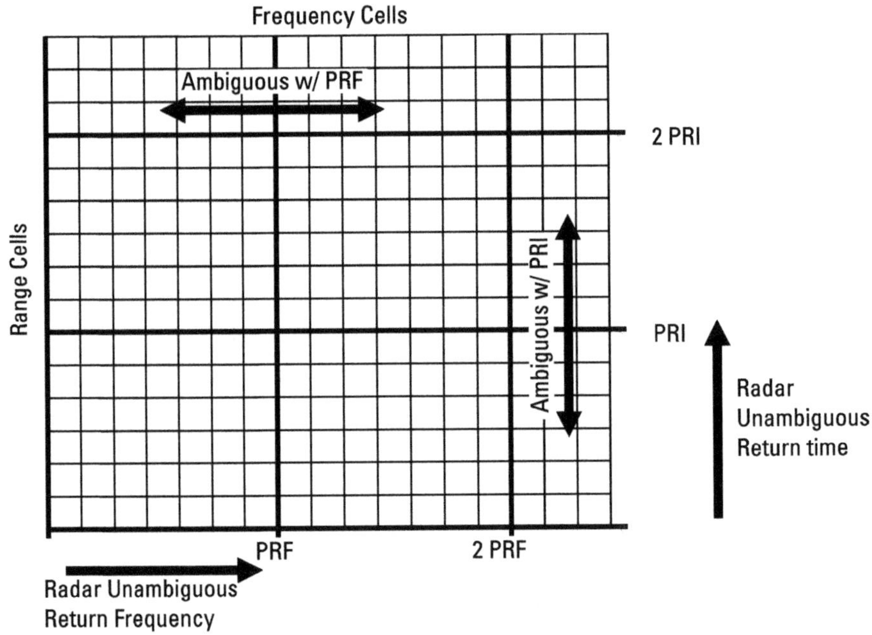

Figure 4.32 The PD radar can be ambiguous in range as a function of its pulse repetition interval and in frequency as a function of its pulse repetition rate.

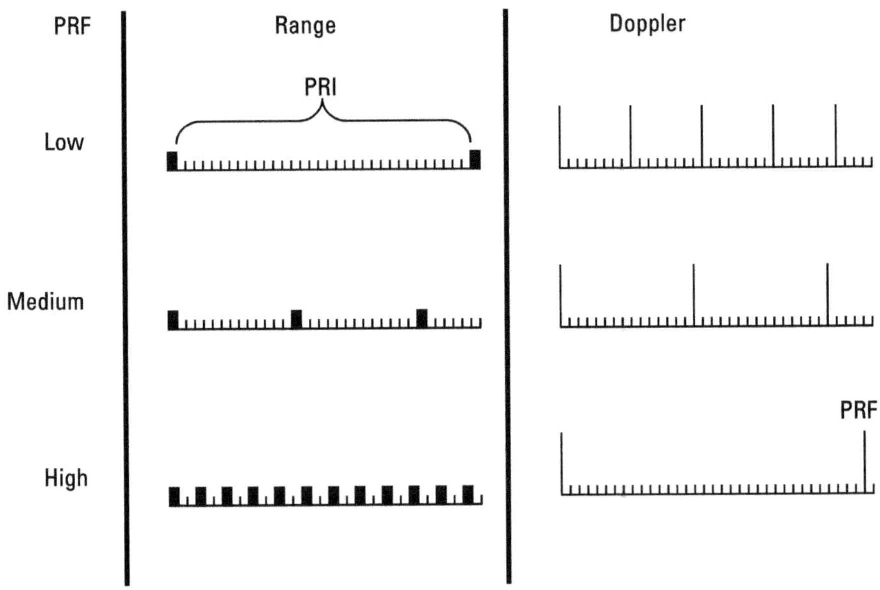

Figure 4.33 Range and frequency cells in low, medium and high PRF Doppler radars.

Low PRF radar is unambiguous in range to a significant target range because of its large PRI; thus, it is very useful for target acquisition. However, its low PRF creates a highly ambiguous Doppler frequency determination. This means that the target radial velocity determination is ambiguous, limiting the radar's ability to make useful range rate/velocity correlation determinations, making it vulnerable to RGPO and RGPI jamming.

High PRF radar is unambiguous in Doppler frequency out to quite high range rates, making it ideal for use in a high speed head-on engagement with a target. Large Doppler frequencies are highly desirable because the target returns are far away from ground returns and internal noise interference. However, the high PRF causes a low PRI, so the high PRF pulse Doppler radar is highly ambiguous in range. This radar may be used in a velocity only mode, or range can be determined by imposing a frequency modulation on the signal as shown in Figure 4.34. Note that a tail chase engagement is characterized by low range rate, so Doppler frequency shifts are much lower than for head-on engagements. This makes the high PRF PD radar less advantageous.

Medium PRF radar is ambiguous in both range and velocity. It was developed to enhance tail chase engagements. The medium PRF PD radar uses several PRFs, each of which creates ambiguity zones in the range/velocity matrix. In processing, it can be determined that some of the PRFs are not ambiguous at the range and velocity of the target being tracked.

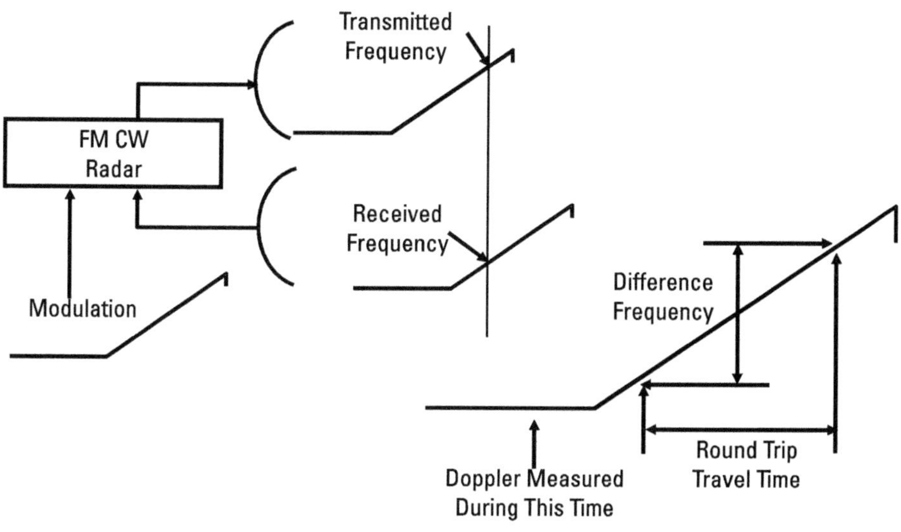

Figure 4.34 If an FM modulation as shown is placed on a radar signal, the difference between the transmitted and received signals will be from the Doppler shift during the linear part and also from the propagation delay (proportional to range) during the ramped part.

4.2.20 Detection of Jamming

Because a PD radar can detect jamming, it will allow any missile system that has a home-on-jam capability to select the home-on-jam operating mode, as discussed in Section 4.2.23.

4.2.21 Frequency Diversity

A radar can have multiple operating frequencies as shown in Figure 4.35. Note that a radar needs an efficient antenna and well-behaved power amplifier, so the range of frequencies used can be expected to be less than 10%. A parabolic antenna can have 55% efficiency if it operates over less than 10% frequency range, but a wider frequency range antenna will have much lower efficiency. For example, a 2–18-GHz EW antenna can be expected to have about 30% efficiency.

The simplest case of frequency diversity is a set of selectable frequencies, with the radar operating at the selected frequency for an extended time. As long as a receiver associated with a jammer can measure the operating frequency, the jammer can be set to the frequency in use and can optimize its jamming bandwidth against that signal. This applies to spot jamming with narrowband noise as well as to deceptive jamming techniques.

A more challenging use of frequency diversity is assignment of one frequency per sweep of the radar antenna. For example, if the radar antenna has a helical scan (one circular azimuth sweep at each of several elevation angles), the radar might change frequencies after each circular sweep. This gives the radar the advantage of a single frequency during its coherent processing interval. When a jammer has a digital radio frequency memory (DRFM), it will be able to measure the frequency (and other parameters) of the first pulse it sees and make accurate copies of all subsequent pulses during the time the radar beam

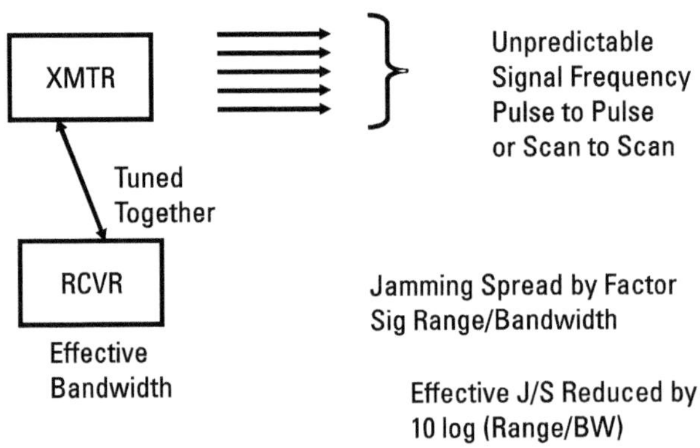

Figure 4.35 Frequency diversity requires a jammer to cover multiple frequencies or an increased frequency range.

is covering the target on which the jammer is located. (Note that we will be discussing DRFMs in detail in Chapter 8.)

The most challenging case of frequency diversity is pulse-to-pulse frequency hopping. In this case, each pulse is transmitted at a pseudo-randomly selected frequency. Because the jammer cannot anticipate the frequency of future pulses, it is impossible to optimally jam the radar. Also note that this type of radar can be expected to avoid frequencies at which jamming is detected, so jamming a few of the frequencies is unlikely to improve the jamming performance. If there are only a few frequencies, it may be practical to set a jammer to each frequency, but more typically, it is necessary to jam the whole frequency hopping range. For example, if the radar operates over a 10% frequency range at about 6 GHz and has a 3-MHz receiver bandwidth:

- The jammer must cover 600 MHz of frequency range.
- The radar only sees the 3 MHz of the jamming signal in its bandwidth.
- Thus, the jamming effectiveness is only 0.05%.
- This reduces the effective J/S (compared to matched jamming) by 23 dB.

4.2.22 PRF Jitter

If a radar has a pseudo-randomly selected pulse repetition interval as shown in Figure 4.36, it is not possible to anticipate the arrival time of radar pulses. Thus, it is not possible to use RGPI jamming. If cover pulses are used to deny the radar range information, they must be extended to cover the full range of possible pulse positions. This requires the jammer to have a longer duty cycle in its cover pulse stream, which reduces the jamming efficiency.

The jamming-to-noise ratio for self-protection jamming is a function of range squared because the radar signal losses power by the square of the range on the way to the target and again on the way back from the target, while the jamming signal only travels from the target location to the radar. As shown in

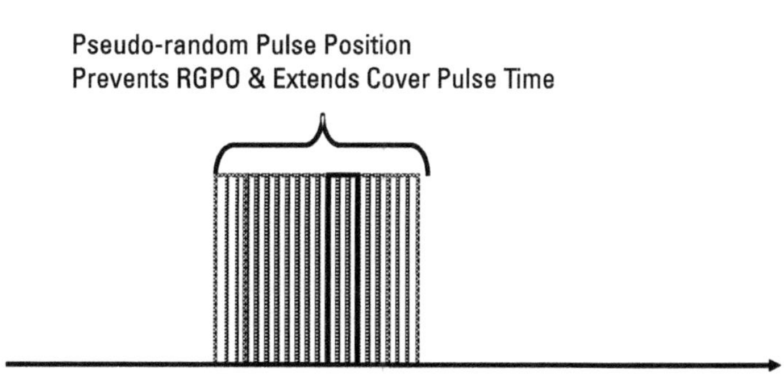

Figure 4.36 Random PRI requires a jammer to cover the full time excursion of pulse times.

Figure 4.37, as the target (on which the jammer is mounted) approaches the radar; the jamming signal in the radar receiver increases by the square of the reducing range while the skin return increases by the fourth power of the reducing range. The range at which the J/S is reduced sufficiently for the radar to reacquire the target is called the burn-through range. Note that the figure shows this range to occur when the jammer and skin return signal are equal. This is a little misleading, because the minimum J/S to protect the target depends on the jamming technique applied and the design of the radar.

Also covered in Chapter 3 is the case of stand-off jamming. The difference is that the jammer is assumed not to move as the target approaches the radar. The range to the target at which the (assumed stationary) stand-off jammer can no longer provide protection is the burn-through range.

The radar range equation, defining the range at which a radar can acquire a target, is given in [1]. The equation is used in several different forms, but all have a time term in the numerator for the time the radar illuminates the target. This is because the radar range depends on the received energy of the skin return signal. The signal energy to noise energy must reach a required level (typically taken as 13 dB) for detection to occur.

Figure 4.38 shows the skin return and jamming signals arriving at the jammed radar. The figure makes the point that the radar is looking for energy while the jammer supplies jamming power. The radar can increase its acquisition range by increasing its effective radiated power or by increasing the duty cycle of its pulse train. Many radars use emission control, outputting only enough radiated power to achieve a good-quality return signal-to-noise ratio. If jamming is detected, the radar can increase its output power to the maximum

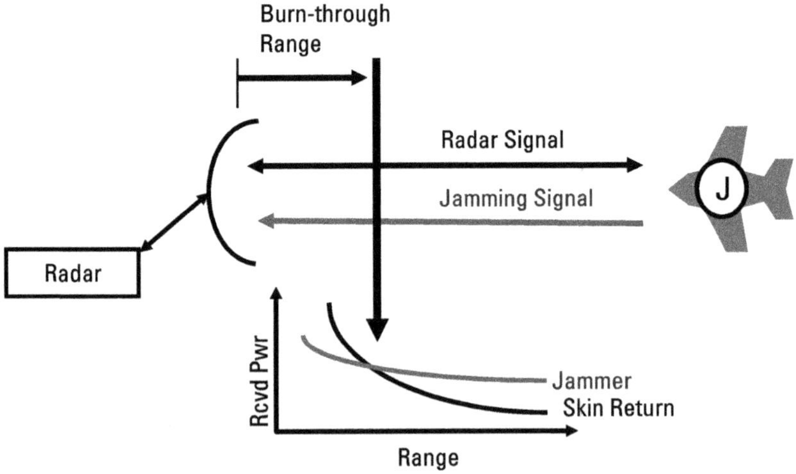

Figure 4.37 A radar's burn through range is the range at which it can reacquire a target in the presence of jamming.

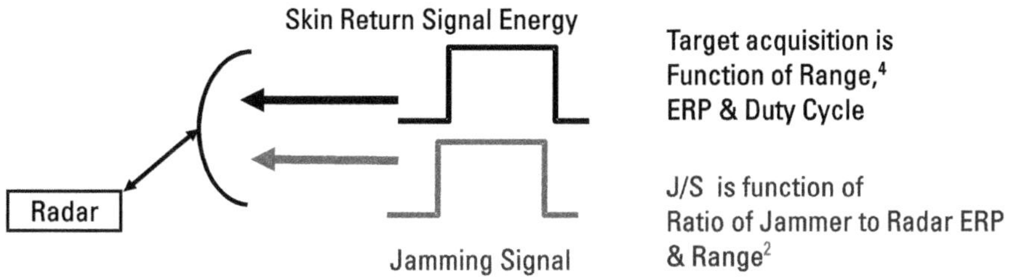

Figure 4.38 Burn-through modes extend the burn through range by increasing either the transmitted power or the duty cycle of the signal.

level. Because the J/S is a function of the ratio of the jammer to radar effective radiated power, any increase in radar power reduces the J/S and thus increases the range over which the radar can overcome the jamming.

Because the radar's acquisition range is directly proportional to time that the target is illuminated, any increase in the radar's duty cycle will increase the acquisition range, hence allowing the radar to acquire (or reacquire) the target at a greater range.

4.2.23 Home on Jam

Many modern missile systems have home-on-jam modes, also called track-on-jam modes. As shown in Figure 4.39, this requires that the missile be able to receive the jamming signal and determine its direction of arrival. If the radar detects jamming, it can then go into a home-on-jam mode, causing the missile to steer itself toward the jammer. This feature makes it very dangerous to use self-protection jamming for terminal protection. Because this mode can also be used against a stand-off jammer, it can threaten this high-value/low-inventory asset if the missile has sufficient range to reach the stand-off jamming location.

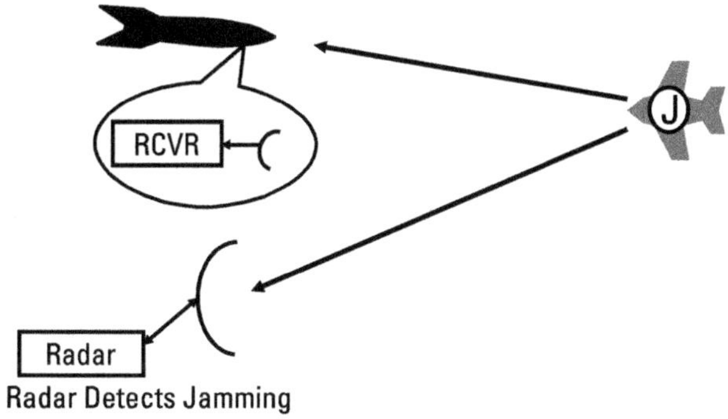

Figure 4.39 Home on jam modes require a passive guidance capability in the missile which allows it to home on the source of jamming energy.

Note that by lofting the missile, it may be practical to achieve more range in the home-on-jam mode.

4.3 Surface-to-Air Missile Upgrades

Figure 4.40 shows the genealogy of Soviet air defense system upgrades. This diagram focuses only on the Russian weapons, although some of the technology has been exported to China, leading to parallel developments that are different in some ways from their Russian roots. In each of the weapon categories shown in the figure, each generation has been designed to overcome countermeasures experienced against legacy systems or shortcomings experienced in operational testing.

The frequency ranges of radars described in open literature are usually given in terms of the NATO radar-frequency bands according to Table 4.2. However, they are sometimes given in terms of the IEEE standard radar-frequency bands as shown in Table 4.3.

The largest portion of this chart relates to the S-300 missile systems, which were developed to overcome the shortcomings of the earlier Soviet missile systems when operated in the presence of countermeasures. This series of SAM systems is shown as flowing from the earlier SA-2, SA-3, SA-4, and SA-5 systems. The design of the S-300 family of systems certainly takes advantage of the features of the earlier systems, but the new systems have significant improvements in capability to avoid the vulnerabilities of the earlier systems.

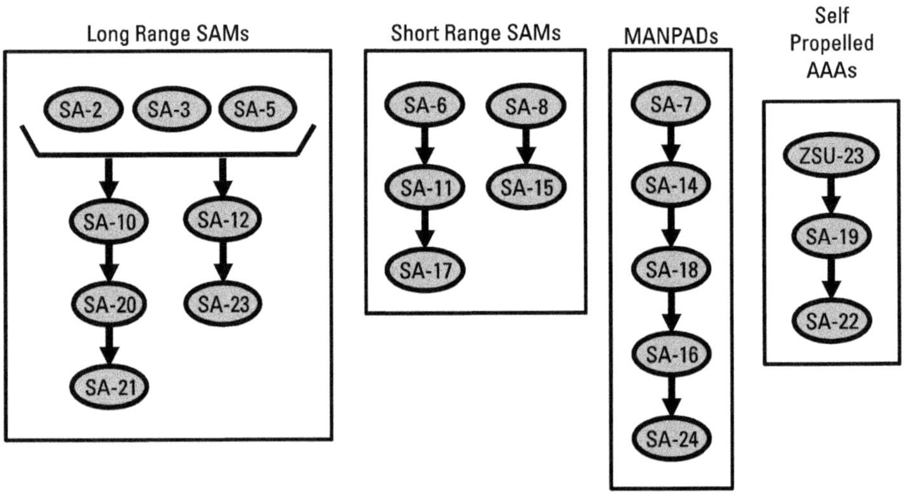

Figure 4.40 There have been many upgrades to threat weapon systems, and the upgrade process is continuing.

Table 4.2
NATO Radar-Frequency Bands

Band	Frequency Range
A	0 to 250 MHz
B	250 to 500 MHz
C	500 to 1,000 MHz
D	1 to 2 GHz
E	2 to 3 GHz
F	3 to 4 GHz
G	4 to 6 GHz
H	6 to 8 GHz
I	8 to 10 GHz
J	10 to 20 GHz
K	20 to 40 GHz
L	40 to 60 GHz
M	60 to 100 GHz

Table 4.3
IEEE Standard Radar-Frequency Bands

Band	Frequency Range
HF	3 to 30 MHz
VHF	30 to 300 MHz
UHF	300 to 1,000 MHz
L	1 to 2 GHz
S	2 to 4 GHz
C	4 to 8 GHz
X	8 to 12 GHz
Ku	12 to 18 GHz
K	18 to 27 GHz
Ka	27 to 40 GHz
V	40 to 75 GHz
W	75 to 110 GHz
mm	110 to 300 GHz

There are also two families of shorter range missile systems that have evolved from the earlier SA-6 and SA-8. The subsequent systems in this family have many of the features of the S-300 system family to overcome specific countermeasure vulnerabilities.

The MANPADS family of weapons is a series of upgrades to the SA-7. These are infrared-guided heat-seeking missiles.

This section deals only with the technical aspects of these systems and does not include descriptions of the various support vehicles or the organization of the force structure in which they are operated. It also omits photographs of the missiles, radars, and vehicles. All of these are thoroughly described (at an unclassified level) in online articles. There are many appropriate Wikipedia references, and there is very good coverage, including many photographs, on the Australia Air Power Web site (www.ausairpower.net).

We will discuss these systems, missiles, and radars in terms of their NATO designators. The above referenced online articles relate all of the NATO designators to their Russian equivalents.

All of these SAM systems and their associated subsystems are developed to support the hide, shoot, and scoot philosophy. The goal is to make the systems as undetectable as possible until the missile is fired and then to move away from the firing location as quickly as possible to avoid destruction of valuable equipment by missiles targeted at the launch site.

Open-source literature provides sketchy details on many of the features of modern missiles. In general, the later the upgrade, the less detail on specific features available. Still, it is useful to gather the information that is provided. At the end of this section, we will discuss the EW implications of the features and upgrades described.

4.3.1 S-300 Series

The S-300 family includes as number of SAM systems. They share the characteristics of vertical cold launch from their packing cases, 5-minute setup times, and 3- to 5-second delays between missile launches. Figure 4.41 shows a vertical cold launch, which is also used in other new generation missiles. The missile is blown out of its packing case or a sealed launch chamber by gas pressure, and then the missile is acquired by a data link and rotated toward the target. Then the missile fuel is ignited. The members of this family of missiles also have significant electronic protection (EP) features.

4.3.2 SA-10 and Upgrades

The land-based SA-10 system with fixed and mobile versions uses the Grumble missile and the FLAP LID fire control system. The missile is described as being able to attack a Mach 4 target. There are two associated acquisition radars for early SA-10s, the TIN SHIELD and the CLAM SHELL. Later versions are supported by the BIG BIRD acquisition radar.

In open literature, the initial lethal range of the SA-10 is stated at 75 km. After several system improvements, the lethal range is described as 150 km. The SA-10 is fired from a truck mounted transporter/launcher (TELAR). The missiles are cold-launched vertically from their cylindrical packing cases using gas pressure. The missile is launched to a few meters of altitude and turned toward

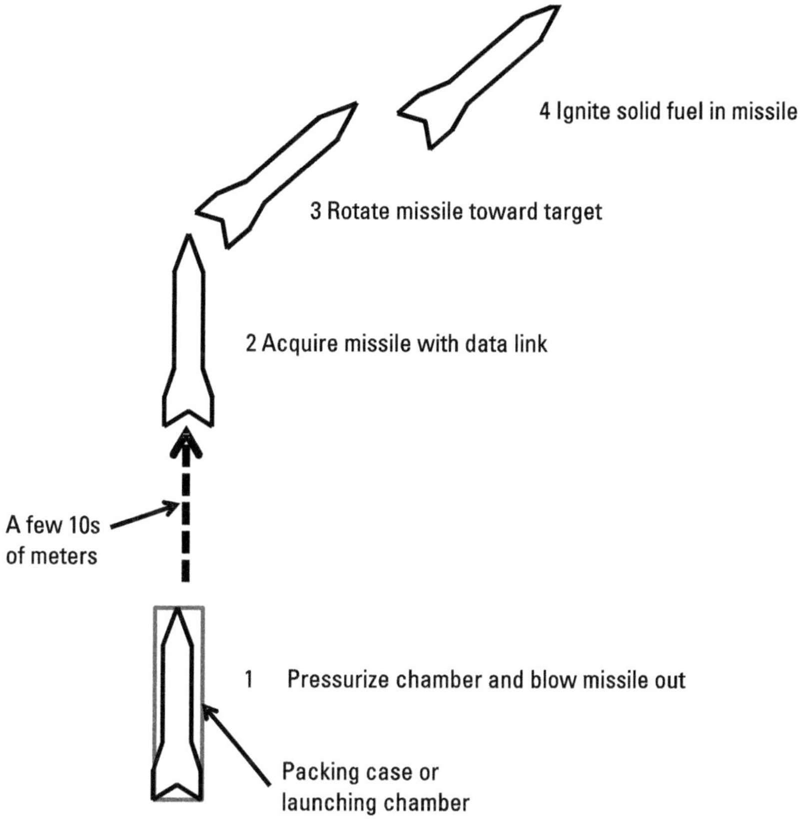

Figure 4.41 In a cold launch sequence, the missile is ejected from a launching chamber or its packing case by cold gas. Then the missile is acquired and rotated toward the target. Then the missile's fuel is ignited.

the target. Then the solid fuel missile engine is lit. This approach gives the SA-10 a very fast reload sequence and greatly simplified operational logistics, supporting the hide, shoot, and scoot philosophy. The tracking radar for this system is the FLAP LID which is an active electronically steered array (AESA) radar incorporating some electronic protection (EP) capabilities. Most of the specific EP capabilities are not identified in open literature, only that the radar has very low antenna side lobes.

4.3.2.1 SA-N-6

The shipboard version of the SA-10 is called the SA-N-6. Open literature describes its lethal range as 90 km. It fires the Grumble missile from rotary launchers. Tracking is provided by the TOP SAIL, TOP PAIR, or TOP DOME radars. It uses command guidance, but also has a terminal semi-active radar homing mode as shown in Figure 4.42.

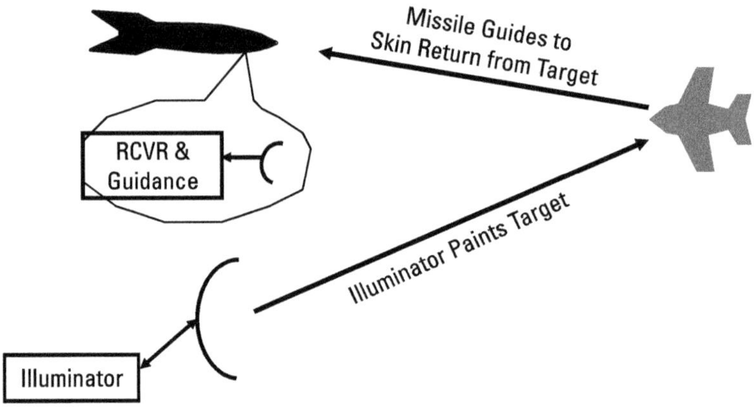

Figure 4.42 Terminal semi-active guidance enabled accurate targeting at long range.

4.3.2.2 SA-N-20

This is described as a MACH 6 missile able to engage targets at a closing up to Mach 8.5. It employs the TOMB STONE tracking radar. And is also described as having a track-via-missile capability as shown in Figure 4.43.

4.3.2.3 SA-20

The SA-10 has been upgraded with a new missile (Gargoyl) and the TOMB STONE tracking radar. It is described as being capable against short- and me-

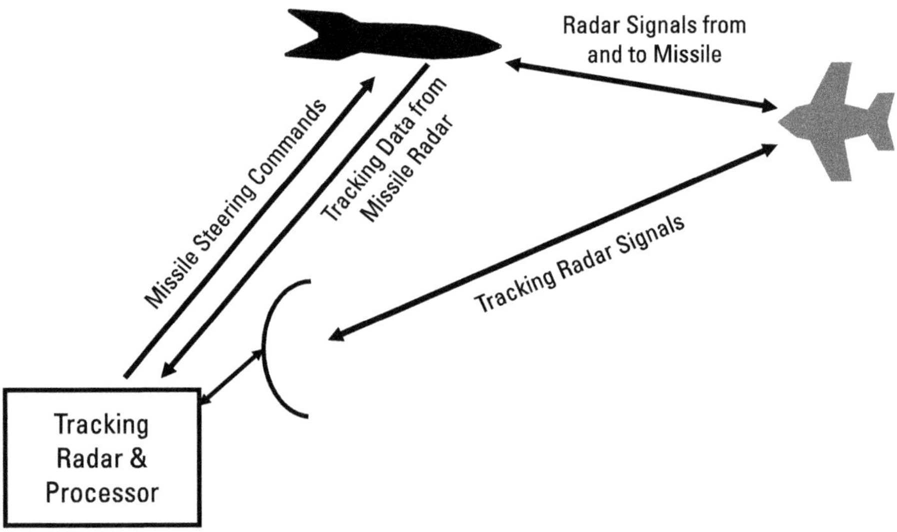

Figure 4.43 When track via missile guidance is employed, a secondary radar on the missile tracks the target and sends tracking information to the primary tracking radar to enhance the total tracking accuracy.

dium-range tactical missiles as well as aircraft. This upgrade has the NATO designation SA-20. It is described as having a 195-km range. The Gargoyl missile is described as having gas-dynamic steering rather than the aerodynamic fins of earlier missiles, giving it greater maneuverability.

4.3.2.4 SA-21

This system has been further upgraded to the SA-21 that uses the GRAVE STONE tracking radar and the TRIUMF missile. It is said to have a range of 240 km with one missile, 396 km with another missile, and 442 km with a third missile. This missile is designed to take out stand-off jammers and combat air traffic control aircraft at extended ranges. It also has smaller missiles with 74-km range with control features that allow very high maneuver rates enabling them to actual impact targets.

4.3.3 SA-12 and Upgrades

The SA-12 SAM system has two types of missiles, the GLADIATOR for aerodynamic targets and the GIANT for ballistic missiles. The Gladiator has an engagement range of 75 km using the GRILL PAN radar and the Giant has an engagement range of 100 km with a maximum altitude of 32 km using the HIGH SCREEN radar.

The GRILL PAN radar is described as having an autonomous search capability.

The SA-12 uses tracked launch and support vehicles for superior cross-country mobility.

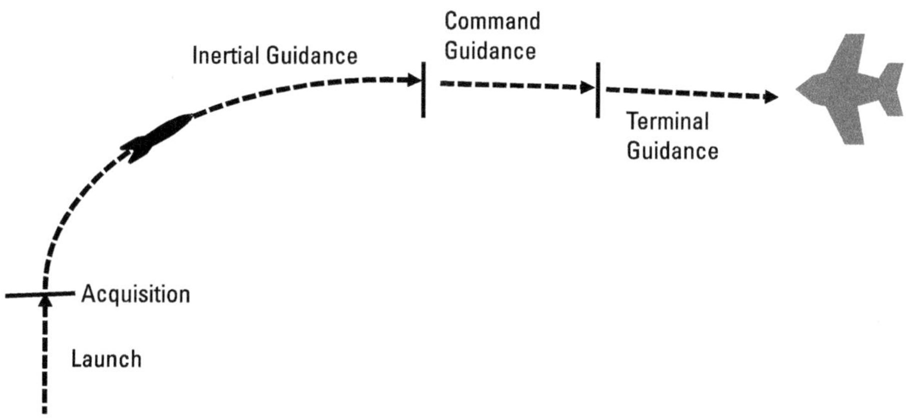

Figure 4.44 Many modern missiles use inertial guidance from the time of missile acquisition, then upgrade with command guidance when the missile approaches the target, then semi-active, passive homing, or TVM guidance for the terminal phase.

The SA-12 has been upgraded to the SA-23, which is described as having a 200-km effective range and advanced radar data processing. It has inertial, command, and semi-active homing guidance as shown in Figure 4.44. It employs a semi-active radar homing (SARH) radar on its TELAR as an illuminator.

4.3.4 SA-6 Upgrades

The SA-6 is a short-range missile system that uses a FIRE DOME radar. It has a range variously described as 20 to 30 km. It can attack a MACH 2.8 target.

This system has been upgraded to the SA-11 which uses the Gainful missile and the STRAIGHT FLUSH AESA tracking radar. Its range has been described as 35 km.

A second upgrade is to the SA-17 system which is described as having a 50-km range.

4.3.5 SA-8 Upgrades

The SA-8 is a low-altitude, short-range system on a wheeled, amphibious platform. It initially had a 9-km range and with later improvements extending the range to 15 km. It uses a J-band frequency agile monopulse tracking radar and a C-band acquisition radar. It also has an electro-optical (EO) tracker.

This system has been upgraded with new radars and missiles to the SA-15. It uses the Gauntlet missile. It has a 12-km range. A feature of the system is that it is autonomous, with surveillance, command and control, missile launch, and guidance all from the same vehicle. It has an IFF function and a phased array PD G/H band tracking radar.

4.3.6 MANPADS Upgrades

The Man Portable Air Defense System (MANPADS) is an optically aimed, IR-guided missile system. These shoulder-fired missile systems are described in open literature as follows. The original missile in this series was the SA-7 STRELLA guided by an uncooled lead sulfide (PbS) sensor. It could attack an aircraft only from the rear. It had a range of 3,700m and a maximum target altitude of 1,500m.

Later upgrades to this system are:

- The SA-14 GREMLIN, which had a better cooled seeker allowing attack from any angle. It had a maximum altitude of 2,300m.
- The SA-16 GIMLET, an improvement on the SA-14 with an all-aspect sensor to counter flares. Its range is 5 km and its maximum altitude is 3,500m.

- The SA-18 GROUSE with a cooled indium antimonide sensor that allows attack from any aspect to 5.2-km range and 3,500-m altitude. It has significantly improved anti-flare protection, including a two-channel tracker.
- The SA-24 GRINCH with standard night vision, and a 6-km range.

4.4 SAM Acquisition Radar Upgrade

Tracking radars for Vietnam era SAM systems were highly dependent on acquisition radars. The acquisition radar, typically in VHF or UHF frequency ranges, would acquire targets and hand them off to the tracking radars. There are two trends developing. One is that some tracking radars have incorporated acquisition modes. The second is that acquisition radars are operating a higher frequency. For example, TIN SHIELD and BIG BIRD operate in S-band and HIGH SCREEN operates at X-band.

In general, the higher operating frequencies allow reduced antenna bandwidths for greater angular resolution and the shorter wavelengths are useful in handling very small radar cross section (RCS) targets. To acquire stealth aircraft, missiles, and UAVs, the ability to acquire low RCS targets is critical. With increasing levels of pulse compression, modern acquisition radars have an increasing target location accuracy and resolution.

Acquisition radars have always had significantly longer range than their associated tracking radars, and that is unchanged. However, these radars are incorporating significant electronic protection features to make them harder to jam by stand-off jammers.

Identification friend foe (IFF) is increasingly incorporated in acquisition radars allowing early identification of potential targets.

4.5 AAA Upgrades

A family of self-propelled anti-aircraft guns started with the ZSU-23-4 SHILKA. It has four 23-mm water-cooled guns mounted on a tracked vehicle. Its range is 2.5 km and is maximum lethal elevation is 1,500m. Later, eight SA-18 or SA-16 heat-seeking missiles were added. It has a GUN DISH radar.

This system has been upgraded to the SA-19 TUNGUSKA, which has two 30-mm guns and eight radar command-guided missiles. The guns have a range of 4 km and elevation of 3 km and the missiles have 8-km range and 3.5-km elevation. It incorporates a HOT SHOT radar with C/D band acquisition function and J-band two-channel monopulse tracking functions.

A further upgrade has been made to the SA-22 GREYHOUND, which has two 30-mm guns and up to 12 command-guided missiles and radar or optical tracking. It has the HOT SHOT radar and integrated IFF. The range of the guns is 4 km with a maximum altitude of 3 km. The missiles have a range of 20 km and a vertical limitation of 10 km.

4.6 EW Implications of Capabilities Described

There are some important implications from each of the enhancements to the modern weapons discussed above. These will be discussed here in terms of their functionality, rather than relating them to specific threat systems; most are present in several different systems and any later enhancements will be incorporated into the continuing flood of missile and threat radar upgrades.

In each case, after describing the impact of a radar improvement feature, the section will have advice on what to do about it.

4.6.1 Increased Lethal Range

Stand-off jamming (SOJ) has been one of the primary techniques for countering threat systems. To review from Chapter 3, SOJ involves flying (typically) two special jamming aircraft in a tight pattern just beyond the lethal range of multiple threat missiles to protect multiple strike force aircraft flying into the lethal range. Because multiple radars are being jammed simultaneously, there is no way to jam into the main beams of the threat radars. Also, the jamming aircraft must distribute its jamming power in multiple directions.

The J/S achieved in stand-off jamming is calculated from the following equation:

$$J/S = 71 + ERP_J - ERP_R + 40\log R_T - 20\log R_J + G_S - G_M - 10\log\sigma$$

where 71 is a constant, ERP_J is the effective radiated power of the jammer in dBm, ERP_R is the effective radiated power of the radar in dBm, R_T is the range from the radar to the target in kilometers, R_J is the range from the jammer to the radar in kilometers, G_S is the radar side-lobe gain (redefined from GR_J above) in decibels, G_M is the radar main beam boresight gain in decibels, and σ is the radar cross section of the target in square meters.

Notice the term −20 log R_J. That means that the J/S is reduced by the square of the additional distance factor. Jamming from the 150 km estimated lethal range of the improved SA-10 as compared to jamming from the 45-km lethal range of the SA-2 reduces the J/S by a factor of 20.5 (13 dB). Moving the jammer out to the estimated lethal range of the 396-km lethal range of the

SA-21 with its most capable missile reduces the J/S by A FACTOR OF 77 (19 dB). Considering that an SOJ needs all of the J/S that it can get to overcome its jamming geometry, this is problematic.

The solution of how to resolve this is to either increase the jamming power or reduce the RCS of the target, but remember that many of the threat radar upgrades enhance performance against low RCS targets. Another approach is to consider stand-in jamming, in which a jammer (unmanned) is placed much closer to the threat radar than the target.

4.6.2 Ultralow Side Lobes

Ultralow side lobes make it harder for an electromagnetic support (ES) system to detect a threat radar and an electronic attack (EA) system to jam it. If you are trying to detect a threat radar in its side lobes, the detection range is reduced by the square of the side lobe reduction.

Likewise, the J/S achieved by an EA system such as an SOJ is reduced by the side lobe reduction factor. See Section 4.2.1.

The solution of how to resolve this is, for the ES system, to optimize the system sensitivity. One way is to scan with a phased array receiving antenna, which will provide additional received signal strength. If the ES system has digital receivers, you may be able to optimize the bandwidth for maximum sensitivity. If the threat radar has a scanning antenna beam, you may be able to get the information you need from the main beam when it scans past your receiving antenna.

If your EA system has an active electronically scanned array (AESA), you can put a beam onto the threat radar that you want to jam to enhance the J/S.

4.6.3 Coherent Side-Lobe Cancelling

Again, consider stand-off jamming. As stated above, the SOJ must jam into the side lobes of the threat radar being jammed. If that radar has coherent side-lobe cancelling (CSLC), it can reduce the received narrow band jamming power of a signal (like typical FM noise) received into its side lobes by a factor of up to 30 dB. This reduces the J/S by that factor, so either the jammer needs 30 dB more power or it must be 32 times closer or the RCS of each protected aircraft needs to be reduced by a factor of 1,000 to achieve the same J/S. See Section 4.2.3.

The solution to resolve this situation is as follows. One technique that is discussed in the literature is to mix pulses with FM noise jamming. The pulses will generate many CW components that took like narrowband jamming signals and will thus tie up all of the coherent side-lobe cancelling channels and thus increase the effectiveness of your jamming.

4.6.4 Side-Lobe Blanking

Whenever a radar with a side lobe blanking capability receives a pulsed signal that is stronger in the output of its special antenna oriented into its side lobes than it is in its main antenna, the radar blanks its main antenna output during the one or few microseconds that those side lobe pulses are present. See Section 4.2.4.

The solution of how to resolve this is, if you time cover pulses from your jammer so that they cover the threat radar's own pulses, the threat radar is essentially jamming itself.

4.6.5 Anti-Cross-Polarization

Anti-cross-polarization is the way that a radar's capability to reduce cross polarization jamming is described. It is stated as some level (in decibels) of anti-cross-polarization. This is accomplished either by the presence of a flat plate phased array antenna without any gain sloping at the edges to reduce side lobes or by the presence of a polarization filter that will not let cross-polarized jamming signals into the radar's receiver. See Section 4.2.6.

The solution of how to resolve this is that you are probably not going to be able to apply cross-polarization jamming to that radar unless you can generate an extremely large J/S. The best answer is to use some other kind of jamming.

4.6.6 Pulse Compression

If we are jamming a threat radar that has pulse compression (PC) and our jamming does not contain the radar's compression waveform (either chirp or Barker code), our J/S will be reduced by the compression factor, which can be up to 30 dB. Again, our J/S can be reduced by up to 30 dB. See Section 4.2.10.

The solution of how to resolve this is to place the compression waveform on jamming signals, whether they are transmitted to the threat radar's main lobe or side lobes. If the compression technique is a linear chirp, you can reproduce that in several ways (sweeping oscillator, direct digital synthesizer, and so forth). However, if it is a nonlinear chirp or a Barker code pulse compression modulation, you will need to incorporate a DRFM into your jammer. This is discussed in detail in Chapter 8.

4.6.7 Monopulse Radar

Monopulse radars are not successfully jammed by some of the jamming techniques described in Chapter 3. Some of the techniques will actually enhance the angle tracking of the monopulse threat radar. See Section 4.2.5.

The solution to this is that the jamming techniques described in Sections 3.4.9 through 3.4.15 are effective against monopulse jamming.

4.6.8 Pulse-Doppler Radar

Pulse Doppler (PD) radar wants to see a coherent signal that falls into one of the channels of its channelized filter. If a jamming signal fills multiple channels or has strong spurious components, the radar will know it is being jammed and can initiate HOJ.

The radar processing will also reduce the J/S achieved by a noise jamming signal that occupies multiple channels. It will also discriminate against signals returned from chaff.

It can also detect separating signals (such as range gate pull off jamming) and track the signal that has a rate of change of range appropriate to the Doppler shifted frequency at which it is received. See Sections 4.2.12, 4.2.14, and 4.2.15.

The solution of how to resolve this is, if you jam with coherent signals, they will fall into a single PD processing filter and will thus be accepted as valid signals, so the jamming will be effective. If the chaff cloud is illuminated by a strong jamming signal, the PD threat radar will accept the chaff cloud as a decoy returning a valid skin return. If you do both range and frequency pull-off jamming, the PD threat radar will accept the jamming signal as a valid return. This is best done using a DRFM.

In earlier conflicts, bulk chaff was dispensed in areas to prevent radars from acquiring aircraft. This was very effective before pulse Doppler radars could discriminate against chaff, but is now of limited use.

4.6.9 Leading-Edge Tracking

If a threat radar has a leading-edge tracking, the latency of a range-gate pull-off (RGPO) jammer will allow the radar to continue tracking the valid skin return because it will never see the delayed RGPO pulses.

The classic solution to this is to use RGPI jamming. However, another solution is to make the latency of the RGPO process short enough to capture the leading-edge tracker. This is typically done using a DRFM, which has extremely short process latency.

4.6.10 Dicke-Fix

The Dicke-fix involves a wideband channel in which strong, low duty cycle pulses are clipped so that they cannot capture the radar's automatic gain control (AGC) in a following narrow band channel. See Section 4.2.10.

The solution to how to resolve this is that there is a special waveform [1] that allows a jamming signal to pass through the Dicke-fix.

4.6.11 Burn-Through Modes

Burn-through modes are modes in which the effective radiated power (ERP) or duty cycle of a radar is increased to extend the burn-through range to the maximum extent practical.

The solution to this is to increase your effective jamming power as much as possible.

4.6.12 Frequency Agility

If a radar has a pseudo-random pulse-to-pulse frequency-hopping signal, it will not be possible to know what the frequency of the next pulse will be. Therefore, a jammer must either jam at each of the radar's frequencies or spread its jamming power over the whole hopping range. This will reduce the J/S achieved by several decibels. See Section 4.2.19.

The solution to this is that the DRFM can once again come to the rescue. If the DRFM and its associated processor measure approximately the first 50 ns of a pulse, it can quickly set the jammer to that frequency. Because modern radars typically have pulses several microseconds long, the jamming energy will be reduced very little by losing this small part of the pulse. This is discussed in Chapter 8.

4.6.13 PRF Jitter

If a threat radar has a pseudo-random pulse repetition interval referred to as a jittered pulse repetition frequency (PRF), it will be impossible to predict the time at which the next pulse will occur. This makes the RGPI jamming technique impossible. Also, cover pulses, which require prediction of the timing of pulses, cannot be efficiently generated. See Section 4.2.20.

The solution to this is that, if cover pulses are to be used against a threat radar with jittered PRF, use extended cover pulses to cover the whole jitter range of the radar's PRI.

4.6.14 Home-on-Jam Capability

Although none of the specific missile systems that we have listed here are identified in open literature as having home-on-jam capability, it is obvious that it will be present in current or near future threats.

Home on jam (HOJ) means that a radar that can detect jamming (clearly including pulse-Doppler radars) can command the missile it is guiding to home on the jamming signal. That means the missile will go directly to any aircraft that is performing self-protection jamming (SPJ).

Also, consider the jamming aircraft performing stand-off jamming. This aircraft is a high-value, low-inventory asset, which is why we deploy it beyond the lethal range of threat missiles. A missile with HOJ can be lofted to maximize its range, beyond the effective range of the guiding radar as shown in Figure 4.45. Then it can home on the SOJ aircraft from above. If the missile has aerodynamic steering capability, it can even attack beyond the range allowed by its fuel. See Section 4.2.21.

Clearly, the solution for self-protection jamming is not to do it. Protect yourself with a decoy to take the missile for you. Chapter 10 describes a number of types of radar decoys that could take a missile for you. Chapter 8 discusses the role of digital RF memories (DRFM) that can be employed to make decoys very sophisticated. Also, consider expendable jammers that can jam from somewhere else. One of these is the miniature air launched decoy J model (the MALDJ). It is a remote jammer that will attract a home-on-jam missile.

4.6.15 Improved MANPADS

The improvements in MANPADS weapons have extended their range and their effective altitude. They are a significant threat to helicopters and other low-flying aircraft.

The solution to this is that it used to be enough to just fly higher to avoid MANPADS, but now it is necessary to consider modern IR jammers such as those described in Chapter 9.

4.6.16 Improved AAA

At the time of Vietnam, it was possible to ignore any report of the presence of an automatic anti-aircraft gun (AAA) if you were flying more than 1,500m above the ground, as that was the maximum vertical envelope of the ZSU-23. Now the AAA upgrades have added heat-seeking missiles with vertical attack envelopes up to 10,000m and 30-mm guns with double the range of the 23-mm guns on the ZSU-23.

Later upgrades have switched from simple heat-seeking missiles to radar-guided missiles. These weapons have become significantly more dangerous.

To overcome these modern AAAs, it is necessary to depend on both IR and radar jammers for protection. Gone are the days when just flying high was enough protection.

References

[1] Schleher, D. C., *Electronic Warfare in the Information Age*, Norwood, MA: Artech House, 1999.

[2] Griffiths, H. G., C. J. Baker, and D. Adamy, *Stimson's Introduction to Airborne Radar*, 3rd ed., New York: SciTech, 2014.

[3] Van Brunt, L. B., *Applied ECM, Vol. 1–3,* Dun Loring, VA: EW Engineering, Inc., 1978, 1982, 1995.

5

Digital Communication

5.1 Introduction

Modern military communication is almost exclusively digital. Modern tactical radios digitize voice before transmission, and much important military command and control communication involves the movement of digital information from on activity to another. A modern integrated air defense network has every element connected by digital data links. In this chapter, we will discuss many aspects of digital communication theory; its advantages and vulnerabilities, specification of digital links, and propagation considerations important to electronic warfare (EW) operations.

This chapter should be considered a background information reference to other chapters, especially Chapters 2, 6, and 7, which discuss subjects in this chapter in less detail, but in various important contexts.

5.2 The Transmitted Bit Stream

As shown in Figure 5.1, a transmitted digital signal must contain more than just the digitized data. A data frame is shown in the drawing.

- There is typically a block of bits that provide frame synchronization.
- In many systems, for example, a command link to an unmanned aerial vehicle (UAV), the information bits may need to be sent to one of several destinations at the receiver location. In a UAV, this could be the

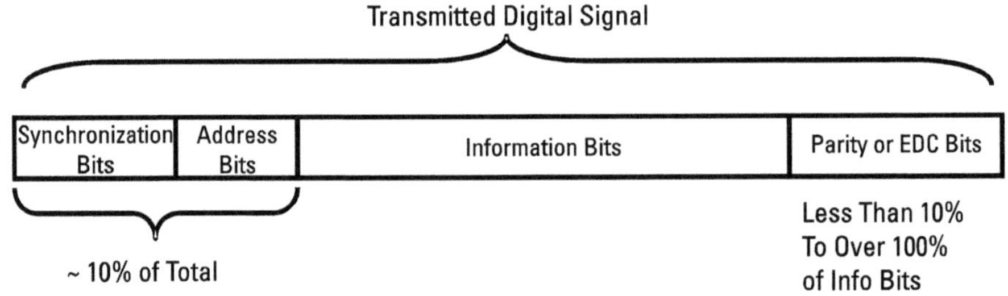

Figure 5.1 A transmitted digital signal contains synchronization, address, information, and parity or EDC bits.

UAV navigation system, one of several payloads, and so forth. Thus, there would need to be a block of address bits.

- The information bits carry the actual transmitted information.
- Because the transmitted data may be corrupted by noise, interference, or jamming in the environment, special bits are added to allow the receiver to either detect and reject bad data blocks or to actually correct erroneous bits in the received signal. The parity or error detection and correction (EDC) block of bits support this function.

5.2.1 Transmitted Bit Rate Versus Information Bit Rate

The transmitted bit rate must be fast enough to send the whole signal frame at the rate that the information in that frame is required at the receiver location. This means that the transmitted data rate could be significantly higher than the required information data rate. The link bandwidth must be wide enough to accommodate this higher bit rate.

5.2.2 Synchronization

There are two aspects of synchronization: bit synchronization and frame synchronization. The digital signal arrives at the receiver as a modulated RF signal with different states for 1 or 0 bits. The receiver demodulates this signal to recover the bits, and then must set a timing circuit (called a bit synchronizer) that outputs a code clock signal aligned with the code clock in the transmitter but delayed by the propagation time of the signal from the transmitter to the receiver (at the speed of light). The bit synchronizer produces a clean digital bit stream with 1s and 0s determined from the demodulated received signal. At this point, some of the bits may be wrong (bit errors) because of degradation in the received RF signal, but the output is a series of bits that can be processed in digital circuitry. As shown in Figure 5.2, the bit synchronizer, in addition to

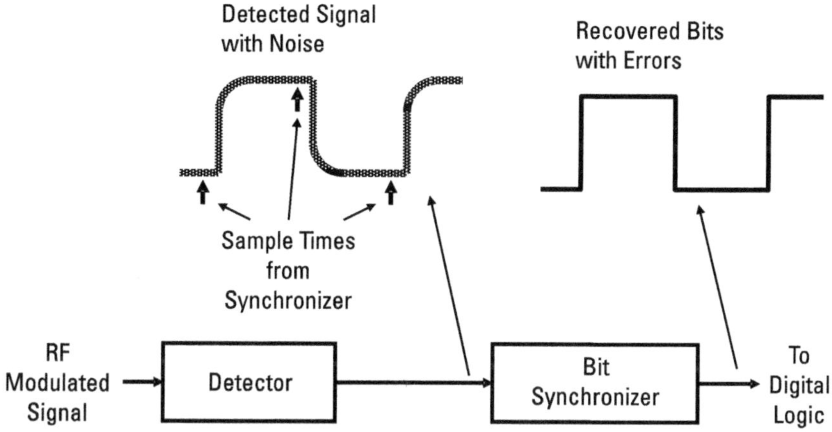

Figure 5.2 A bit synchronizer circuit creates binary bits from the demodulated output of the receiver's discriminator.

generating the code clock, also determines when the RF signal is sampled to decide whether a received bit is a 1 or a 0.

When information is transmitted digitally, the transmitter sends a typically continuous series of bits (1s and 0s) that is meaningless unless the receiver can determine the function of each bit. The information is organized into frames of many bits, and the receiver must be able to determine the beginning of each frame. The position of each bit in the frame then identifies its function. This process is called synchronization. In some data transfer systems, there is a separate modulation value for a synchronization pulse at the start of the data frame. However, typically, there is a unique series of bits in the digital bit stream that the receiver can compare against a stored bit sequence to identify the beginning of the frame.

Figure 5.3 shows the thumb-tack correlation of a series of bits. A digital signal will have approximately the same number of 1s and 0s, and they will be close to randomly distributed. The correlation value of the two signals is determined by comparing their states. If, at any instant, the two signals are equal (e.g., both 1s) the correlation is one. If they are not equal (i.e., a one and a zero), the correlation is zero. Because the bits are randomly distributed, averaging the correlation value over a block of bits will yield 0.5 correlation value. If the received code is moved in time against the reference code [by changing the frequency of the code clock (slightly) to slide one signal against the other], the correlation of the two signals will start to increase as soon as the received code is within one bit period of the reference code and will have a correlation value of 1 (100% correlation) when the two bit streams are exactly aligned. The receiver will store (as the reference code) the unique series of random 1s and 0s

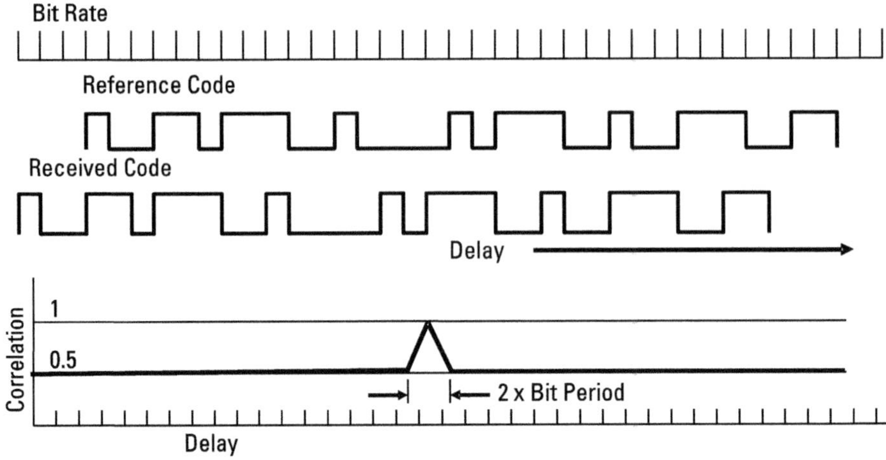

Figure 5.3 A received digital signal must be synchronized so that the information in its bits can be recovered.

in the synchronization block (refer to Figure 5.1). It will average the correlation over a series of bits as long as the synchronization block, and stop delaying the received code when the average correlation pops up to 100%. Then the receiver can identify the function of each received bit from its position in the frame.

Note that there is no reason synchronization bits must be in a contiguous group at the beginning of the frame. They could just as well be distributed pseudo-randomly through the frame to make it more difficult for a hostile receiver or jammer to recover the frame or interfere with frame synchronization. Because preventing synchronization is a highly efficient method of jamming digital communication, an important communication link can be expected to have a very robust synchronization scheme.

5.2.3 Required Bandwidth

The thumb-tack synchronization diagram in Figure 5.3 shows a very sharp correlation triangle that is two bit periods wide. This requires that the bits be square, which, in turn, requires an infinite bandwidth. When the link bandwidth is narrowed, the bits become rounded, which dulls the correlation as shown in Figure 5.4. Dixon stated that the 3-dB bandwidth of the main lobe of the digital signal frequency spectrum is adequate to support the recovery of the transmitted digital signal [1] (see Figure 5.5).

The 3-dB bandwidth is also given in [1] as 0.88 × the transmitted bit rate for most digital RF modulations, but is only 0.66 × the transmitted bit rate for minimum shift keying (MSK). MSK is an efficient modulation that is widely

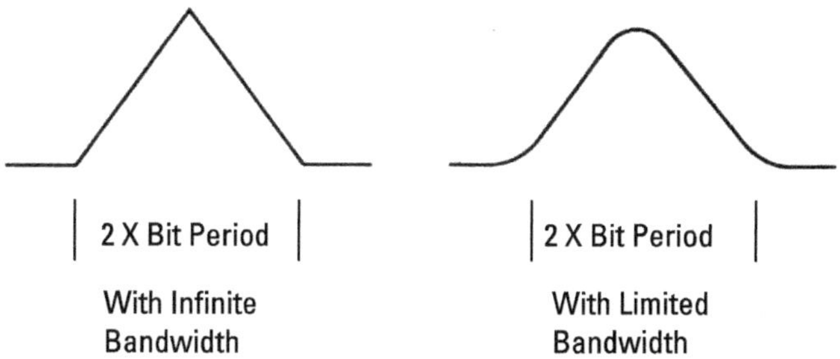

Figure 5.4 The shape of the correlation curve is dependent on the bandwidth of the link over which the digital signal is carried.

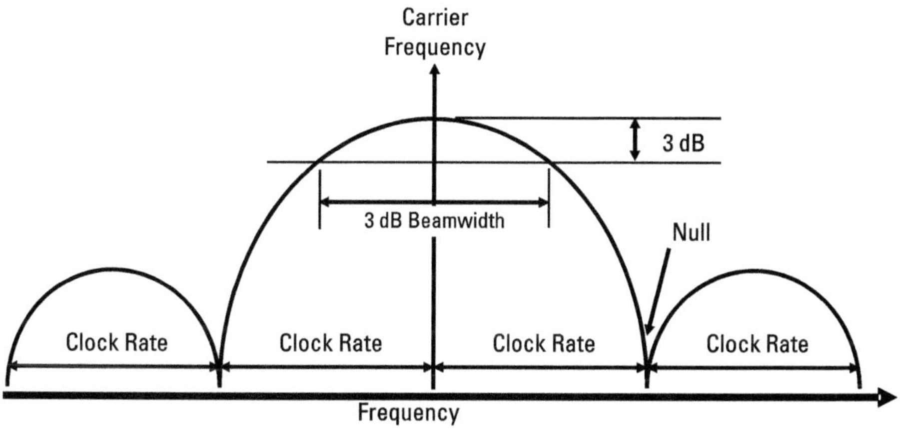

Figure 5.5 The digital signal spectrum includes a main lobe and side lobes with clearly defined nulls spaced at multiples of the clock rate from the carrier frequency.

used in digital links because this reduced bandwidth versus bit rate allows improved receiver sensitivity. In Section 5.4, we will be discussing a number of modulations and their implications in detail.

5.2.4 Parity and EDC

The final block of bits in the frame of Figure 5.1 is to preserve information fidelity by detecting or correcting bit errors. For systems designed to operate in very hostile environments, these bits, or other techniques for fidelity preservation, can significantly increase the bandwidth required to pass a given amount of data in the required amount of time.

5.3 Protecting Content Fidelity

One very important requirement for networking is that correct information arrives at a remote location. Because most information is sent digitally, this means that the bit error rate must be low enough to allow proper function of the networked activity.

5.3.1 Basic Fidelity Techniques

There are several approaches to assuring the fidelity of information sent over a transmission link. You will note that each technique makes trade-offs among data rate, latency, level of fidelity assurance, and system complexity.

You can use majority encoding, which involves sending the data multiple times as shown in Figure 5.6. Let us assume that each data block is sent three times. At the receiver, the received data blocks are compared. If all three agree, the data is passed to an output register. If two of the three agree, their version of the data is passed to the output. If none agree, the data can either be rejected or some arbitrary decision can be made. The fidelity is improved, but the throughput data rate is reduced by a factor of 3 and the output data is delayed by three times the duration of a data block. Sending more repetitions would increase the fidelity in a hostile environment, but would further reduce the throughput rate and increase the latency.

You can also send the data blocks multiple times, but with multiple parity bits added to each block as shown in Figure 5.7. As discussed next, the parity bits for each data block can be checked and any block containing bit errors can be rejected. The first data block received without errors is passed to the output register. In this case, the data throughput rate is reduced and the latency increased by both the percentage of parity bits per block and the number of block repetitions sent. For example, if each block were sent five times and there were 10% parity bits in each block, the throughput data rate would be reduced by a factor of 5.5 and a latency of 5.5 times a data block duration introduced. However, this approach improves the data fidelity.

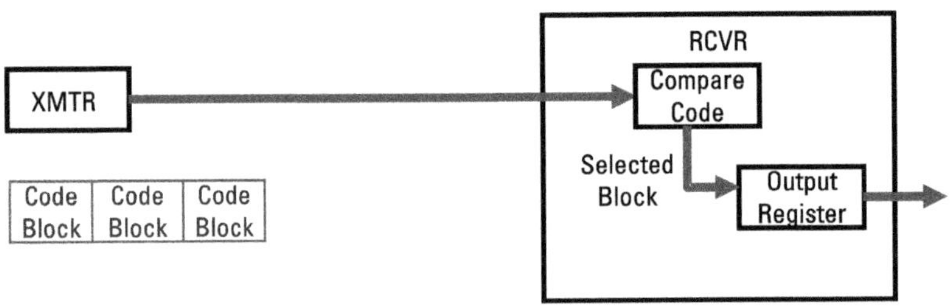

Figure 5.6 Majority encoding requires multiple transmissions of a block of code and the receiver selects the block that is received the same the most times for output.

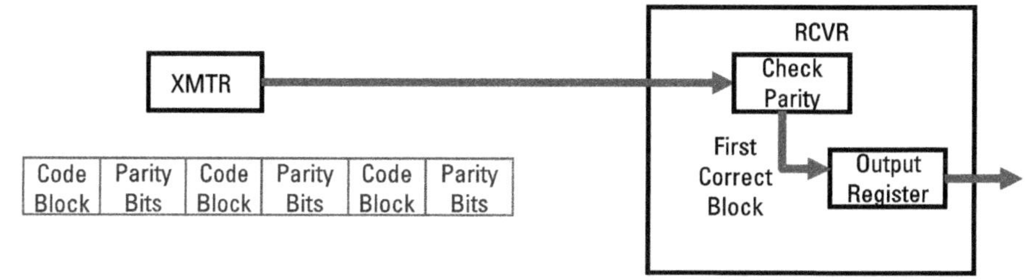

Figure 5.7 Repetitive transmission with many parity bits requires that each information code block be sent with enough parity bits that a block with errors can be dependably detected. The receiver rejects any block that does not pass parity check and outputs the first error free information block received.

You can retransmit the received data back to the transmitter, check the returned data bit for bit, and repeat the data block if there were any errors as shown in Figure 5.8. Only correct data blocks are placed into the output register, and the transmitter is authorized to send the next data block. If a data block has errors, that block is resent until an error free block is received. This approach assures that every data block will be correctly transmitted (eventually). However, the complexity of a return transmission link is added. Consider a wideband data link from a remote sensor to a control station. Typically, the command link from the control station to the remote sensor has far less bandwidth than the data link; a command link may not even be required. If this fidelity protection approach is used, there must be a link from the control station to the remote sensor, and it must be as wide as the data link. We will be discussing the impact of link bandwidth on network operation later. If the environment does not have significant interference, there will be very little reduction of the data throughput rate or latency with this approach. When there is

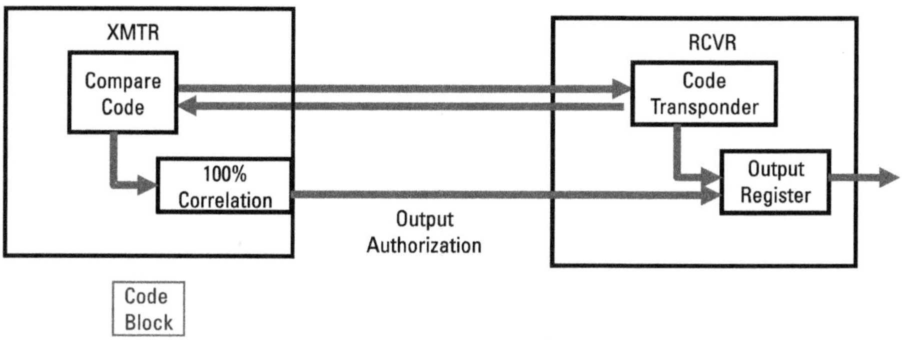

Figure 5.8 Retransmission data validation requires that each information code block be retransmitted to the transmitter where it is compared with the originally transmitted data. If it is correct an authorization signal is sent to the receiver to allow the code block into the output register.

significant interference or jamming, lots of bit errors can be expected and thus more data block resends, causing decreased data throughput rate and increased latency as a function of the level of hostility of the environment.

You can add an error detection and correction (EDC) code to each data block as shown in Figure 5.9. If there are errors in a data block, the EDC corrects those errors. This approach is called forward error correction. It provides error free data transmission up to some maximum correctable bit error rate. No return link is required, and the throughput rate and latency are not changed by the level of hostility of the environment. The amount of reduction of throughput rate and increase of latency depend on the percentage of each data block dedicated to the EDC code; the higher the percentage of code bits, the greater the number of bit errors that can be corrected.

The final approach is to simply increase the transmitter power so that signals are received at a higher signal-to-noise ratio (SNR) and desired signal-to-interference ratio. You can get the same effect by reducing the transmitted bit rate which allows a reduced receiver bandwidth. Either of these measures will reduce the received bit error rate, improving the information fidelity. Increased transmitter power can be a significant increase in system complexity, and the reduced data rate will reduce the data throughput rate.

5.3.2 Parity Bits

As discussed above, extra bits are added to the transmitted digital data to protect the information fidelity. This is particularly important in hostile environments with interference including jamming. These extra bits can either be parity bits or an error detection and correction code. Parity bits check that the proper information has been received. The more parity bits provided, the higher the confidence that, if all of the parity bits are received correctly, there were no errors in the received data block.

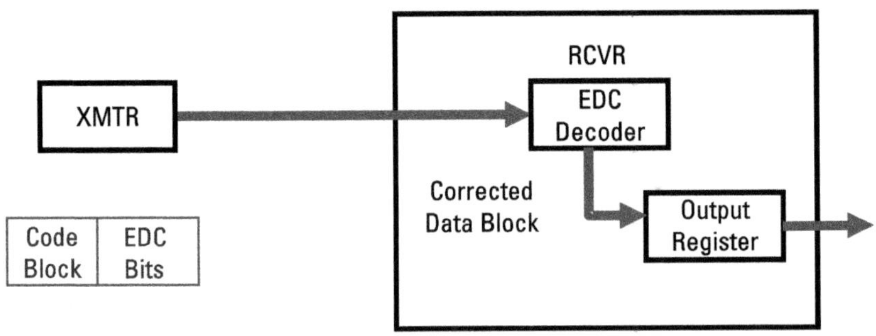

Figure 5.9 Forward error correction requires that an error detection and correction code be placed on each code block. The EDC code is decoded to correct bit errors and the corrected code is output.

5.3.3 EDC

However, an EDC code provides forward error correction. Such a code will detect bad bits (or bytes) and correct them in the received data stream up to some bit (or byte) error rate. The power of the code increases with the number of extra bits or bytes that are added to the data block.

There are two classes of EDC codes. A convolution code is most efficient for randomly spread bit errors. It corrects individual bits. The power of a convolutional code is stated as (n/k), which indicates that there are a total of n output code bits for k information bits. That is, $n - k$ additional EDC code bits are added for each k information bits.

The second class of EDC codes comprises block codes. Block codes correct whole data bytes and are generally more efficient when bit errors come in groups. An example of such a case is a frequency-hopping signal (which you will recall must be digital). If the transmitter hops to a frequency at which there is a strong interfering signal, all of the bits sent at that frequency will be wrong. Actually, there will most likely be close to 50% bit errors. Thus, several contiguous bytes will have many errors.

Partial-band jamming is a technique in which some (but not all) of the hopping slots is a technique often used to jam frequency-hopping communication systems. If encountered, it will also cause groups of erroneous bytes when the hopper hops to one of the jammed channels.

The power of a block code is stated as (n, k) meaning that there are n bytes (or symbols) sent for each k information symbol. Thus, $n - k$ extra bytes are added for each n information symbols sent.

An example of a block code is the (31,15) Reed-Solomon code that is used in Link 16, which provides real-time interconnection among airborne, shipboard, and ground military assets. (Note that this code is also used in space broadcasting of compressed television signals.) This specific code can correct $(n - k)/2$ bad symbols in each n symbol sent. It can also correct one fewer and give an indication to about 10^{-3} accuracy whether there are any additional uncorrected errors. Because this code sends 31 total bytes for each 15 information bytes, the digital bit transmission rate is more than twice the rate that information bits are sent. In general, this means that over twice the bandwidth is required to send information at any given rate. The advantage is that all of the received bytes will be corrected as long as not more than 8 of 31 bytes contain errors.

5.3.4 Interleaving

When using a block code to protect a frequency hopping link, it is common to transmit a whole block of bytes (i.e., 31 bytes for a 31,15 code) during a single hop. Remember that an occupied hop (i.e., at the frequency of an interfering signal) will cause all of the received bits to be bad. To overcome this problem, the transmitted bytes are interleaved so that not more than 8 of 31 bytes (in this case) will be transmitted at one frequency. Figure 5.10 shows a linear interleaving scheme in which the second 8 bytes are delayed into the next hop, the next into the following hop, and so forth. Thus, no more than 8 contiguous bytes will be lost during an occupied hop. Note that pseudorandom interleaving over a somewhat longer series of bytes is common. Any interleaving approach will cause some increase in latency.

5.3.5 Protecting Content Fidelity

One very important requirement for networking is that correct information arrives at a remote location. Because most information is sent digitally, this means that the bit error rate must be low enough to allow proper functioning of the networked activity.

5.4 Digital Signal Modulations

5.4.1 Single Bit per Baud Moduatlions

A digital waveform cannot be directly transmitted; it must be modulated onto an RF carrier using one of several modulations. Some of the modulations carry one bit per transmitted baud and some carry multiple bits per baud. The choice of modulation impacts the amount of bandwidth required to carry a given number of bits per second of information, and the percentage of bit errors that will be caused by the SNR in the transmitting link. This discussion will require 2 months.

Figure 5.11 shows three of the waveforms that carry 1 bit per baud. These are pulse amplitude modulation (PAM), frequency shift keying (FSK), and on-off keying (OOK). PAM generates one modulating amplitude for a 1 and another for a 0. FSK carries a 1 at one frequency and a 0 at another frequency. OOK is shown with a signal present for a digital 1 and no signal for a 0; these can be reversed. Generally, the bandwidth required to send one of these

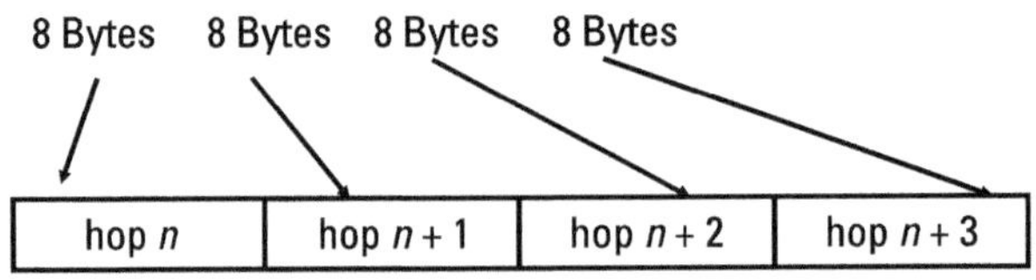

Figure 5.10 Interleaving places adjacent data into other parts of the signal stream to protect against systematic Interference or jamming.

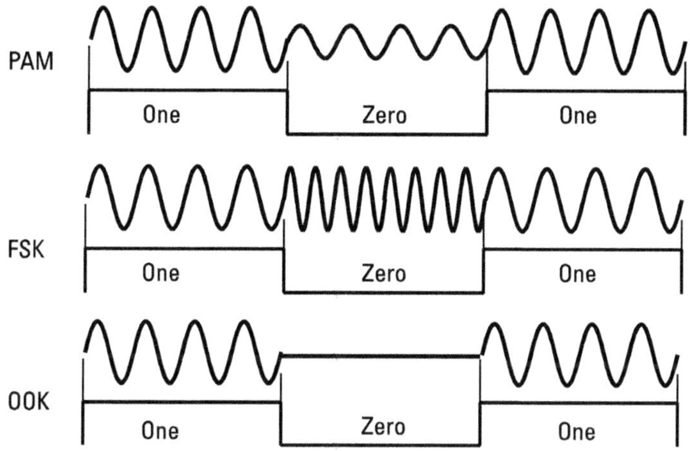

Figure 5.11 Digital information can be carried by number of modulations, including pulse amplitude modulation, frequency shift keying, and on off keying. Each has a unique modulation condition for a one and another for a zero.

codes is 0.88 × the bit rate. This is the width of the frequency spectrum of the modulated signal 3 dB down from the peak of the curve shown in Figure 5.5.

Figure 5.12 shows two waveforms that carry digital information by phase modulating a carrier. Binary phase-shift keying (BPSK) is shown with a zero phase shift when a 1 is carried and a 180° phase shift when a 0 is carried. These can be reversed. Quadrature phase shift keying (QPSK) has four defined phases, 90° apart. Each of these phase conditions defines two bits of information. As shown in the figure, a 0° phase shift indicates a "zero, zero" digital signal, 90° indicates a "zero, one" signal, and so on. Obviously, any two binary values can be assigned to any of the four phase states. Also shown in this figure are signal vector diagrams for each of these modulations. In a signal vector diagram, the length of the arrow indicates the signal amplitude and the angle of the arrow indicates its phase. The arrow rotates counterclockwise 360° during each RF cycle of the transmitted signal. In this case, the phases shown are relative to a reference signal.

5.4.2 Bit Error Rates

Figure 5.13 shows the signal with noise. The noise vector will have some statistically defined amplitude and phase pattern. The received signal is the vector sum of the transmitted signal vector and the noise vector. Thus, the shaded circle is the locus of the ends of the signal and noise vector.

Figure 5.14 shows the decision process in the receiver when a signal with noise is received. The abscissa of this diagram is the modulation dimension. The ordinate is the probability that the received signal (with noise) will be at

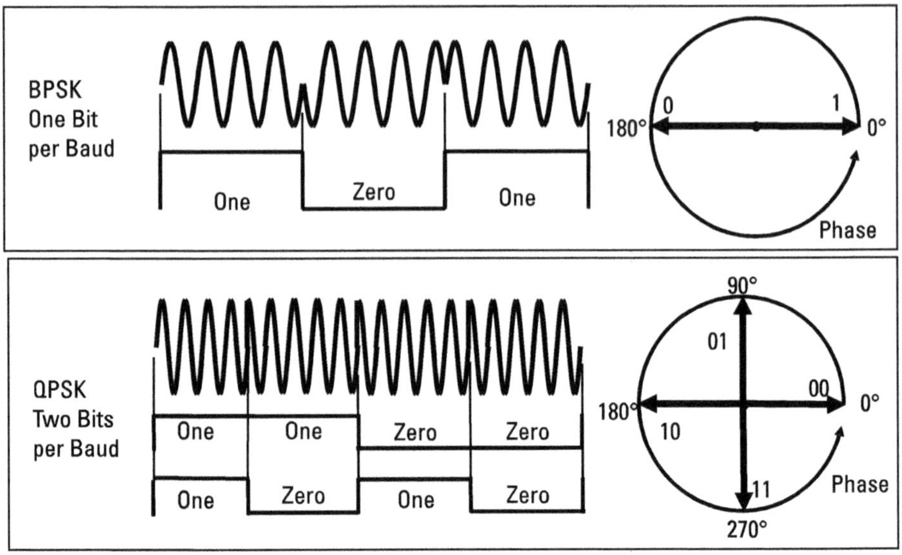

Figure 5.12 Two common digital modulations carry information in the phase of the transmitted signal. Binary phase shift keying has two phase positions and one bit per transmitted baud. Quadrature phase shift keying has four phase positions and two bits per transmitted baud.

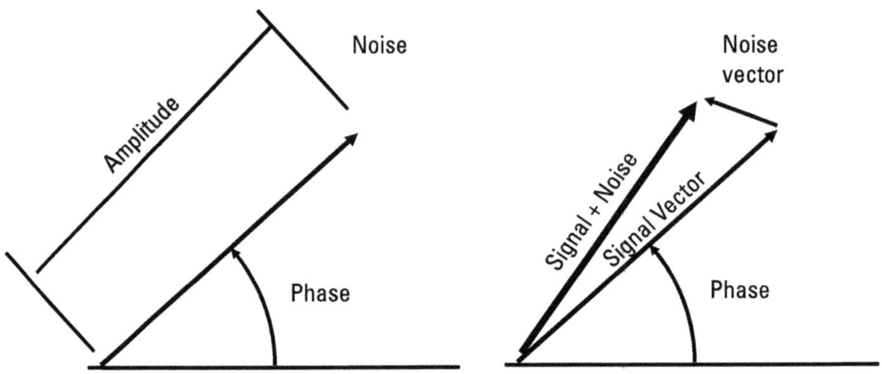

Figure 5.13 The received signal + noise has the noise vector statistically distributed at the end of the signal vector. The received signal is the vector sum of the transmitted signal vector and the noise vector.

each modulation value. The modulation dimension is frequency in FSK modulation, amplitude in PAM modulation and phase in PSK modulation. If the noise is Gaussian, the modulation value of the received signal (for example, a 0) will have the probability distribution of the Gaussian curve centered on the value transmitted for a zero. Likewise for a transmitted 1, the probability that

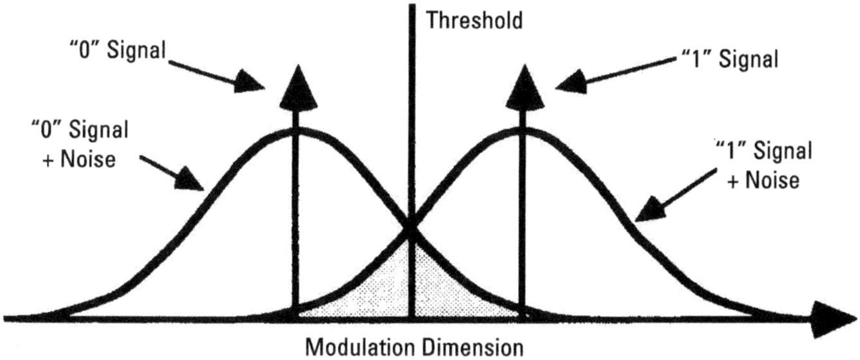

Figure 5.14 The receiver has a threshold in the modulation dimension (amplitude, frequency or phase) that determines whether a received signal with noise is to be declared a "one" or a "zero."

the received frequency will have any modulation level is defined by the Gaussian curve centered on the 1 value. There is a threshold value that determines whether a 0 or 1 is received. If the received signal is to the left of the threshold, a 0 is output. If it is to the right, a 1 is output. The shaded area under both Gaussian curves represents the incorrectly received bits. The greater the predetection SNR, the narrower this Gaussian curve will be. The bit error rate is the number of incorrectly received bits divided by the total number of received bits. It is inversely proportional to the predetection SNR. Consider that the bit error area under the two curves is smaller if the predetection SNR is greater, because the Gaussian curves are tightened around the noise free one and zero values.

A graph of bit error rate versus E_b/N_0 is shown in Figure 5.15. Note that E_b/N_0 is the predetection SNR adjusted for the bit rate (in bits per second) to bandwidth (in hertz) ratio. This graph will have a different curve for each type of modulation. The more coherent the waveform, the farther to the left the curve moves. For this modulation, an E_b/N_0 value of 11 dB will produce a bit error rate of 10^{-3} (i.e., one of each thousand received bits will be incorrect).

Figure 5.15 shows the probability that the received signal will be at any given modulation value. If the noise caused the received signal to be on the wrong side of the 1 versus 0 threshold, a bit error occurred. The solid line curve in Figure 5.16 is the same as Figure 5.15. Now let us see what happens to the diagram if the SNR is increased. The diagram changes to that represented by the dash line curves. Notice that the dashed probability curves are much tighter to the transmitted modulation values and that the area under the two curves when the received signal (with noise) is on the wrong side of the threshold is significantly smaller. Thus, the bit error rate is reduced.

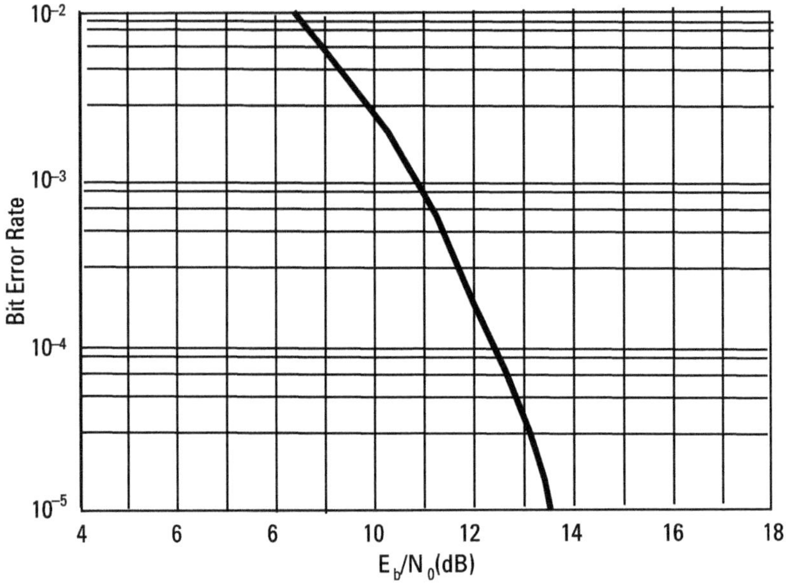

Figure 5.15 The bit error rate in a received signal is an inverse function of E_b/N_0.

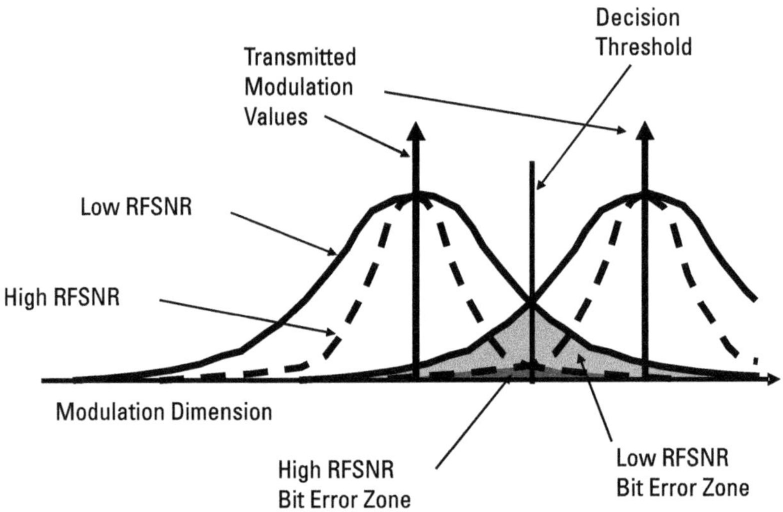

Figure 5.16 As the signal to noise ratio in a received digital signal increases, the bit error rate decreases.

5.4.3 m-ary PSK

Figure 5.17 shows a digital waveform that carries more bits per transmitted baud. This is called an m-ary phase shift-keyed signal. In this case, m is 16

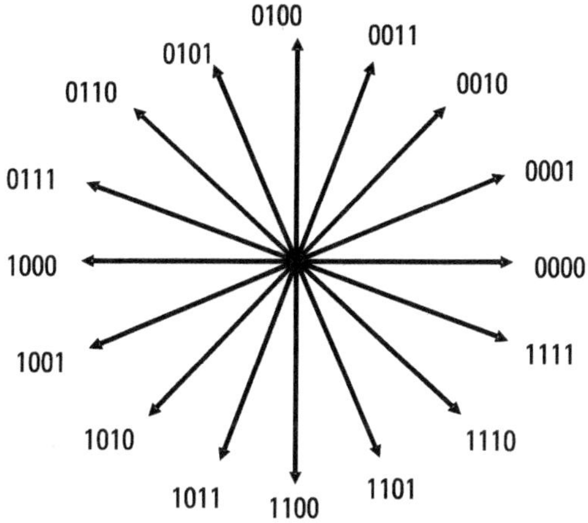

Figure 5.17 An m-ary phase shift keyed modulation has m phase positions. In this case there are 16 phase positions and each phase value defines four bits of information.

because there are 16 defined phases. The radial vectors in the diagram show each of the transmitted phase vectors (without noise). There are four bits represented by the transmitted phase of each baud as indicated in the diagram. This is a highly efficient modulation because four bits are sent in each baud. Thus, the transmitted bandwidth is only one-fourth of that required for transmission of any given data bit rate using one of the modulations discussed in Section 5.3.1. The shown 16-ary PSK requires approximately 7.5 dB greater predetection SNR to provide the same bit error rate achieved by BPSK. This is because phase noise on the received signal causes each of the signal and noise vectors in Figure 5.7 to move away from their transmitted phases. The closer the assigned phase angles are to each other, the greater the vulnerability to noise. Thus, the requirement for greater SNR for any required level of bit error rate.

5.4.4 I&Q Modulations

Figure 5.18 shows an I&Q modulation. I&Q refers to in-phase and quadrature and is used to describe this family of modulations because the location of the end of the signal vector (in I&Q space) for each transmitted baud identifies it. Each of the 16 locations shown in this diagram is a transmitted signal state defined by the phase and the amplitude of the carrier. Because there are 16 locations, each represents four binary bits. The advantage of I&Q modulation over m-ary PSK is that the locations can be more widely separated in parametric space and are thus less subject to bit errors caused by noise on the received signal.

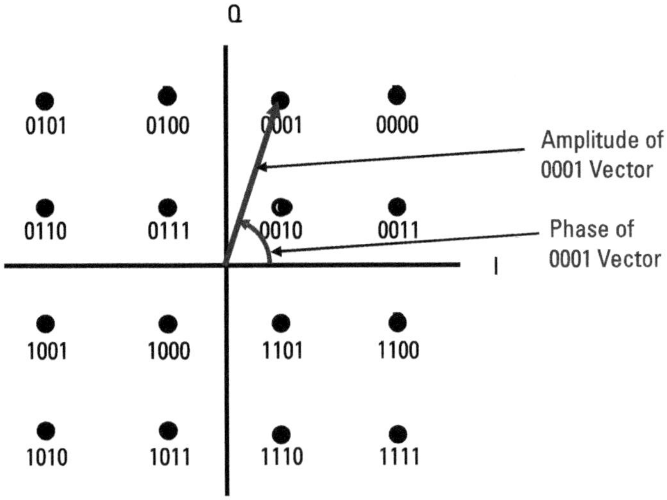

Figure 5.18 This I&Q modulation has sixteen amplitude & phase conditions so each condition defines four bits of information.

5.4.5 BER Versus E_b/N_0 for Various Modulations

Figure 5.19 directly compares the bit error rate versus E_b/N_0 for three types of modulation. The left curve is for the family of modulations that carry one bit of data in each transmitted baud. The middle curve uses a particularly efficient waveform to move between one and zero modulation values, and the right curve is for a modulation that carries multiple bits per transmitted baud. Note that the shapes of the three curves are the same, but they are offset horizontally. It is important to note that the bandwidth required to carry the information by each of these modulations also varies. The left curve is the least frequency efficient and the right curve is most frequency efficient.

5.4.6 Efficient Bit Transition Modulation

Figure 5.20 shows two frequency efficient modulations. The top curve makes transitions between 1 and 0 along a sinusoidal path. The bottom curve shows a minimum shift keyed (MSK) modulation. This modulation is very efficient because the waveform moves between the zero and one positions in the most energy efficient way. Table 5.1 shows the null-to-null and 3-dB bandwidth for minimum shift keying versus that for the less frequency efficient waveforms. Because the 3-dB bandwidth is typically taken as the required transmission bandwidth, an MSK signal requires only three-fourths of the bandwidth.

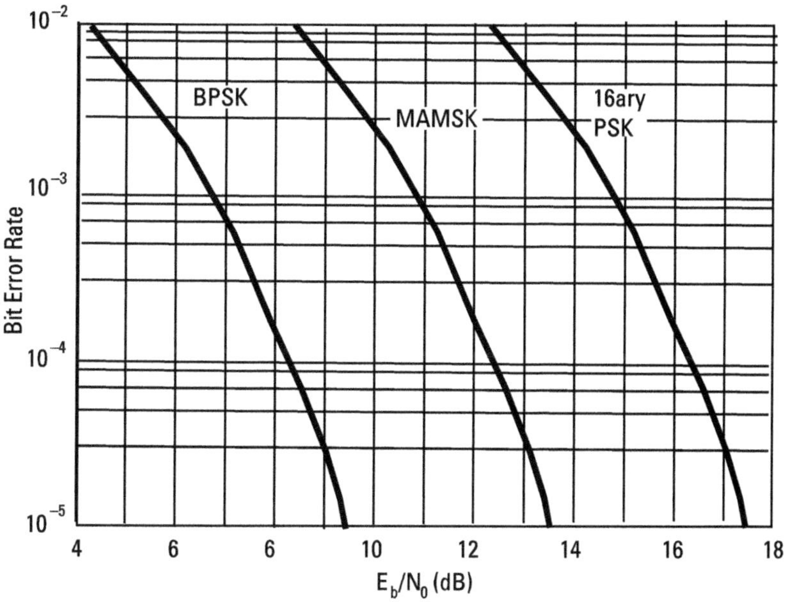

Figure 5.19 The bit error rate in a received signal is an inverse function of E_b/N_0.

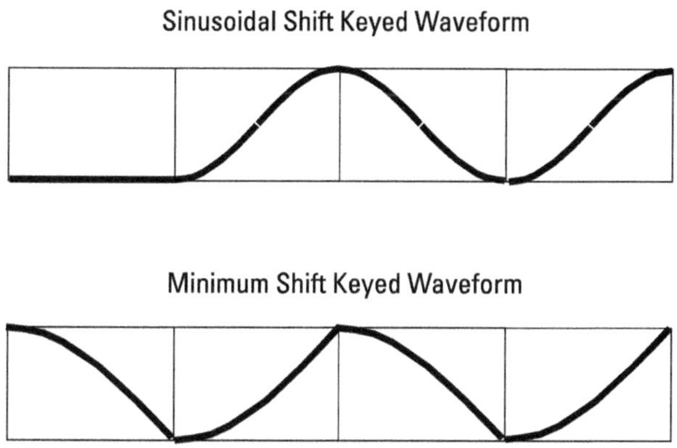

Figure 5.20 Shaped waveforms move between zero and one values in such a way that the transmission bandwidth is reduced.

5.5 Digital Link Specifications

To pass data from one location to another, the digital data link must have adequate link margin. This margin includes some elements that are clearly measurable, like link distance and system gains and losses. It also includes some elements that are statistical (like weather). The link availability is related to the

Table 5.1
Bandwidth Versus Waveform of Digital Signal

Waveform	Null-to-Null Bandwidth	3-dB Bandwidth
BPSK, QPSK, PAM	2 × code clock	0.88 × code clock
MSK	1.5 × code clock	0.66 code clock

link margin. The greater the margin, the higher the probability that the link will be performing up to full specifications at any given time.

The link, including a few elements that have not been discussed earlier, is shown in Figure 5.21.

5.5.1 Link Specifications

Typical specifications for an overall digital link are shown in Table 5.2.

5.5.2 Link Margin

The link margin is the amount that the received signal power exceeds the receiver sensitivity.

$$M = P_R - S$$

where M is link margin (in decibels), P_R is signal strength at the receiver system input (dBm), and S is receiver system sensitivity at output of receiving antenna, including the effects of any cable losses from the antenna (dBm).

The received signal power is a function of the ERP, propagation losses, and receiving antenna gain.

$$P_R = ERP - L + G_R$$

where ERP is the effective radiated power from the transmitting antenna (dBm) including adjustments for transmitting antenna pointing error gain reduction

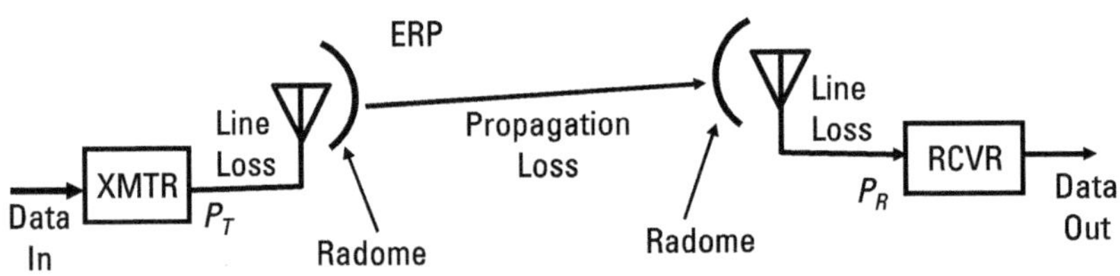

Figure 5.21 The received power in a data link receiver is a function of all of the gains and losses between the transmitter and receiver.

Table 5.2
Typical Link Specifications

Specification	Definition
Maximum range	Maximum operating range of link
Data rate	Transmission data bit or symbol rate
Bit error rate	Ratio of bits incorrectly received
Angular tracking rate	Maximum angular tracking rate and angular acceleration of transmit or receive antennas
Weather	Rain conditions under which the link will meet its other specifications
Antijam capability	The jamming to received signal ratio under which the link will meet full performance specifications
Antispoof capability	The authentication measures of the system to prevent hostile insertion of false data

and radome loss; L is the propagation loss between the transmitting and receiving antennas, including line-of-sight or two-ray propagation loss, diffraction loss, atmospheric loss, and rain loss (all in decibels); and G_R is the receiving antenna gain including radome loss and antenna and gain reduction caused by pointing error.

The three important propagation loss models used to predict general future performance of systems in dynamic conditions are discussed in Chapter 6.

Figure 5.22 shows the antenna pointing error in the transmitting antenna. This same geometry applies to the receiving antenna not perfectly pointed at the transmitter. In our previous radio propagation discussions related to intercept and jamming situations, we discussed transmitting antenna gain toward the receiver and receiving antenna gain toward the transmitter. This gain has been used in jamming and intercept equations. In that case, we were typically talking about jamming or intercepting into or out of radar main beam versus side lobes. In this case, we are generally in the main lobe of the link antennas, but away from the antenna boresight by a small angle. The gain reduction relative to boresight can be calculated with reasonable accuracy, but it is normally more practical to get the gain patterns of the antennas from the manufacturers and determine the gain reduction at the angle from boresight equal to the specified maximum antenna pointing error.

5.5.3 Sensitivity

The receiver system sensitivity, as discussed in Chapter 6 is:

$$S(\text{dBm}) = kTB(\text{dBm}) + NF(\text{dB}) + RFSNR(\text{dB})$$

where kTB is the internal noise in the receiver, referenced to the receiver input.

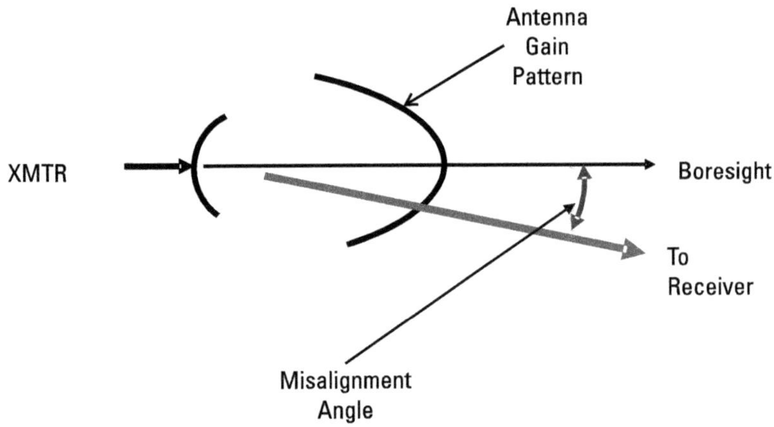

Figure 5.22 The transmit antenna gain in the direction of the receiver is reduced from the boresight gain by a factor determined from the offset angle.

Within the atmosphere, a common expression for kTB is –114 dBm + 10 log(bandwidth/1 MHz). This assumes that the receiver is at 290K.

NF, the system noise figure, is the amount of noise above kTB added by the receiver system, referred back to the receiver input.

RFSNR is the predetection SNR. In some literature, this is called the CNR (the carrier-to-noise ratio) to differentiate it from the output SNR. Note that the signal power used in the calculation is the total predetection signal power, not just the carrier power, which is why we use RFSNR in the *EW 101* series.

In digital links, the RFSNR is related to the bit error rate as a function of a ratio called E_b/N_0 as shown in Figure 5.23. There are two typical curves shown in this figure; however, the actual curve for a specific link is determined by the digital modulation used to carry the data.

5.5.4 E_b/N_0 Versus RFSNR

E_b/N_0 is the energy per bit divided by the noise density (i.e., the noise per hertz of noise equivalent bandwidth).

$$E_b = S/R_b$$

where S is the received signal power (PR in Figure 5.1) and R_b is the bit rate (bits per second). Note that this refers to the data bits rather than all of the bits sent (i.e., not the synchronization and error correction bits).

$$N_0 = N/B$$

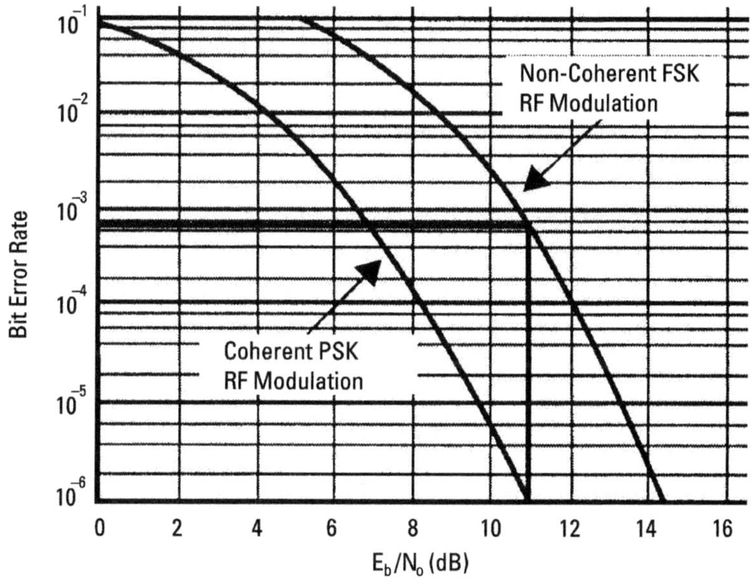

Figure 5.23 The bit error rate in a demodulated digital signal is a function of E_b/N_0.

where N is the noise in the receiver (i.e., kTB + Noise figure) and B is the noise equivalent bandwidth that can be approximated as equal to the symbol rate.

Thus, E_b/N_0 is related to RFSNR by the equation:

$$E_b/N_0 = SB/NR_b$$

In decibel form, this equation is:

$$E_b/N_0\,(\text{dB}) = RFSNR\,(\text{dB}) + \left[B/R_b\right](\text{dB})$$

5.5.5 Maximum Range

The maximum range is the distance at which the received signal is equal to the sensitivity plus the specified operating margin. Note that there is a trade-off between margin and maximum range and that, for the moment, we are ignoring any weather related losses. To determine the maximum range, start with the received power formula in Section 5.5.1. Then expand the loss term (L) for the appropriate propagation model. In most data link cases, this will be the line-of-sight model, making the received power formula:

$$P_R = ERP - 32 - 20\log(d) - 20\log(F) + G_R$$

where P_R is the signal strength into the link receiver (in dBm), d is the link distance (in kilometers), F is the operating frequency (in megahertz), and G_R is the receiving antenna gain (decibels).

Both the ERP and the G_R values are reduced by the appropriate antenna pointing losses. Then set P_R equal to the sensitivity (S) in dBm + the required link margin (M) in decibels.

The above equation is now:

$$S + M = ERP - 32 - 20\log(d) - 20\log(F) + G_R$$

Solving for the range term:

$$20\log(d) = ERP - 32 - 20\log(F) + G_R - S - M$$

Then solve the 20 log(d) term for distance, which is the maximum range in kilometers:

$$d = \text{antilog}\{[20\log(d)]/20\} \text{ or } 10^{\{[20\log(d)]/20\}}$$

5.5.6 Minimum Link Range

The minimum link range must also be considered. This is impacted by the dynamic range of the link's receiving system and by the angular tracking rate. The dynamic range is the range of received power over which the receiver can operate properly without saturation. In Chapter 6, dynamic range is discussed as it applies to EW and reconnaissance systems. These systems must have a wide instantaneous dynamic range to allow reception of weak signals in the presence of strong interfering signals, and cannot typically include automatic gain control (AGC). However, a data link receiver is designed to receive only its intended data signal, so it can use AGC to allow operation over a very wide range of received signal strength levels. The link angular tracking rate is discussed in Section 5.5.9.

5.5.7 Data Rate

The data rate is the number of data bits per second that can be carried by the link. Note that this is not the total number of transmitted bits per second, as there will be synchronization, address, and parity or error correction bits as shown in Figure 5.24. This relates to bandwidth. Typically the transmission bandwidth would be the 3-dB bandwidth of the digital spectrum shown in Figure 5.25. Relating this to the sensitivity discussion in Section 5.1.2, this bandwidth is the "B" in kTB.

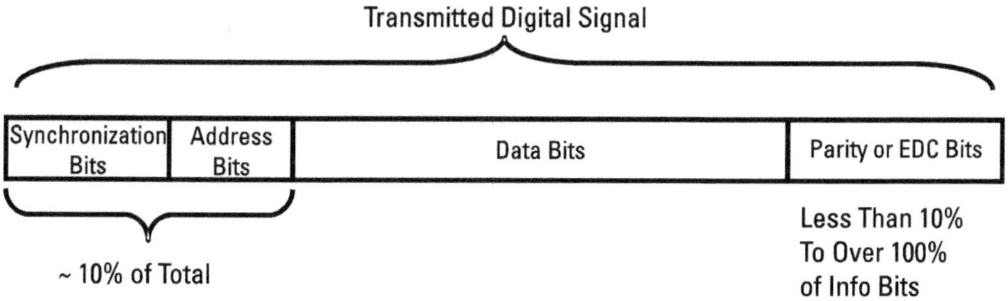

Figure 5.24 A transmitted digital signal contains synchronization, address, information, and parity or EDC bits in addition to the data bits.

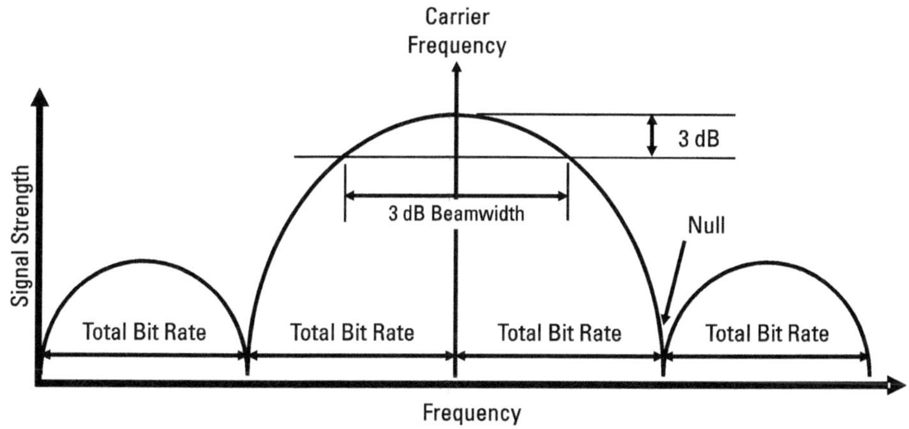

Figure 5.25 The typical transmission bandwidth for a digital signal is the 3 dB bandwidth of the digital signal spectrum of the full digital signal.

5.5.8 Bit Error Rate

The bit error rate is the ratio of incorrectly received bits to the total number of bits sent. In Section 5.1.2, we covered the definition of E_b/N_0. The predetection SNR (RFSNR) defined in this discussion is part of the sensitivity calculation.

5.5.9 Angular Tracking Rate

The link angular tracking rate specification relates to the geometry of the link application. If one or both of the ends of the link are on moving platforms and have narrow beam antennas, the pedestals on which those antennas are mounted must be able to track the other link terminal at the maximum cross-range velocity at the minimum specified range as shown in Figure 5.26. This diagram

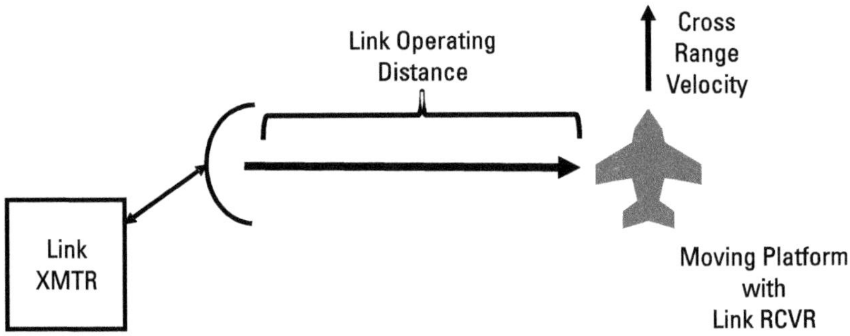

Figure 5.26 The required angular tracking rate for a link is a function of the maximum cross range velocity of the other link terminal and the minimum operating range.

illustrates a fixed link transmitter and a moving link receiver. It could as well use a fixed receiver with a moving transmitter, or both elements could be moving.

5.5.10 Tracking Rate Versus Link Bandwidth and Antenna Types

One of the important factors in the selection of links connecting moving platforms is the requirement for narrow-beam antennas. As the transmitted data rate dictates the required transmission bandwidth and receiver sensitivity varies inversely with bandwidth, wide bandwidth links may require significant antenna gains at the transmitting or receiving ends (or both) to achieve adequate link performance. Increased antenna gain implies reduced antenna beamwidth, which increases the criticality of antenna pointing accuracy.

In general, a low data rate link can be implemented with a simple dipoles or similar antennas on moving platforms and relatively wide-beam antennas on fixed link terminals. This minimizes antenna pointing problems. However, a wideband link may require directional antennas at both ends. This can make antenna pointing requirements a significant issue.

5.5.11 Weather Considerations

First, consider atmospheric attenuation. Figure 5.27 shows the atmospheric attenuation per kilometer as a function of frequency. Note that there are two curves in this figure. One is for standard atmospheric conditions. This curve assumes a humidity level that supports 7.5 grams of water content per cubic meter of air. The other curve is for dry atmospheric conditions (i.e., 0 grams of water per cubic meter). Note that in extremely dry air, the loss at low frequencies is significantly lower than for standard air. To use either curve, trace up from the frequency to the appropriate line, then left to the loss per kilometer scale. The atmospheric link loss is this number multiplied by the maximum specified link operating range.

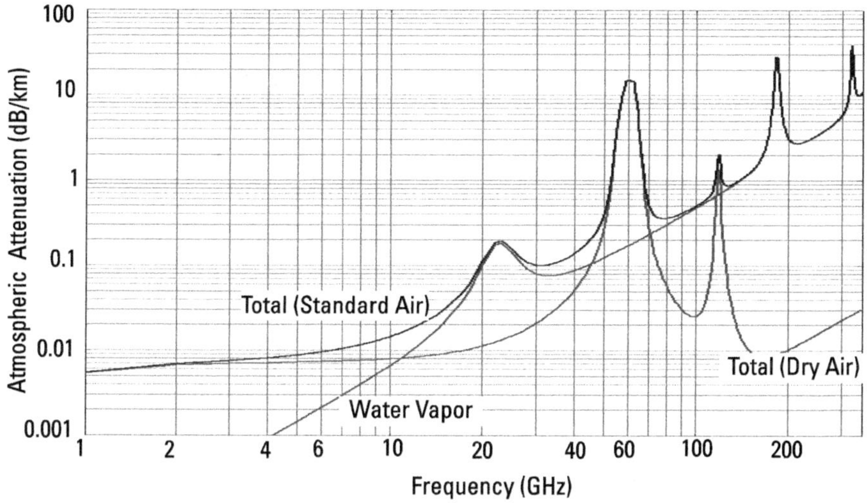

Figure 5.27 Atmospheric attenuation is a function of frequency and humidity.

If the link is from a platform on or near the ground to a satellite, Figure 5.28 applies. This is the loss through the whole atmosphere as a function of elevation angle to the satellite.

Now consider rain. Figure 5.29 shows the loss per kilometer for various rain rates. Again, start from the frequency and go up to the appropriate curve. Then go left to the loss per kilometer that the link passes through rain at that rate.

When specifying rain loss margin for a link, we still have the problem of estimating what the rain rate should be. A common approach is to start with the link availability specification. For example, the link may be specified to be available 99.9% of the time, that is, to have an unavailability rate of 0.1% (1.44 minutes per day). There is a wealth of online data about the percentage of time that a particular rain rate can be expected in just about any region on the Earth. A fairly common number for 0.1% of the time is around 20 mm/hr. If this number were to apply to the part of the Earth where your link is to operate, you could use line D (or slightly above curve D) in Figure 5.29. The loss per kilometer would then be multiplied by the maximum link distance.

If the link goes to a satellite, it is necessary to calculate the path length from the terminal on or near the Earth up to the elevation at which the temperature can be expected to be 0°C. Tables and graphs of this elevation can be found online. For our 99.9% availability case, the 0°C isotherm is about 5 km high at latitudes below 25° and then sweeps down to 1 km at 70° latitude. The path length in the rain is calculated by the formula:

$$D_{RAIN} = \Delta E1/\sin(E)$$

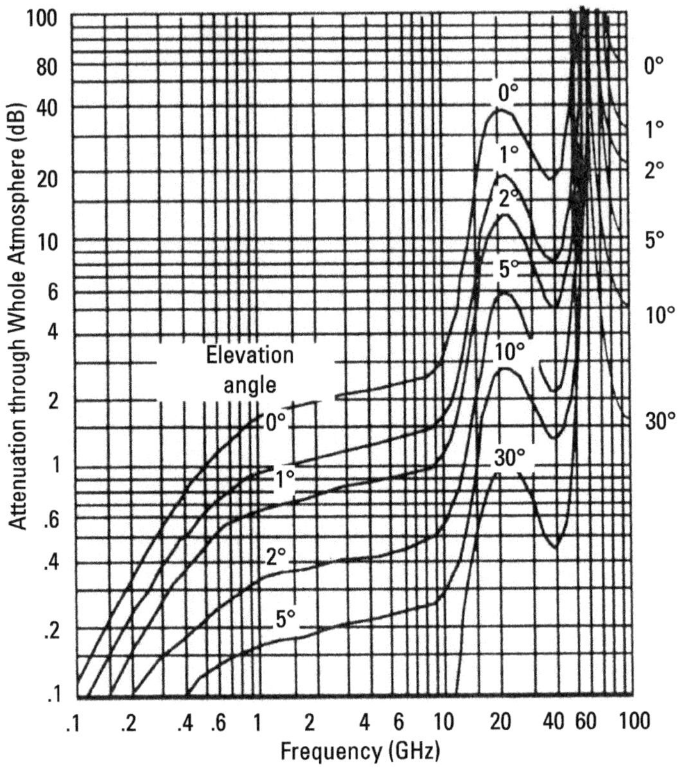

Figure 5.28 Atmospheric loss in a satellite to ground link is a function of frequency and satellite elevation angle.

where D_{RAIN} is the path length to which rain attenuation applies, ΔEl is the difference in elevation between the lower platform and the 0°C isotherm and E is the elevation angle to the satellite from the lower platform.

Once D_{RAIN} is calculated, multiply the rain attenuation per kilometer determined from Figure 5.9 by this distance.

5.5.12 Antispoof Protection

It is very important that an enemy not be able to enter your data link to pass false information. The general answer to this problem is to require authentication. In even the simplest voice links, a password is required before a user can enter information into the network. This also applies to manual entry onto digital links. For multiple user high-duty cycle digital networks, this same approach can be used. However, there is a great risk of compromise if an enemy is able to determine the password.

One very common and very effective type of authentication is encryption. If a high-level encryption is used, it is extremely unlikely that an enemy

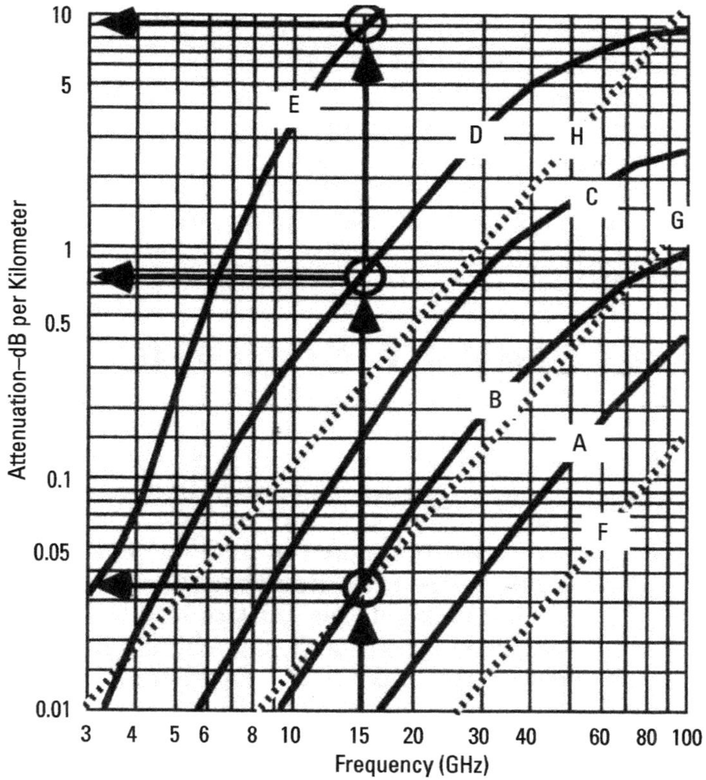

Rain				
	A	0.25 mm/hr	.01 in/hr	Drizzle
	B	1.0 mm/hr	.04 in/hr	Light Rain
	C	4.0 mm/hr	.16 in/hr	Moderate Rain
	D	16 mm/hr	.64 in/hr	Heavy Rain
	E	100 mm/hr	4.0 in/hr	Very Heavy Rain

Fog			
	F	0.032 gm/m^3	Visibility greater than 600 meters
	G	0.32 gm/m^3	Visibility about 120 meters
	H	2.3 gm/m^3	Visibility about 30 meters

Figure 5.29 Rain loss is a function of frequency and the rate of rainfall.

can enter the net at all. This approach also provides the important feature of message security.

5.6 Antijam Margin

Protection against jamming of a data link can be provided in many ways, including:

- Maximizing the transmitter ERP;
- Using narrow beam antennas;
- Nulling signals received from directions other than from the link transmitter;
- Using spread spectrum modulations;
- Employing error correction codes.

Jamming effectiveness is measured in terms of the jammer-to-signal ratio (J/S). The higher the J/S (commonly stated in decibels), the better the jamming. Maximizing the ERP reduces the J/S by increasing "S." Narrow-beam antennas increase the probability that a jamming signal will have to enter the receiver through antenna side lobes that have significantly less gain than the main beam, which is presumably oriented toward your link transmitter (i.e., reducing "J").

As shown in Figure 5.30, a side-lobe canceller has an antenna with its gain in side-lobe directions. Any signal that is received stronger in this special antenna than from the regular link antenna causes a phase-reversed copy to be added to the received link signal, thereby cancelling (or significantly reducing) the jamming signal. Reduction of received jamming signals can also be implemented with a phased array antenna that can generate multiple nulls in selected directions.

Spread spectrum signals are discussed in Chapter 2 and in more detail in Chapter 7. Each of the three techniques described (frequency hopping, chirp, and direct sequence spread spectrum) causes the transmitted signal to be spread (pseudo-randomly) over a much wider frequency range than required to carry link information. The receiver reverses the pseudo-random spreading of the received link signal and thus provides a processing gain. This gain enhances the received link signal but does not enhance jamming signals because they do not have the pseudo random function that has been applied to your transmitted link signal. This process reduces the J/S, because the jamming signals exit the despreader with significant attenuation. The formula for the jamming margin provided by processing gain is:

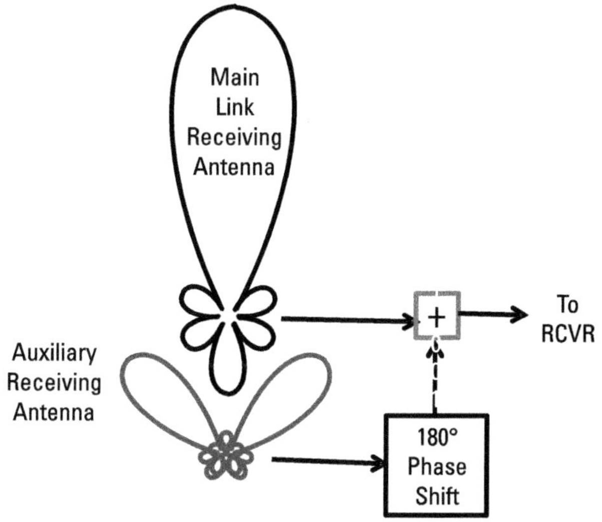

Figure 5.30 The received power of a jamming signal can be reduced by a sidelobe canceler.

$$M_J = G_P - L_{SYS} - SNR_{RQD}$$

where M_J is the jamming margin (in decibels), G_P is the processing gain (in decibels), L_{SYS} is system losses (in decibels) note these are often set to zero, SNR_{RQD} is the SNR (i.e., above the jamming) that the link requires for proper operation

5.7 Link Margin Specifics

Link margin is the difference between the minimum signal level in the receiver for proper link connectivity and the actual signal level received as the link is configured.

Table 5.3 shows the items that need to be considered in calculating the link margin. This table was adapted from a similar table in [2].

The subtotal items in this table are related by the following two formulas:

$$RSP = ERP - TPL + TRG$$

where RSP is the received signal power, ERP is the effective radiated power, TPL is the total path loss, and TRG is the total receiver gain.

$$NLM = RSP - RSS$$

Table 5.3
Link Budget

Link Margine Inputs	**Link Margin Subtotals**	**TSP & RSS**	**Net Margin**
+Transmitter power (NLM) TSP - RSS -Transmitter loss +Transmit antenna gain -Transmitter Radome loss	Effective Radiated Power (ERP) Add left column items	Total Signal Power (TSP) Add second column values	Net Link Margin (NLM) TSP - RSS
-Path loss -Transmit antenna pointing error -Rain loss (0.999) -Multipath -Atmospheric loss	Total Path Losses (TPL) Add left column items		
-Receiving radome loss +Receiving antenna Gain -Receiving Polarization loss -Receiver loss -Receiving antenna pointing error	Total Receiver Gain (TRG) Add left column items		
+Receiver Noise Figure +kTB +Predection Signal to Noise Ratio	Receiver Sensitivity (RSS) Add left column items	Second column value (RSS)	

where *NLM* is the net link margin, *RSP* is the received signal power, and *RSS* is the receiver system sensitivity.

5.8 Antenna Alignment Loss

The most accurate way to assign a link budget loss for antenna misalignment is to get the antenna gain pattern from the manufacturer and read the gain loss relative to boresight gain for the angle equal to the pointing accuracy specification. This is still a good idea, but it is very handy to have a formula for the loss versus pointing error for an ideal parabolic antenna. The following formula gives the 3-dB beamwidth as a function of wavelength and antenna diameter

$$\alpha = 70\lambda/D$$

where α is the 3-dB bandwidth in degrees, λ is the wavelength in meters, and D is the diameter of the antenna in meters.

If it is more convenient to input operating frequency than wavelength, the formula becomes:

$$\alpha = 21{,}000/DF$$

where α is the 3-dB bandwidth in degrees, F is the operating frequency in megahertz, and D is the diameter of the antenna in meters.

The formula for the gain reduction as a function of the error angle and the 3-dB beamwidth (for relatively small offset angles) is:

$$\Delta G = 12(\theta/\alpha)^2$$

where ΔG is the gain reduction in decibels because of antenna misalignment, θ is the antenna pointing accuracy in degrees, and α is the 3-dB beamwidth.

A convenient decibel formula for the gain reduction as a function of frequency, antenna diameter, and antenna pointing accuracy is:

$$\Delta G = -0.565 + 20\log(F) + 20\log(D) + \theta^2$$

where ΔG is the gain reduction in decibels because of antenna misalignment, θ is the antenna pointing accuracy in degrees, F is the operating frequency in megahertz, and D is the antenna diameter in meters.

5.9 Digitizing Imagery

An important issue in net-centric warfare is the transportation of imagery from the point of origin to the point at which an operator or other decision-maker needs to access the information carried in the imagery. The imagery can be from a large part of the electromagnetic spectrum: visible light, infrared (IR), or ultraviolet (UV).

There are two basic approaches to the capture of imagery. One way is to scan an area using a raster scan as shown in Figure 5.31. In this technique, a single sensor (IR, UV, or visible light) (or set of sensors) is directed through the angular area of interest. The spacing of the lines in the raster is close enough to provide the required resolution of the picture in the vertical dimension. The horizontal resolution is determined by the angular movement between the samples of the data from the sensor.

In analog video, this sampled data has a frame synchronization pulse at the beginning of each picture captured and a line synchronization pulse at the beginning of each line in the raster pattern. For commercial television (in the United States), there are 575 lines in the raster and 575 samples taken per line. Every second line (alternating) is sent 60 times per second. This captures 30 full pictures per second. In Europe, there are 625 raster lines and 625 samples per line. Every second line is sent 50 times per second, yielding 25 full pictures per second. In either case, this allows full-motion video because the human eye can only see a new picture 24 times per second. This analog video signal requires a bandwidth of just under 4 MHz in full color. By digitizing the output of the scanned sensor, a digital video signal is produced.

Figure 5.32 shows the other approach to capturing imagery data. In this case, there are a number of imagery sensors in an array. Each sensor captures one pixel of the picture. The outputs of these sensors are sequentially sampled and digitized to form a serial digital signal suitable for transmission.

The bit rate of the digital signal is determined by the formula:

$$\text{Bit Rate} = \text{Frames per Second} \times \text{Pixels per Frame} \times \text{Bits per Pixel}$$

A standard, full resolution digitized video signal has 720 by 486 pixels per picture with 16 bits for each pixel. This makes 720 × 486 × 16 bits per picture.

In the United States, with 30 frames per second, this requires a bit rate of 167,961,600 bits/sec.

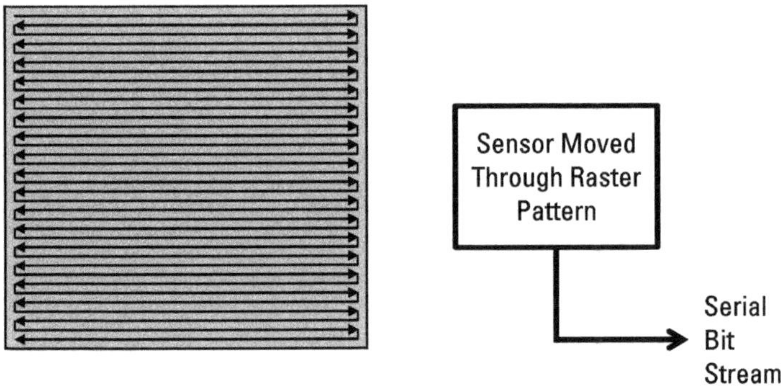

Figure 5.31 If imagery is sensed using a raster scan, the intensity of each color in each pixel is digitized into a serial bit stream.

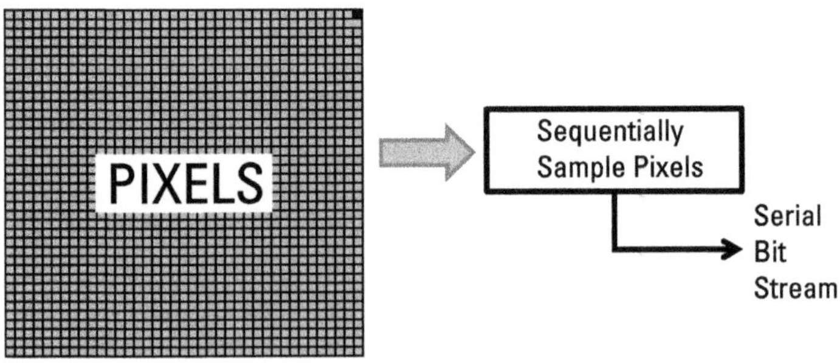

Figure 5.32 If the imagery sensor has a sensor array, the intensity of each color is digitized for each pixel and output as a serial bit stream.

In Europe, with 25 frames per second, the required bit rate is 139,968,000 bits/sec.

The type of modulation carrying this digital data could require a great deal of link bandwidth. We will discuss various ways of reducing this data rate.

5.9.1 Video Compression

There are various basic measures that will reduce the required bandwidth. One way is to transmit analog video. Unfortunately, this option has the disadvantages that analog signals are very difficult to securely encrypt and their quality can be severely reduced if transported over long distances requiring multiple transmissions. If digital video is used, the required data rate (hence, bandwidth) can be reduced using several techniques:

- Reduce the frame rate.
- Reduce the data density (i.e., reduce the resolution).
- Reduce the angular area of coverage (with the same resolution).
- Take advantage of the fact that the eye sees luminance (brightness) at twice the resolution of chrominance (color). This allows full color with 8-bit resolution per color to be captured with only 16 bits per pixel.
- Use digital data compression software.

There are three basic digital compression techniques:

- *Direct cosine transform compression* (DCT) writes a digital word to describe an 8×8 section of the picture captured. This is a very mature technique. As the SNR of the received digital signal degrades, the picture breaks into square blocks. A single bit error will take out 64 pixels and under some circumstances can take out a whole picture, which can require multiple frames to resynchronize. Therefore, systems using DCT compression must usually incorporate forward error correction.
- *Wavelet compression* performs a series of highpass filter operations on the picture, replacing a series of 1s with a single 1. After repeating this operation 10 or 12 times, a compressed digital representation of the whole picture is generated. With this approach, each bit error has the effect of slightly blurring the whole picture. This means that, in general, forward error correction is not advantageous.
- *Fractal compression* is a process in which the picture is divided into geometric shapes and a digital bit stream is generated to describe the density, color, and placement of each shape. This technique requires a great deal

of memory and processing power. The performance of this compression technique is comparable to that of DCT and wavelet compression but has the advantage of allowing significant enlargement.

Each of these techniques reduces the data rate that must be transmitted, thereby reducing the required link bandwidth. All three techniques compress each frame of video, which allows efficient editing and analysis to recover information from the digital data. The compression ratio depends on the required quality of the recovered video, but ratios of 30 to 50 are usually discussed.

Temporal compression involves removing redundant data from frame to frame.

It is possible to achieve very high compression ratios with this compression approach. The disadvantage is that digital editing becomes very difficult.

5.9.2 Forward Error Correction

By encoding transmitted digital signals with additional bits, it is possible to detect bit errors up to some limit and to correct those bit errors at the receiver. The more additional bits are incorporated, the more bit errors can be corrected. These additional bits increase the transmitted bit rate, hence the required link bandwidth.

5.10 Codes

Codes are widely used in modern communication and EW, including:

- Encryption;
- Frequency-hopping sequences;
- Pseudo random synchronization of chirp signals;
- Direct sequence chip generation.

In these applications, the codes appear to be randomly generated. They involve maximal length binary sequences which have the following characteristics:

- $2^n - 1$ bits before repeating, where n is the number of shift registers required to generate the code.
- When synchronized, the number of bit agreements is equal to the number of bits in the code.

- When unsynchronized, the number of agreements less the number of disagreements is –1.

Table 5.4 shows various numbers of shift register stages versus the length of the code before it will repeat. Note that the security of a code is related to the length of the code. A rule of thumb for military systems and applications is that a code should not repeat for two years in normal operation if security is important.

Figure 5.33 shows the shift register configuration to generate the linear seven-digit Barker code, 1110100. Note that there are three shift register stages and one feedback loop with a modulo 2 adder. There can be more feedback loops in any desired configuration. The use of binary adders in all feedback loops is characteristic of linear codes, which are used when security is not an issue.

For nonlinear codes, the feedback loops use such devices as digital AND gates, OR gates, and so forth. These are used when security is important.

When operation is initiated, all of the shift registers in Figure 5.33 are in the 1 state. Figure 5.34 shows the state of each of the shift register stages with each clock cycle. The code repeats itself after seven cycles. After each clock cycle, the state of stage 3 is shifted to stage 2 and the binary sum of states 1 and

Table 5.4
Number of Shift Register Stages Versus Length of Code

Stages	Code Length
3	7
4	15
5	31
6	63
7	127
31	2,147,483,647

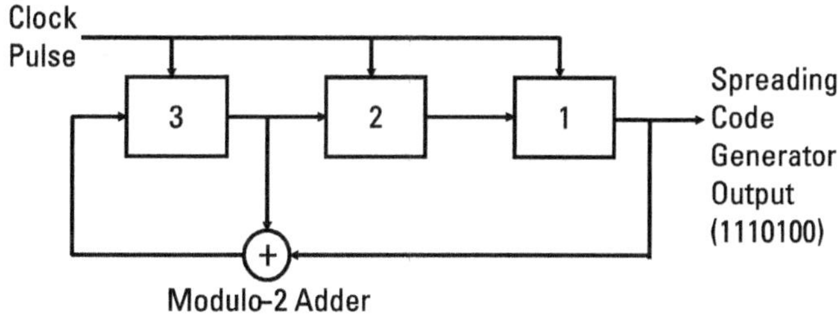

Figure 5.33 A sift register generator with 3 stages will generate a seven bit code sequence.

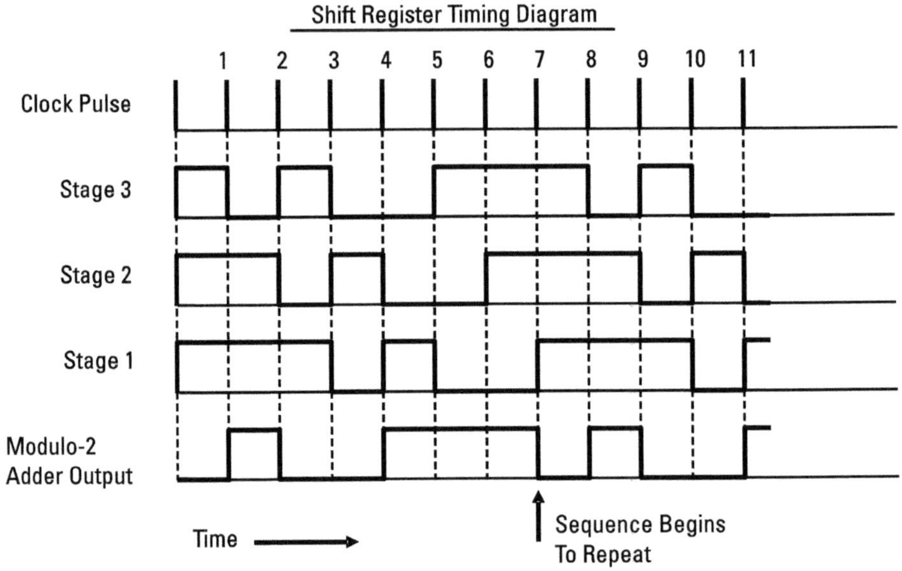

Figure 5.34 At each clock cycle, the status of each stage and the modulo 2 adder is moved to the next stage.

3 are input to stage 3. In this process, there is no carrying, that is, $1 + 1 = 0$ and a 1 is not carried to the next register stage.

Now consider Figure 5.35, which shows the status of each of the three stages. These three bits form an octal binary number. Note that in the right column of this table, the first seven clock cycles (which produce our 1110100 code at the output of stage 1) are the octal bits to generate a random sequence of numbers between 1 and 7. This series of octal codes could be used to set the

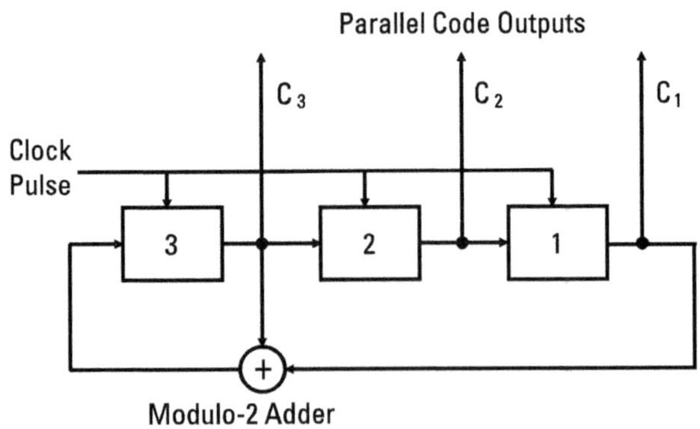

Figure 5.35 The condition of each of the three stages at each clock cycle forms an octal word describing a series of pseudo randomly selected numbers.

Hop Time	Binary Code C_3 C_2 C_1			Frequency Step
1	1	1	1	7
2	0	1	1	3
3	1	0	1	5
4	0	1	0	2
5	0	0	1	1
6	1	0	0	4
7	1	1	0	6
8	1	1	1	7
⋮	Code Repeats			⋮

Figure 5.36 The right hand column of this table is a pseudo random sequence between 1 and 7 formed by the octal codes in the middle column.

count-down values in a frequency hopping radio synthesizer. This would cause a pseudo-random selection of the hopping frequencies.

References

[1] Dixon, R., *Spread Spectrum Systems with Commercial Applications*, New York: Wiley-Interscience, 1994.

[2] Seybold, J., *Introduction to RF Propagation*, New York: Wiley, 1958.

6

Legacy Communication Threats

6.1 Introduction

The main focus of this chapter is on the basics of radio propagation and how it applies to communications electronic warfare (EW). This material is referenced in many other places in the book.

Other material in this chapter relates to the intercept, emitter location, and jamming of normal communication signals. The same EW functions against more complex signals, mainly low probability of intercept signals that will be covered in Chapter 7.

6.2 Communications Electronic Warfare

EW is the art and science of denying an enemy the benefits of the electromagnetic spectrum while preserving those benefits for friendly forces. This means the whole spectrum. In this series, we will be focusing on part of the spectrum most commonly used for tactical communication. In this book, we take tactical communication to be more than military point-to-point radio communication; it also includes command and data links between base stations and remote military assets, broadcast transmissions to multiple receivers, and remote detonation of weapons.

We will start with a brief review of radio propagation in very high frequency (VHF), ultrahigh frequency (UHF), and low microwave bands, and then we will cover some principles and examples of electronic support (ES), electronic attack (EA), and electronic protection (EP) in those bands.

6.3 One-Way Link

The most dramatic difference between EW against radars and EW against communications is that radars typically use two-way links, that is, the transmitter and receiver are generally (not always) in the same location with transmitted signals reflecting from targets. In communication, the transmitter and receiver are in different locations. The purpose of communication systems of all types is to take information from one location to another. Thus, communication uses the one-way communication link as shown in Figure 6.1.

The one-way link includes a transmitter, a receiver, transmit and receive antennas, and everything that happens to the signal between those two antennas. Figure 6.2 is a diagram that represents the one-way link equation. The abscissa of this diagram is not to scale; it merely shows what happens to the level of a signal as it passes through the link. The ordinate is the signal strength (in dBm) at each point in the link. The t ransmitted power is the input to the transmit antenna. The antenna gain is shown as positive, although in practice any antenna can have positive or negative gain (in decibels). It is important to add that the gain shown here is the antenna gain in the direction of the receiving antenna. The output of the transmit antenna is called the effective radiated power (ERP) in dBm. Note that the use of dBm units is not really correct; in fact, the signal at this point is a power density, properly stated in microvolts per meter. However, if we were to place a theoretical ideal isotropic antenna next to the transmit antenna (ignoring the near-field issue), the output of that antenna would be the signal strength in dBm. Using the artifice of this assumed ideal antenna allows us to talk about signal strength through the whole link in dBm without converting units, and is thus commonly accepted practice. The formulas to convert back and forth between signal strength in dBm and field density in microvolts per meter are:

$$P = -77 + 20\log(E) - 20\log(F)$$

where P is the signal strength arriving at the antenna in dBm, E is the arriving field density in microvolts per meter, and F is the frequency in megahertz.

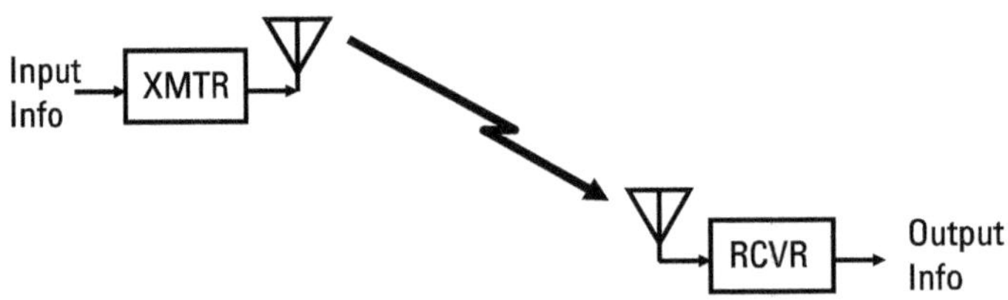

Figure 6.1 A one way communication link includes a transmitter, a receiver, two antennas, and everything that happens between those antennas.

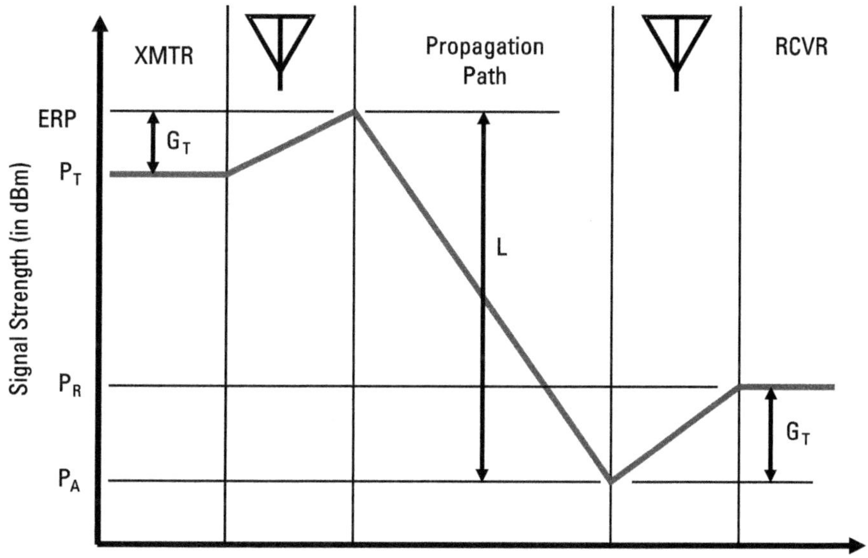

Figure 6.2 The one way link equation calculates the received power as a function of all other link elements.

Conversely, the arriving signal strength can be converted to field density by the formula:

$$E = 10^{\left[P+77+20\log(F)\right]/20}$$

where E is the field density in microvolts per meter, P is signal strength in dBm, and F is the frequency in megahertz.

Between the transmit and receive antennas, the signal is attenuated by the propagation loss. We will discuss the various types of propagation loss in detail. The signal arriving at the receiving antenna does not have a commonly used symbol, but we will call it P_A for convenience in some of our later discussions. Because P_A is outside the antenna, it should really be in microvolts per meter, but using the same ideal antenna artifice, we use the units of dBm. The receiving antenna gain is shown as positive, although it can be either positive or negative (in decibels) in real-world systems. The gain of the receiving antenna shown here is the gain in the direction of the transmitter.

The output of the receiving antenna is the input to the receiver system in dBm. We call it the received power (P_R). The one-way link equation gives P_R in terms of the other link components. In decibels units, it is:

$$P_R = P_T + G_T - L + G_R$$

where P_R is the received signal power in dBm, P_T is the transmitter output power in dBm, G_T is the transmit antenna gain in decibels, L is the link loss from all causes in decibels, and P_R is the transmitter output power in dBm.

In some literature, the link loss is dealt with as a gain, which is negative (in decibels). When this notation is used, the propagation gain is added in the formula rather than subtracted. In this book, we will consistently refer to loss as a negative number in decibels and therefore subtract loss in link equations.

In linear (i.e., nondecibel) units, this formula is:

$$P_R = (P_T G_T G_R)/L$$

The power terms are in watts, kilowatts, and so forth and must be in the same units. The gains and losses are pure (unitless) ratios. Because the link loss is in the denominator, it is a ratio greater than 1. In subsequent discussions, the loss formulas both in decibels and in a linear form will consider loss to be a positive number.

Figures 6.3 and 6.4 show important cases of the use of one-way links in electronic warfare. Figure 6.3 shows a communication link and a second link from the transmitter to an intercept receiver. Note that the transmit antenna gain to the desired receiver and to the intercept receiver may be different. Figure 6.4 shows a communication link and a second link from a jammer to the receiver. In this case, the receiving antenna may have different gain toward the

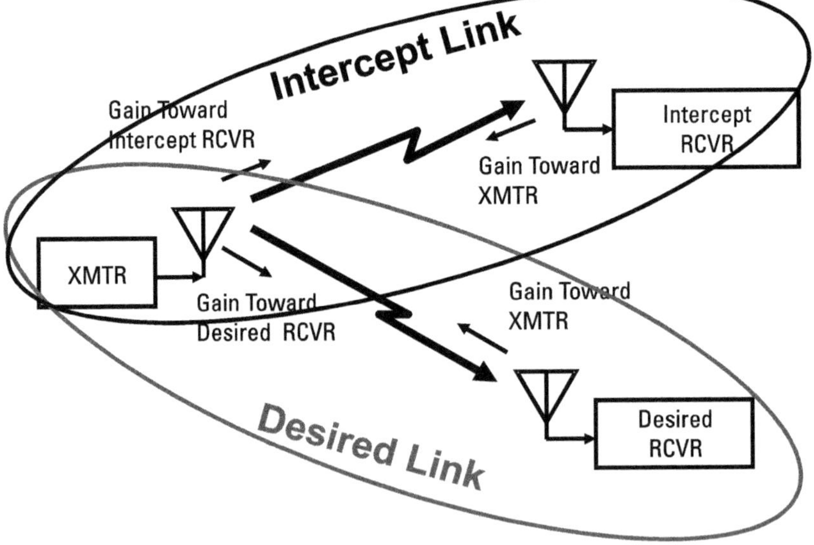

Figure 6.3 When a communication signal is intercepted, there are two links to consider; the transmitter to intercept receiver link and the transmitter to desired receiver link.

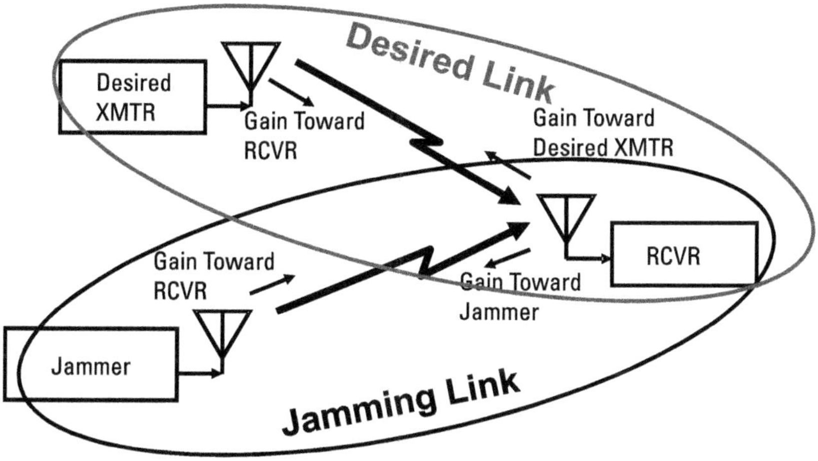

Figure 6.4 When a communication signal is jammed, there is a link from the desired transmitter to the receiver and a link from the jammer to the receiver.

desired transmitter and the jammer. Each of the links (in both figures) have the elements shown in the diagram of Figure 6.2.

6.4 Propagation Loss Models

In the description of the link, we clearly separated the transmitting and receiving antenna gains from the link losses. This implies that the link loss is between two unity gain antennas. By definition, an isotropic antenna has unity gain or 0-dB gain. All of the discussion of link losses in this section will be for propagation losses between isotropic antennas.

There are a number of widely used propagation models, including the Okumura and Hata models for outdoor propagation and the Saleh and SIRCIM models for indoor propagation. There is also small scale fading, which is short-term fluctuation caused by multipath. These models are discussed in [1]. These detailed models all require computer models of the environment to support analysis of each reflection path in the propagation environment.

Because EW is dynamic by nature, it is common practice not to use these detailed computer analyses, but rather to use three important approximations to determine the appropriate propagation loss models in practical applications. These three models are line of sight, two-ray, and knife-edge diffraction.

Reference [1] also discusses these three propagation models to some extent. Table 6.1 summarizes the conditions under which these three models are used.

Table 6.1
Selection of Appropriate Propagation Loss

Clear propagation path	Low frequency, wide beams, near ground	Link longer than Fresnel zone distance	Use two-ray model
		Link shorter than Fresnel zone distance	Use line-of-sight model
	High frequency, narrow beams, far from ground		
Propagation path obstructed by terrain	Calculate additional loss from knife-edge diffraction		

6.4.1 Line-of-Sight Propagation

Line-of-sight (LOS) propagation loss is also called free space loss or spreading loss. It applies in space and between transmitters and receivers in any other environment in which there are no significant reflectors and the ground is far away in comparison with the signal wavelength (see Figure 6.5).

The formula for LOS loss comes from optics, in which propagation loss is calculated by projecting the transmitting and receiving apertures on a unit sphere with its origin at the transmitter. This is converted to radio frequency propagation by considering the geometry of two isotropic antennas. As shown in Figure 6.6, the isotropic transmitting antenna propagates its signal spherically, with its total energy spread over the surface of the sphere. The sphere expands at the speed of light until its surface touches the receiving antenna. The area of the surface of a sphere is:

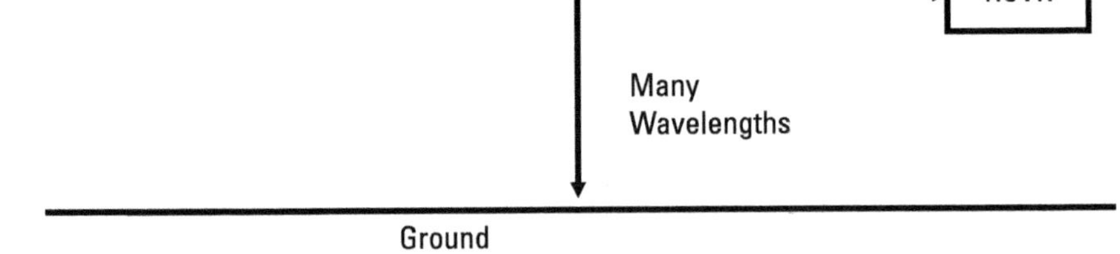

Figure 6.5 If both the transmitter and receiver are many wavelengths above the ground or if the antenna beams are narrow enough to exclude significant energy to and from the ground, The line of sight propagation model is appropriate

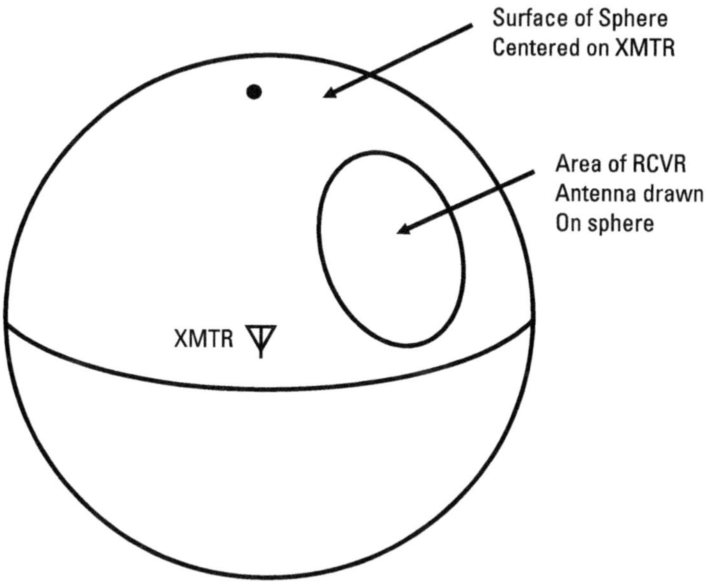

Figure 6.6 Line of sight loss is the ratio of the surface of a sphere centered on the transmitter with radius equal to the transmission distance—and the effective area of the receiving antenna.

$$4\pi R^2$$

where *R* is in this case the distance from transmitter to receiver.

The effective area of the isotropic (i.e., unity gain) receiving antenna is:

$$\lambda^2/4\pi$$

where λ is the wavelength of the transmitted signal.

We want the loss to be a number larger than one, so we can divide the transmitted power by the loss to get the receive power. Thus, we determine the loss ratio by dividing the surface area of the sphere by the area of the receiving antenna:

$$\text{Loss} = (4\pi)^2 R^2/\lambda^2$$

where both the radius and the wavelength are in the same units (typically meters).

Note that some authors treat this as a gain by which the transmitted signal is multiplied. This inverts the right side of the formula.

If we convert from wavelength to frequency, the loss formula becomes:

$$\text{Loss} = (4\pi)^2 R^2 F^2 / c^2$$

where R is the transmission path distance in meters, F is the transmitted frequency in hertz, and c is the speed of light (3×10^8 m/s).

Allowing distance to be input in kilometers and frequency in megahertz requires a conversion factor term. Combining terms and converting to decibel form gives the loss in decibels as:

$$L(\text{dB}) = 32.44 + 20\log_{10} R + 20\log_{10} F$$

where R is the link distance in kilometers and F is the transmit frequency in megahertz. The 32.44 term combines the conversion factors and the c and π terms, converted to decibels. By using this constant, we can input link parameters in the most convenient units.

Alternate forms of this equation change the constant to 36.52 if the distance is in statute miles and to 37.74 if the distance is in nautical miles. The formula is often used in applications to 1-dB accuracy. In this case the constants are simplified to 32, 37, and 38, respectively.

There is a widely used nomograph that gives the line of sight loss in decibels as a function of the distance and the frequency. This is shown in Figure 6.7. To use this nomograph, draw a line between the frequency in megahertz and the link distance in kilometers. Your line crosses the center axis at the LOS loss in decibels. In this figure, the loss at 1 GHz and 10 km is shown as just under 113 dB. Note that the above formula calculates the value at 112.44 dB.

6.4.2 Two-Ray Propagation

When the transmitting and receiving antennas are close to a single dominant reflecting surface (i.e., the ground or water) and the antenna patterns are wide enough to allow significant illumination of that surface, the two-ray propagation model must be considered. As we will see, the transmitted frequency and the actual antenna heights determine whether the two-ray or LOS propagation model applies.

Two-ray propagation is also called 40 log(d) or d^4 attenuation because the loss varies with the fourth power of the link distance. The dominant loss in two-ray propagation is the phase cancellation of the direct wave by the signal reflected from the ground or water as shown in Figure 6.8. The amount of attenuation depends on the link distance and the height of the transmitting and receiving antennas above the ground or water. You will note that (unlike LOS attenuation) there is no frequency term in the two-ray loss expression. In nonlogarithmic form, the two-ray loss is:

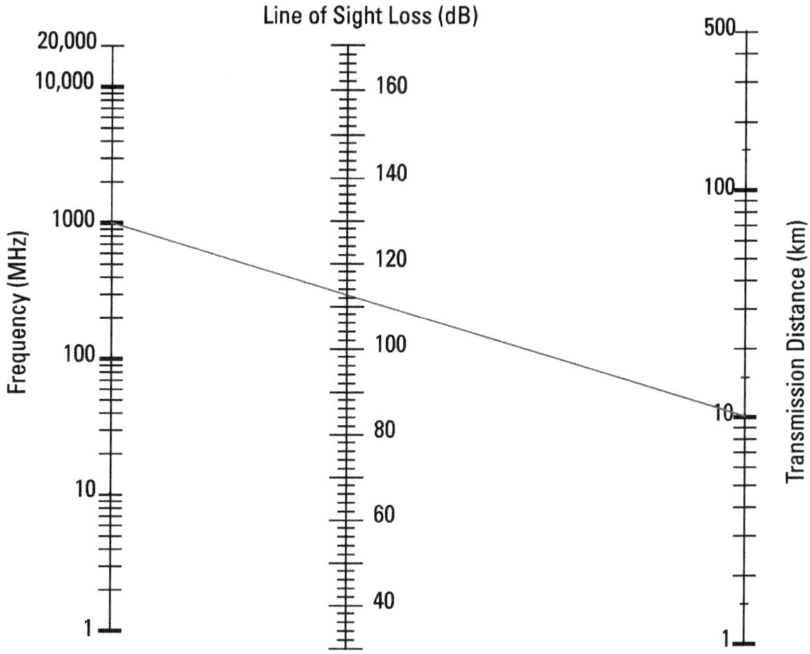

Figure 6.7 A line drawn from the frequency value to the transmission distance value passes through the line of sight loss value.

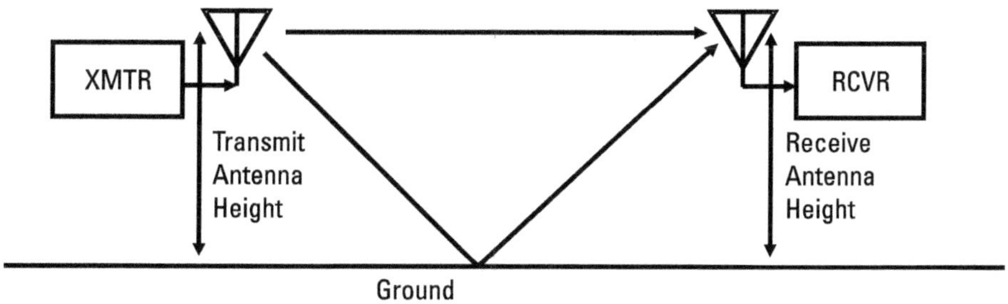

Figure 6.8 In two-ray propagation, the dominant loss effect is the phase cancellation between the direct and reflected signals.

$$L = d^4 / \left(h_T^2 \times h_R^2\right)$$

where d is the link distance, h_T is the transmitting antenna height, and h_R is the receiving antenna height.

The link distance and antenna heights are all in the same units.

The decibel formula for the two-ray propagation loss is:

$$L = 120 + 40\log(d) - 20\log(h_T) - 20\log(h_R)$$

where d is the link distance in kilometers, h_T is the transmitting antenna height in meters, and h_R is the receiving antenna height in meters.

Figure 6.9 gives a nomograph for the calculation of two-ray loss. To use this nomograph, first draw a line between the transmitting and receiving antenna heights. Then draw a line from the point at which the first line crosses the index line through the path length to the propagation loss line. In the example, two 10-m-high antennas are 30 km apart, and the attenuation is a little less than 140 dB. If you calculate the loss from either of the above formulas, you will find that the actual value is 139 dB.

6.4.3 Minimum Antenna Height for Two-Ray Propagation

Figure 6.10 shows minimum antenna height for two-ray propagation calculations versus transmission frequency. There are five lines on the graph for:

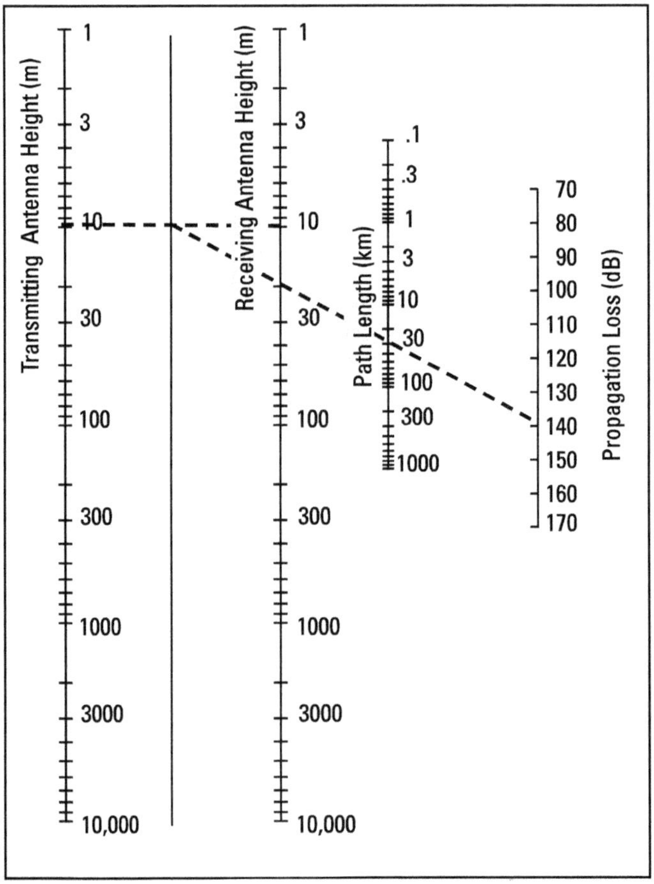

Figure 6.9 Two ray propagation loss can be determined as shown on this nomograph.

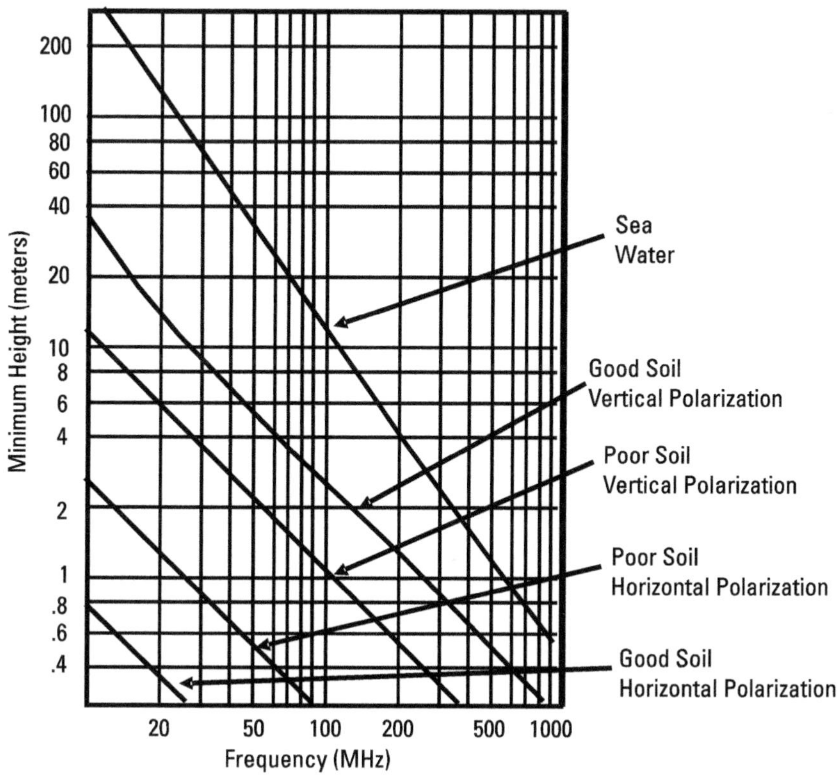

Figure 6.10 If antennas are below the minimum height shown in this graph, use the indicated minimum height in the two-ray propagation loss calculation.

- Transmission over sea water;
- Vertically polarized transmission over good soil;
- Vertically polarized transmission over poor soil;
- Horizontally polarized transmission over poor soil;
- Horizontally polarized transmission over good soil.

Good soil provides a good ground plane. If either antenna height is less than the minimum shown by the appropriate line in this graph, the minimum antenna height should be substituted for the actual antenna height before completing the two-ray attenuation calculation. Please note that if one antenna is actually at ground level, this chart is highly suspect.

6.4.4 A Note About Very Low Antennas

In the communication theory literature, discussions of very low antennas all seem to be constrained to antenna heights at least a half wavelength above the ground. A recent, far from complete, test gives some insight into the performance of antennas lower than that. A 400-MHz, vertically polarized, 1-m-high transmitter was moved various distances from a matched receiver while the receiver was lowered from 1m high to the ground. Over level, dry ground, the received power reduced by 24 dB when the receiving antenna was at the ground. With a 1-m-deep ditch across the transmission path (near the receiver), this loss was reduced to 9 dB.

6.4.5 Fresnel Zone

As mentioned above, signals propagated near the ground or water can experience either LOS or two-ray propagation loss, depending on the antenna heights and the transmission frequency. The Fresnel zone distance is the distance from the transmitter at which the phase cancellation becomes dominant over the spreading loss. As shown in Figure 6.11, if the receiver is less than the Fresnel zone distance from the transmitter, LOS propagation takes place. If the receiver is farther than the Fresnel zone distance from the transmitter, two-ray propagation applies. In either case, the applicable propagation applies over the whole link distance.

The Fresnel zone distance is calculated from the following formula:

$$FZ = 4\pi h_T h_R / \lambda$$

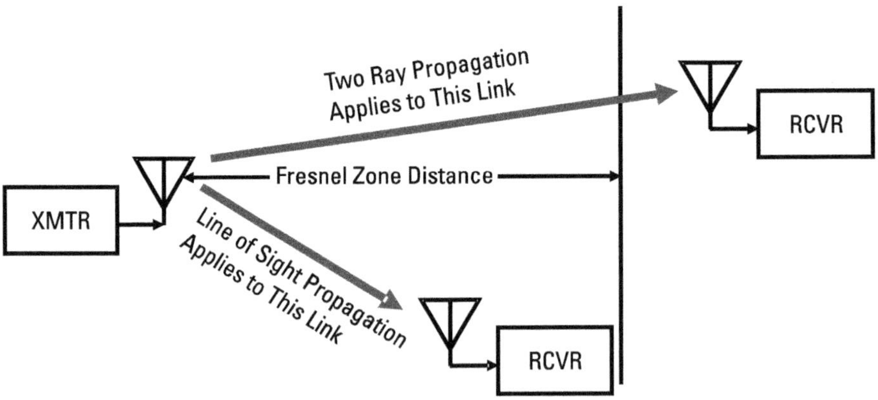

Figure 6.11 If the link is shorter than the Fresnel zone distance, is uses line of sight propagation. If it is longer than the Fresnel zone distance, it uses two ray propagation.

where FZ is the Fresnel zone distance in meters, h_T is the transmitting antenna height in meters, h_R is the receiving antenna height in meters, and λ is the transmission wavelength in meters.

Note that several different formulas for Fresnel zone are found in literature. This one is chosen because it yields the distance at which LOS and two-ray attenuation are equal. A more convenient form of this equation is:

$$FZ = \left[h_T \times h_R \times F\right]/24{,}000$$

where FZ is the Fresnel zone distance in kilometers, h_T is the transmitting antenna height in meters, h_R is the receiving antenna height in meters, and F is the transmission frequency in megahertz.

6.4.6 Complex Reflection Environment

In locations with very complex reflections, for example, when transmitting down a valley as shown in Figure 6.12, it is suggested in the literature that the LOS propagation loss model will give a more accurate answer than the two-ray propagation model.

6.4.7 Knife-Edge Diffraction

Non-LOS propagation over a mountain or ridge line is usually estimated as though it were propagation over a knife edge. This is a very common practice and many EW professionals report that the actual losses experienced in terrain closely approximate those estimated by equivalent knife edge diffraction estimation.

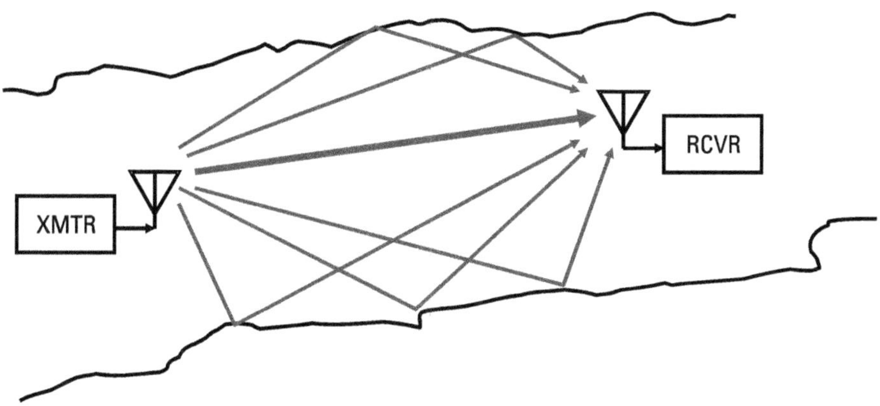

Figure 6.12 In a very complex reflection environment, like transmission down a valley, the actual propagation loss can be expected to be closer to line of sight than two ray.

The knife-edge diffraction (KED) attenuation is added to the LOS loss as it would be if the knife edge were not present. Note that the LOS loss rather than the two-ray loss applies when a knife edge (or equivalent) is present (see Figure 6.13).

The geometry of the link over a knife edge is shown in Figure 6.14. *H* is the distance from the top of the knife edge to the LOS as though the knife edge were not present. The distance from the transmitter to the knife edge is called d_1 and the distance from the knife edge to the receiver is called d_2. For KED to take place, d_2 must be at least equal to d_1. If the receiver is closer to the knife edge than the transmitter, it is in a blind zone in which only tropospheric scattering (with significant losses) provides link connection.

As shown in Figure 6.15, the knife edge causes loss even if the LOS passes above the peak, unless the line of sight path passes several wavelengths above. Thus, the height value *H* can be either the distance above or below the knife edge.

Figure 6.16 is a KED calculation nomograph. The left scale is a distance value *d*, which is calculated by the following formula:

$$d = \left[\sqrt{2}/(1 + d_1/d_2)\right] d_1$$

Table 6.2 shows some calculated values of *d*.

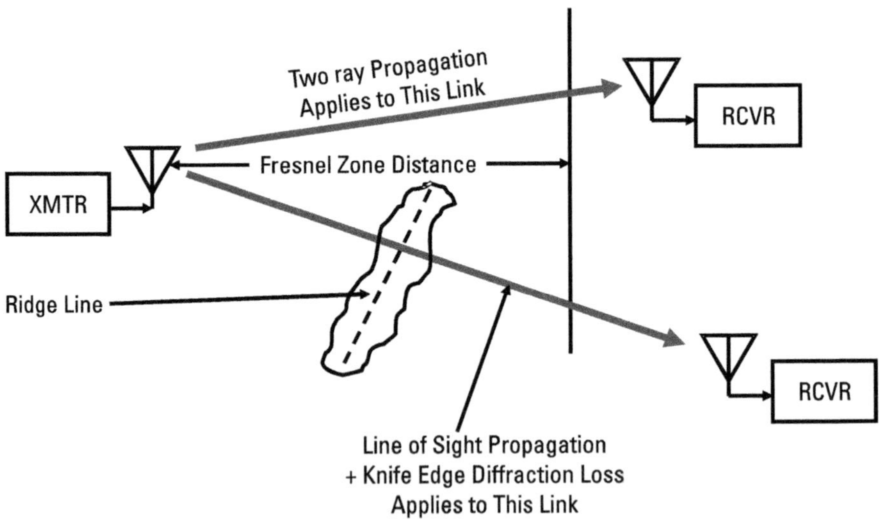

Figure 6.13 Even if the link distance is greater than the Fresnel Zone distance, line of sight propagation applies if there is an intervening ridge line.

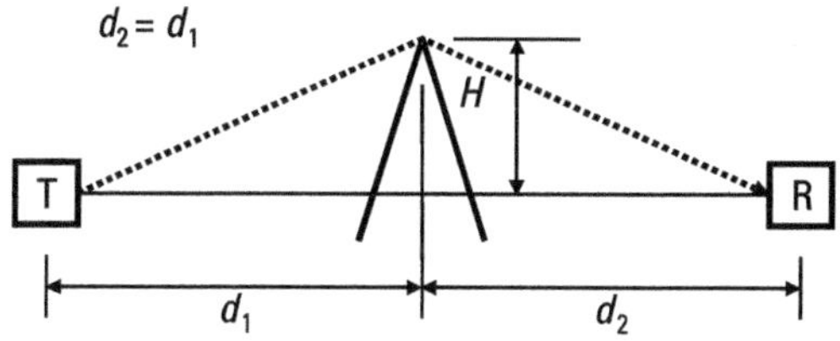

Figure 6.14 The knife edge diffraction geometry is set by the distance to the knife edge, the distance past the knife edge, and the height of the knife edge relative to the line of sight path if there were no knife edge.

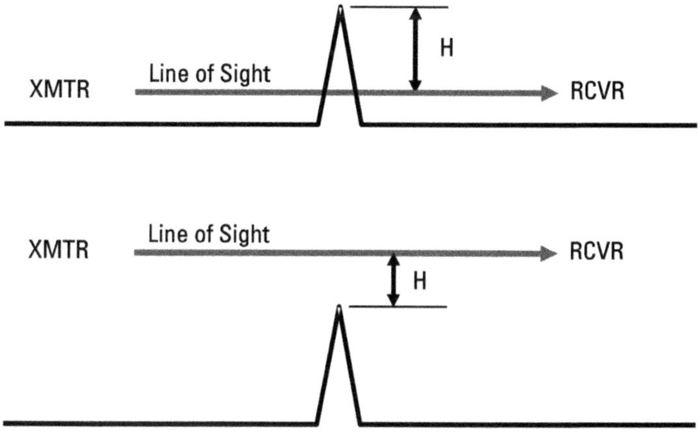

Figure 6.15 The line of sight path can pass above or below the knife edge. If it is not too far above, knife edge diffraction loss will still occur.

If you skip this step and just set $d = d_1$, the KED attenuation estimation accuracy will only be reduced by about 1.5 dB.

Returning to Figure 6.16, the line from d (in kilometers) passes through the value of H (in meters). At this point, we do not care whether H is the distance above or below the knife edge. Extend this line to the center index line.

Another line passes from the intersection of the first line with the center index through the transmission frequency (in megahertz) to the right scale, which gives the KED attenuation. At this point, we identify whether H was above or below the knife edge. If H is the distance above the knife edge, the KED attenuation is read on the left scale. If H is the distance below the knife edge, the KED attenuation is read on the right scale.

Consider an example (which is drawn onto the nomograph): d_1 is 10 km, d_2 is 24.1 km, and the LOS path passes 45m below the knife edge. d is 10 km (from Table 6.1) and H is 45m. The frequency is 150 MHz. If the LOS path were 45m above the knife edge, the KED attenuation would have been 2 dB.

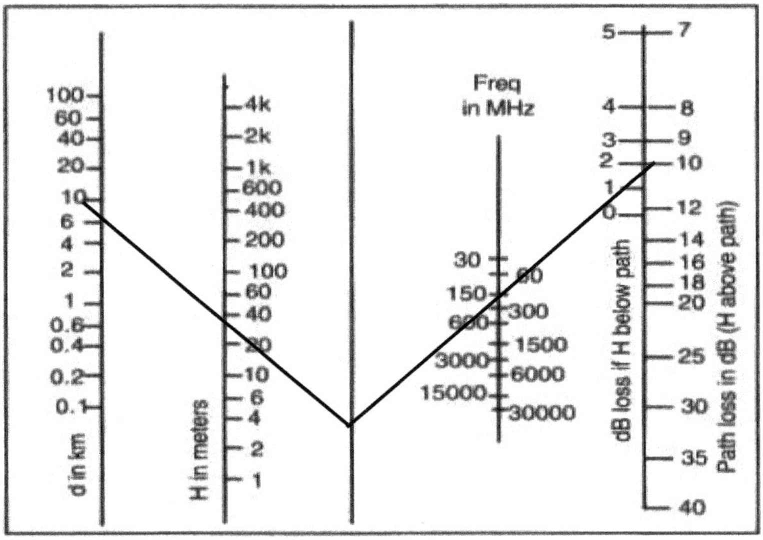

Figure 6.16 Knife edge diffraction can be determined graphically from the values of d and H and the frequency.

However, because the LOS path is below the knife edge, the KED attenuation is 10 dB.

The total link loss is then the LOS loss without the knife edge and the KED attenuation:

$$\text{LOS loss} = 32.44 + 20\log(d_1 + d_2) + 20\log(\text{frequency in megahertz})$$
$$= 32.44 + 20\log(34.1) + 20\log(150) = 32.44 + 30.66 + 43.52$$
$$= \text{approximately } 106.6\,\text{dB}$$

So the total link loss is 106.6 + 10 = 116.6 dB.

6.4.8 Calculation of KED

The math for calculation of KED is very complex, so apiece-wise approximation is suggested in [1].

First, you must calculate an intermediate value v from the formula:

$$v = H\sqrt{\frac{2(d_1 + d_2)}{\lambda d_1 d_2}}$$

Table 6.2
Values of d

	d
$d_2 = d_1$	$0.707\ d_1$
$d_2 = 2\ d_1$	$0.943\ d_1$
$d_2 = 2.41\ d_1$	d_1
$d_2 = 5\ d_1$	$1.178\ d_1$
$d_2 >> d_1$	$1.414\ d_1$

where the d_1, d_2, and H values are the same as in Figure 6.2 and λ is the transmission wavelength.

Table 6.3 then gives the KED gain as a function of the variable v. Note that the KED loss in decibels is the negative of the gain in decibels.

This piecewise solution can be set up in an Excel or Mathcad file or similar software, but for manual calculations, the nomograph in Figure 6.16 is recommended.

6.5 Intercept of Enemy Communication Signals

6.5.1 Intercept of a Directional Transmission

The situation shown in Figure 6.17 is the intercept of a data link by a hostile receiver. The transmitter has a directional antenna pointed toward the desired receiver, and the hostile receiver is not in the main lobe of the transmitting antenna pattern. The transmitter and receiver are both located on elevated terrain, so the receiving antenna is not illuminated by significant reflections from local terrain. This means that the propagation loss is determined from the LOS model discussed in Section 6.4.1.

The received power in the intercept receiver is the transmitter power, increased by the transmit gain in the direction of the intercept receiver, reduced by the propagation loss and increased by the receiving antenna gain in the direction of the transmitter. Thus, the received power is calculated from the formula:

Table 6.3
KED Gain Versus v

V	*G* (in decibels)
$v < 1$	0
$0 < v < 1$	$20 \log_{10} (0.5 + 0.62\ v)$
$-1 < v < 0$	$20 \log_{10} (0.5 \exp [0.4 - [0.95\ v])$
$-2.4 < v < -1$	$20 \log_{10} (0.4 - \mathrm{sqrt} [0.1184 - (0.1\ v + 0.38)^2)]$
$v < -2.4$	$20 \log_{10} (0.225/v)$

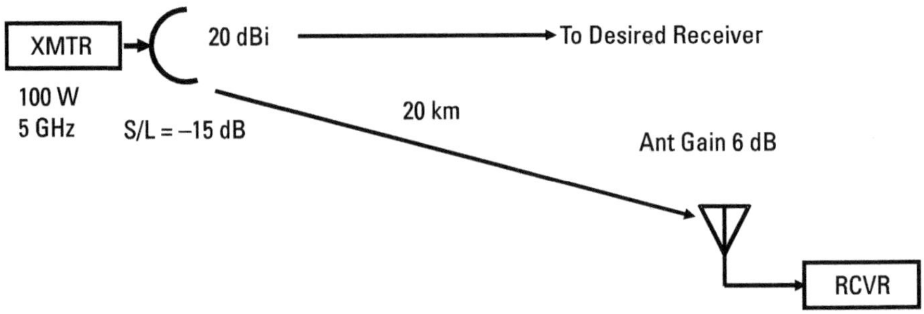

Figure 6.17 Analysis of the intercept link from a hostile transmitter to an intercepting receiver determines the quality of the intercept.

$$P_R = P_T + G_T - [32.44 + 20\log(d) + 20\log(f)] + G_R$$

where P_R is the received power, G_T is the transmit antenna gain (toward the receiver), d is the link distance (in kilometers), f is the transmitted frequency (in megahertz), and G_R is the receiving antenna gain (toward the transmitter).

The link transmitter outputs 100W (i.e., 50 dBm) into its antenna at 5 GHz. The transmitting antenna has 20-dBi boresight gain and the receiver is located 20 km away in a −15-dB side lobe (i.e., 15-dB lower gain that the peak of the main beam), the transmit antenna gain for the intercept link is 5 dB. The receiving antenna is oriented toward the transmitter and has 6-dBi gain. The link transmitter outputs 100-W at 5 GHz. The received power in the intercepting receiver is calculated as:

$$P_R = +50\,\text{dBm} + 5\,\text{dBi} - [32.44 + 26 + 74\,\text{dB}] + 6\,\text{dBi} = -71.4\,\text{dBm}$$

6.5.2 Intercept of a Nondirectional Transmission

In the intercept situation shown in Figure 6.18, the transmitter and receiver both are near the ground and have wide angular coverage antennas; therefore, they may be subject to either LOS or two-ray propagation. The proper propagation mode is determined by calculation of the Fresnel zone distance by the formula (from Section 6.4.5):

$$FZ = (h_T \times h_R \times f)/24{,}000$$

where FZ is the Fresnel zone distance (in kilometers), h_T is the transmit antenna height (in meters), h_R is the receiving antenna height(in meters), and f is the transmitted frequency.

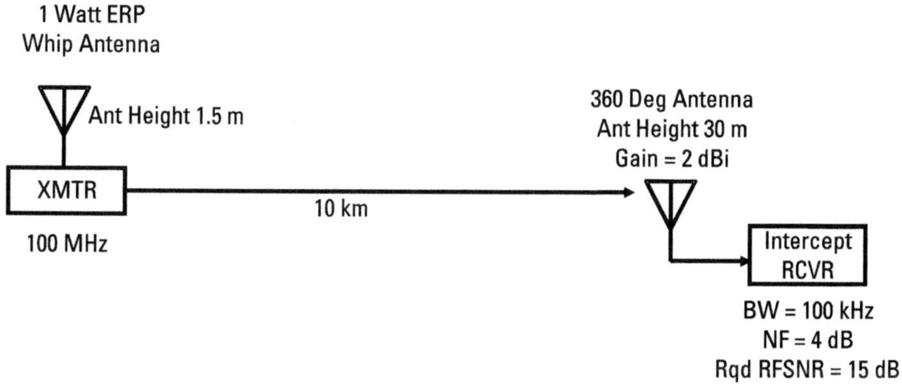

Figure 6.18 A signal from a ground transmitter to a ground based intercept system is subject to either line of sight or 2 ray propagation loss, depending on the link geometry.

If the transmitter to receiver path length is shorter than the Fresnel zone distance, LOS propagation applies. If the path is longer than the Fresnel zone distance, two-ray propagation applies.

The target emitter is a handheld push to talk system with a whip antenna, 1.5m from the ground. Note that the effective height of a whip antenna is at the bottom of the whip. The receiving antenna has a 2-dBi gain. The power transmitted from the target emitter has effective radiated power of 1W (30 dBm) at 100 MHz. The Fresnel zone distance is:

$$(1.5 \times 30 \times 100)/24{,}000 = 188\text{m}$$

The Fresnel zone distance is far less than the 10-km path distance, so two-ray propagation applies.

The propagation loss, from the formula in Section 6.4.2, is:

$$120 + 40\log(d) - 20\log(h_T) - 20\log(h_R)$$

Thus, the received power in the intercept receiver is calculated to be:

$$P_R = ERP - \left[120 + 40\log(d) - 20\log(h_T) - 20\log(h_R)\right] + G_R$$

Plugging in the values from Figure 6.18,

$$P_R = 30\,\text{dBm} - [120 + 40 - 3.5 - 29.5] + 2\,\text{dB} = -95\,\text{dBm}$$

This intercept problem has a different complication in that the intercept receiver has a relatively wide bandwidth. If the transmitter has a typical 25-kHz bandwidth, the receiver bandwidth is four times as wide to allow a more rapid frequency search.

To determine whether or not the signal is successfully intercepted, we must calculate the sensitivity of the receiver using the following formula:

$$Sens = kTB + NF + Rqd\ RFSNR$$

where *Sens* is the receiver sensitivity in dBm, *NF* is the receiver noise figure in decibels, and *Rqd RFSNR* is the required predetection signal-to-noise ratio in decibels.

Remember that the sensitivity is the minimum signal strength that a receiver can receive and still do its job.

$$kTB = -114\,\text{dBm} + 10\log(\text{bandwidth}/1\ \text{MHz}) = -124\,\text{dBm}$$

The receiver system noise figure is given as 4 dB and the required RFSNR is given as 15 dB, so:

$$Sens = -124 + 4 + 15 = -105\,\text{dBm}$$

Because the signal is received at a level 10 dB above the receiver system's sensitivity level, the intercept receiver has achieved a 10-dB performance margin.

6.5.3 Airborne Intercept System

In Figure 6.19, the intercept system is located in a helicopter that is 50 km from the enemy transmitter at an altitude above local terrain of 1,000m. The target emitter is a handheld 400-MHz transmitter transmitting 1-W ERP at 400 MHz. The bottom of its whip antenna is 1.5m above the local terrain.

First, we need to calculate the Fresnel zone distance for the intercept link using the formula given above:

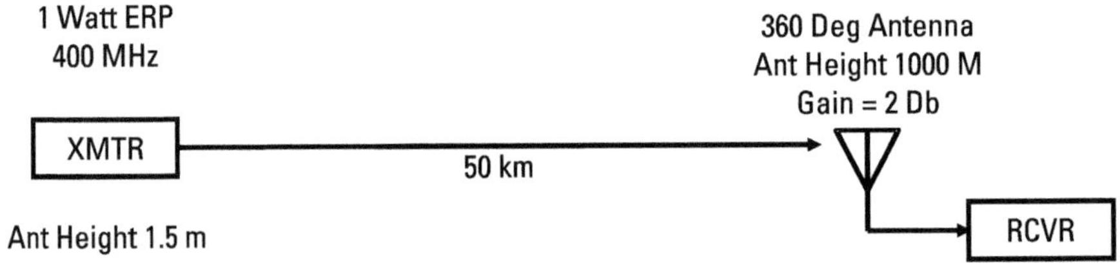

Figure 6.19 An airborne intercept system can achieve significant performance because of the impact of the receiver elevation on propagation loss.

$$FZ = (h_T \times h_R \times f)/24{,}000$$
$$= (1.5 \times 1{,}000 \times 400)/24{,}000 = 25\,\text{km}$$

Because the transmission path is longer than the Fresnel zone distance, two-ray propagation occurs, so:

$$P_R = ERP - [120 + 40\log(d) - 20\log(h_T) - 20\log(h_R)] + G_R$$

The received intercept signal strength is:

$$P_R = 30\,\text{dBm} - [120 + 68 - 3.5 - 60]\,\text{dB} + 2\,\text{dBi} = -94.5\,\text{dBm}$$

6.5.4 Non-LOS Intercept

Figure 6.20 shows intercept of tactical communication emitter across a ridge line 11 km from the emitter. In this problem, the direct line distance from the transmitter to the intercept receiver is 31 km, the transmit antenna height is 1.5m, and the intercept antenna height is 30m. The transmit signal has 1-W ERP at 150 MHz and the receiving antenna has 12 dBi gain (G_R).

As discussed in Section 6.4.7, the link loss is the LOS loss, (ignoring the terrain interference) plus a KED loss factor. If the ridge rises 210m above the local ground (assuming flat earth), it will be 200m above the LOS between the two antennas.

The LOS loss, using the formula from Section 6.4.1, is:

$$32.4 + 20\log D + 20\log f$$

Note that we use a capital D here for full link distance to avoid confusion with the lowercase d used in the KED loss determination.

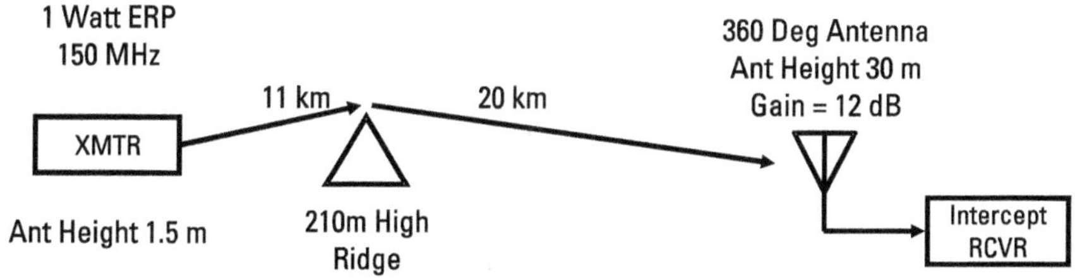

Figure 6.20 If an intercept system is over a ridge line from its target emitter, but farther from the ridge than the target transmitter, the propagation loss will be line of sight plus a knife edge diffraction factor.

$$LOS\ los = 32.4 + 20\log(31) + 20\log(150) = 32.4 + 29.8 + 43.5 = 105.7\,\text{dB}$$

We will round this to 106 dB.

To determine the KED loss, we first calculate d from the formula:

$$d = \left[\text{sqrt}(2)/\left(1 + \left(d_1/d_2\right)\right)\right]d_1$$

where d is the distance term entered into the KED loss nomograph, d_1 is the transmitter to ridge distance, and d_2 is the ridge-to-receiver distance.

For this problem, d= [sqrt(2)/1.55] 11 = 10, but remember we could also just set $d = d_1$ for a slightly less accurate KED determination.

Figure 6.21 is the nomograph from Section 6.4.7 used to calculate KED loss with the values from this problem drawn in. It shows that the values from this problem (d= 10 km, H= 200m, f= 150 MHz) will cause 20-dB KED loss. Thus, the total link loss is:

$$LOS\ loss + KED\ loss = 106\,\text{dB} + 20\,\text{dB} = 126\,\text{dB}$$

The received power in the intercept receiver is then:

$$P_R = ERP - Loss + G_R = 30\,\text{dBm} - 126\,\text{dB} + 12\,\text{dB} = -84\,\text{dBm}$$

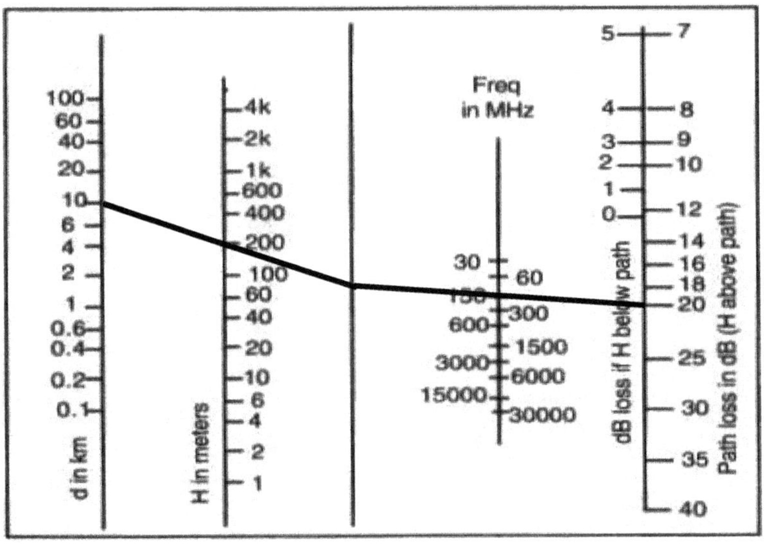

Figure 6.21 If the derived value of d is 10 km, a ridge line rises 200 m above the direct signal path and the signal is at 150 MHz, the knife edge diffraction loss is 20 dB.

6.5.5 Intercept of Weak Signal in Strong Signal Environment

Figure 6.22 is the block diagram of an intercept receiver system. The effective system bandwidth is 25 kHz, its receiver has an 8-dB noise figure, and it has a preamplifier with a 20-dB gain and a 3-dB noise figure. There is a 2-dB loss between the antenna and the preamplifier and 10 dB circuit losses between the preamplifier and the receiver.

The system must receive weak signals [providing 16-dB predetection signal-to-noise ratio (SNR)] in the presence of a large number of strong signals that may be in band, so we will determine its dynamic range.

First, we need to determine the system sensitivity using the techniques discussed in Section 6.5.2. Remember that the sensitivity is the sum of kTB, system noise figure, and required predetection SNR.

$$kTB = -114\,\text{dBm} + 10\log(\text{effective bandwidth}/1\text{ MHz}) = -130\,\text{dBm}$$

The system noise figure is determined from the chart in Figure 6.23. Draw a vertical line from the receiver noise figure (8 dB) on the abscissa and a horizontal line from the value *preamp noise figure +preamp gain – loss before the receiver* (13 dB) on the ordinate. The two lines cross at the degradation factor, which is 1 dB. The system noise figure is the sum of: *loss before preamp + preamp noise figure + degradation factor*, so the system noise figure is 2 dB + 3 dB + 1 dB = 6 dB.

The system sensitivity is then:

$$-130\,\text{dBm} + 6\,\text{dB} + 16\,\text{dB} = -108\,\text{dBm}$$

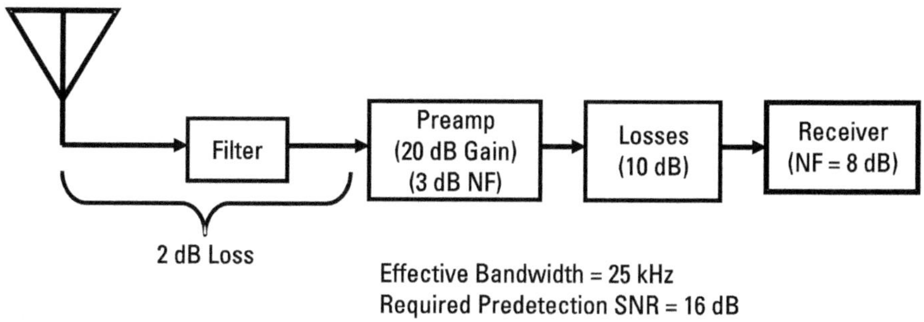

Figure 6.22 The intercept system has a front end filter to prevent second order spurious responses. After the preamplifier, there is a signal distribution network to feed multiple receivers. This diagram shows only the path to one of the receivers.

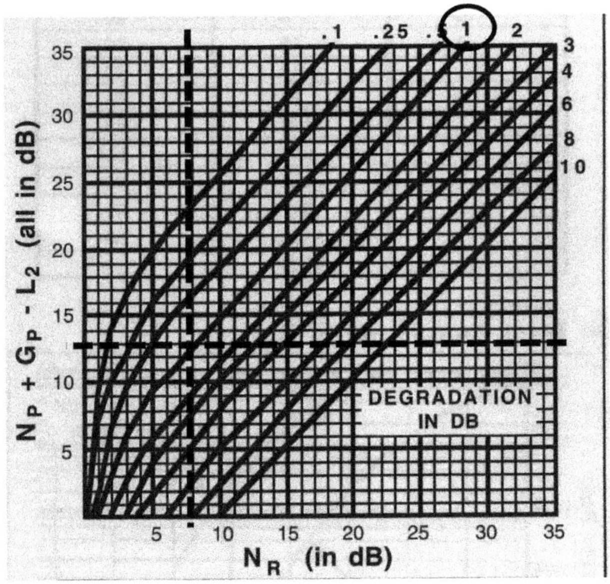

Figure 6.23 This diagram shows that the degradation of the preamplifier noise figure is 1 dB.

Note that a signal entering the system from the receiving antenna at –108 dBm will be –90 dBm at the output of the preamplifier (2 dB for loss before the preamp + 20 dB preamp gain).

The system design is such that second order spurious responses are filtered out, so the third-order response of the preamplifier determines the dynamic range. The third-order intercept point for the chosen preamplifier is +20 dBm.

Figure 6.24 is a diagram from with which we determine the receiver system dynamic range. Draw a line at a 3-to-1 slope through the fundamental preamp output line at +20 dBm. Then draw a horizontal line across the chart at –90 dBm (–108 dBm – 2 dB loss before preamp + 20 dB preamp gain). This is the preamp output level for a sensitivity level input signal from the antenna). The vertical distance from the crossing of the third-order spur line and the sensitivity line to the fundamental output level line is 70 dB. The dynamic range of the receiver system is 70 dB.

This means that the system can intercept –108-dBm signals in the presence of –38-dBm signals.

6.5.6 Search for Communications Emitters

Military organizations do not publish their operating frequencies openly and go to great trouble to keep those frequencies from being known by enemies. However, in general, it is necessary to know the frequency at which an enemy is operating to perform the various EW operations. Thus, frequency search is

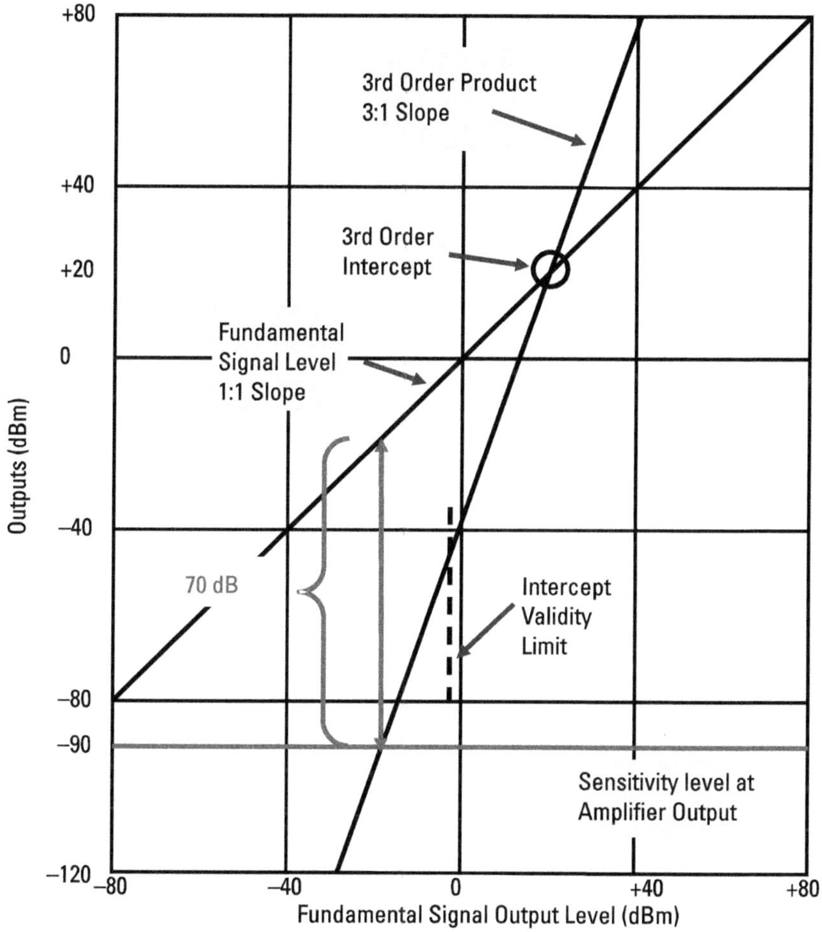

Figure 6.24 This figure from the March 2007 EW101 shows that the system dynamic range is 70 dB.

an important EW function. In this section, we will discuss the basic principles of frequency search to highlight the trade-offs that must be made. When using wideband receivers, these same trade-offs are required, unless the whole range of interest is covered instantaneously.

6.5.7 About the Battlefield Communications Environment

Modern warfare, which requires a great deal of mobility for almost all assets, is highly dependent on radio communication. This includes large numbers of both voice and data links. The tactical communication environment is often described as having 10% channel occupancy. This is a bit misleading because it refers the likelihood that at any microsecond you would expect 10% of all available RF channels to be active. If you stay on each channel for a few seconds, the occupancy rate is much higher, closer to 100%. This means that any search for a specific emitter must find it among a thick forest of nontargeted emitters.

6.5.8 A Useful Search Tool

Figure 6.25 shows a commonly used tool to assist in the development or evaluation of a frequency search approach. It is a graph of frequency versus time on which the characteristics of target signals and the time versus frequency coverage of one or more receivers can be plotted. The frequency scale should cover the whole frequency range of interest (or some part of that range) and the time scale should be long enough to show the search strategy. The signal depictions show each signal bandwidth versus expected duration of the signal. If the signals are periodic or change frequency in some predictable way, these characteristics can be shown on the graph. The receiver is shown tuned to a particular frequency with its bandwidth and the time during which it covers that specific frequency increment.

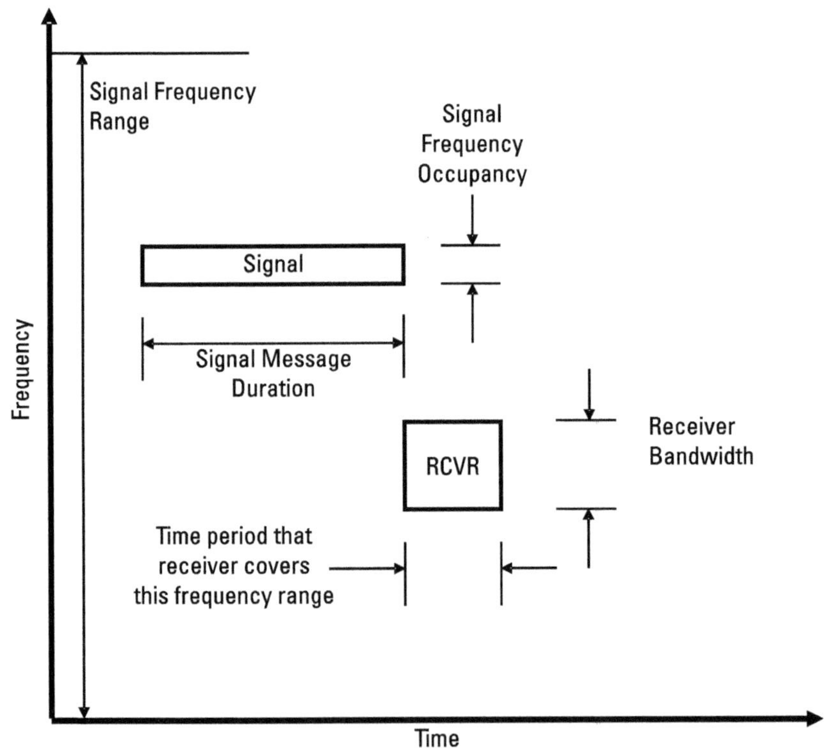

Figure 6.25 A graph of frequency vs time showing both receivers and target signals is a useful search analysis tool.

A typical sweeping receiver strategy is shown in Figure 6.26. The parallelograms show the frequency versus time coverage of the swept receiver. The receiver bandwidth is the height of the parallelogram at any frequency and the slope is the receiver's tuning rate. Note that Signal A is optimally received (its whole bandwidth over its whole duration). Signal B is received if you do not need to see its whole bandwidth and Signal C is received if you do not need to see its whole duration. You can set the rules to fit the nature of the signals and the purpose of your search.

6.5.9 Technology Issues

Years ago, military intercept receivers were mechanically tuned, so it was necessary to search by manually tuning or by automatically tuning across the whole covered band in a single sweep that was more or less linear. This approach was commonly called garbage collection because you needed to look at every signal in the environment and then pick the few signals of interest from a typically large collection of signals that were not of interest. Identification of signals of interest required rather complex analysis by a trained operator. Remember that 50 years ago, computers were rooms full of vacuum tubes requiring large quantities of forced air cooling and that their capabilities were miniscule compared to modern computers.

With the availability of digitally tuned receivers and one-chassis (and eventually one-chip) computers with massive memory and blinding speed, much more civilized search approaches became practical. Now you can store

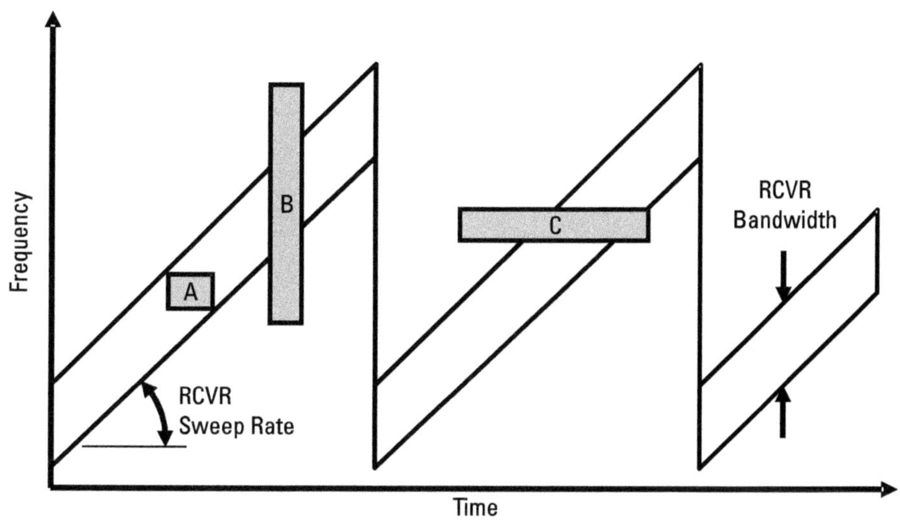

Figure 6.26 The receiver search plan and targets of interest assists in the analysis of probability of intercept.

the frequencies of known signals of interest and automatically check each one before looking for new signals of interest. A fast Fourier transform (FFT) can be performed on each potential signal of interest to allow computerized spectral analysis. An interest/no-interest determination can be made from the results of the spectral analysis; perhaps enhanced by a quick look at the general emitter location (if the system has direction finding or emitter location capability).

6.5.10 Digitally Tuned Receiver

Figure 6.27 shows a digitally tuned superheterodyne receiver. A digitally tuned, receiver has a synthesizer local oscillator and an electronically tuned preselector to allow very fast selection of any signal frequency within the tuning range. Tuning can be either by operator or computer control.

Figure 6.28 shows a phase-lock-loop synthesizer block diagram. Note that the voltage tuned oscillator is phase locked to a multiple of the frequency of an accurate and stable crystal oscillator. This means that the tuning of a digitally tuned receiver is accurate and repeatable, making the search approach described above practical. Note that the bandwidth of the feedback loop in the synthesizer is set at the optimum compromise between low noise signal output (i.e., narrow loop bandwidth) and high tuning speed (i.e., wide loop bandwidth). In the search mode, time must be allowed for the synthesizer to settle before beginning the analysis of any signals in the selected instantaneous receiver bandwidth coverage.

When a digitally tuned receiver is used in the search mode, it is tuned to discrete frequency assignments as shown in Figure 6.29. The search need not move linearly through the whole band of interest, but can check specific frequencies or scan frequency sub-bands of high interest in any order desired. It is often desirable to provide 50% overlap of receiver tuning steps. This prevents a

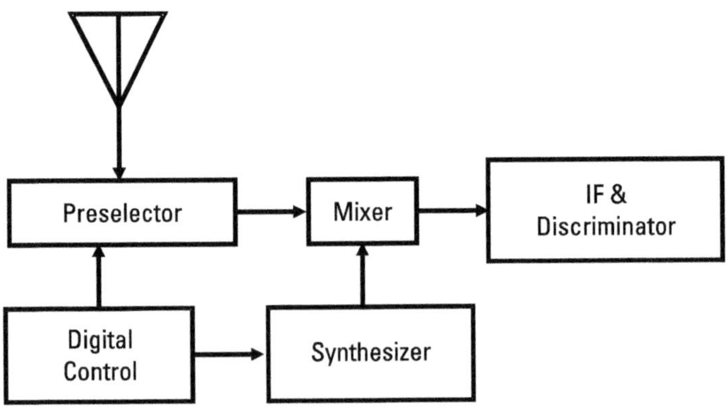

Figure 6.27 A digitally tuned receiver can be quickly tuned to any part of the band at any time.

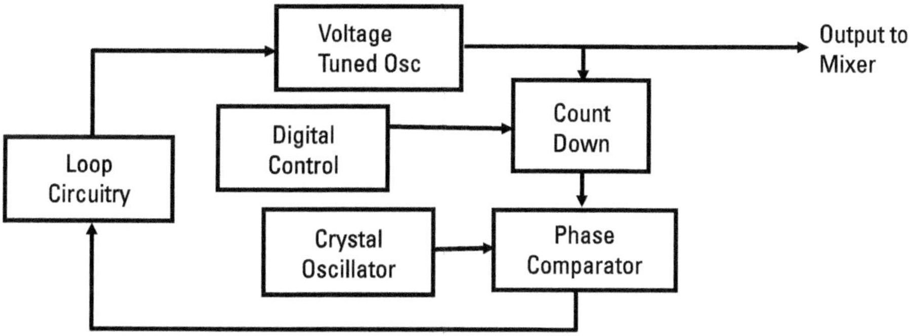

Figure 6.28 A phase-lock-loop synthesizer tunes a voltage controlled oscillator to the frequency which will allow a counted down signal to come in phase with a crystal oscillator. The countdown ratio is digitally selected to determine the synthesizer output frequency.

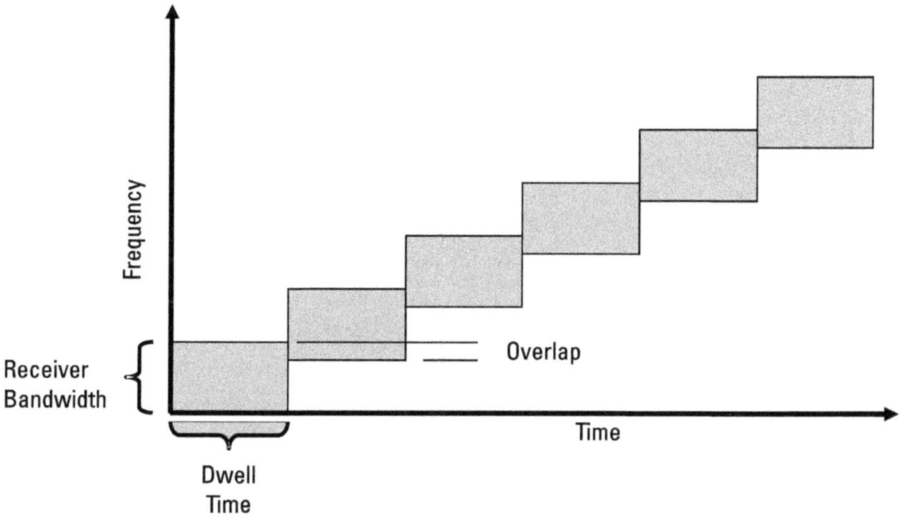

Figure 6.29 A digitally tuned receiver moves to discrete frequency assignments

band-edge intercept of a signal of interest. However, a 50% overlap will require twice as much time to cover the signal range of interest. The amount of overlap is a trade-off that must be made to optimize the search in a specific situation.

As shown in Figure 6.30, an intercept system will often use multiple monitor receivers controlled by a special search receiver. When the search receiver encounters a signal, a rapid analysis of that signal determines whether or not this is a signal of interest and of high enough priority to be allocated one of the monitor receivers. If so, a monitor receiver is tuned to the frequency of that signal and its operating parameters (for example, demodulation mode) are set ap-

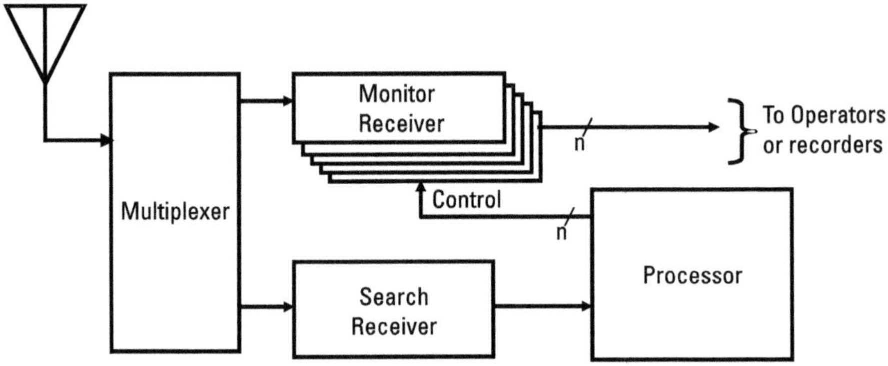

Figure 6.30 A search receiver can be used to determine the frequencies of signals of interest, allowing monitor receivers to be quickly tuned to the highest priority signals. The search receiver can be a wideband receiver type or an optimally swept narrowband receiver.

propriately. The output of each monitor receiver can go to an operator position or to an automatic recording or content analysis location.

6.5.11 Practical Considerations Effecting Search

Theoretically, a receiver can sweep at a rate which allows a signal to remain within the receiver bandwidth for a time equal to the inverse of the bandwidth (e.g., 1 μs in a 1-MHz bandwidth). However, system software requires time to determine if a signal is present. This might require as much as 100 to 200 μs, which can be significantly more than a time equal to 1/bandwidth.

Performing processing on each signal present (such as modulation analysis or emitter location) will typically take longer to identify the signal as a signal of interest. This level of processing may take as much as one or more milliseconds for each signal found. Remember that there may be a signal in 10% of the available channels. For example, in the 30- to 88-MHz band, there are 2,320 25-kHz channels, so you would expect to find 232 occupied channels in a full band search.

6.5.12 A Narrowband Search Example

Here is a narrowband search example. We want to find a 25-kHz-wide communication signal that is between 30 and 88 MHz. We will assume that the signal is up for 0.5 second. Note that a signal this short is probably a key click, which at one time was the shortest signal an intercept system had to worry about. In this example, times will be rounded to the nearest millisecond.

Our receiving antenna covers 360° of azimuth, and the search receiver bandwidth is 25 kHz. The receiver must dwell at each tuning step for a time equal to the inverse of the bandwidth. To avoid band edge intercepts, we will overlap our tuning steps by 50%.

$$\text{Dwell} = 1/\text{bandwidth} = 1/25\,\text{kHz} = 40\,\mu s$$

Figure 6.31 shows the search problem in the diagram format we discussed in Section 6.5.8. Note the overlap of the receiver coverage, which causes us to change frequency only 12.5 kHz with each tuning step.

For 100% probability of finding the signal of interest, the receiver must cover the whole 58 MHz in 0.5 second. The number of bandwidths required to cover the signal range is:

$$58\text{ MHz}/25\text{ kHz} = 2{,}320$$

With 50% overlap, the 58-MHz frequency range requires 4,640 tuning steps.

At 40-μs dwell per step, 4,640 steps require 186 ms.

This means that the receiver can find the signal of interest in less than one-half of the assumed minimum signal duration, so 100% probability of intercept is easily achieved.

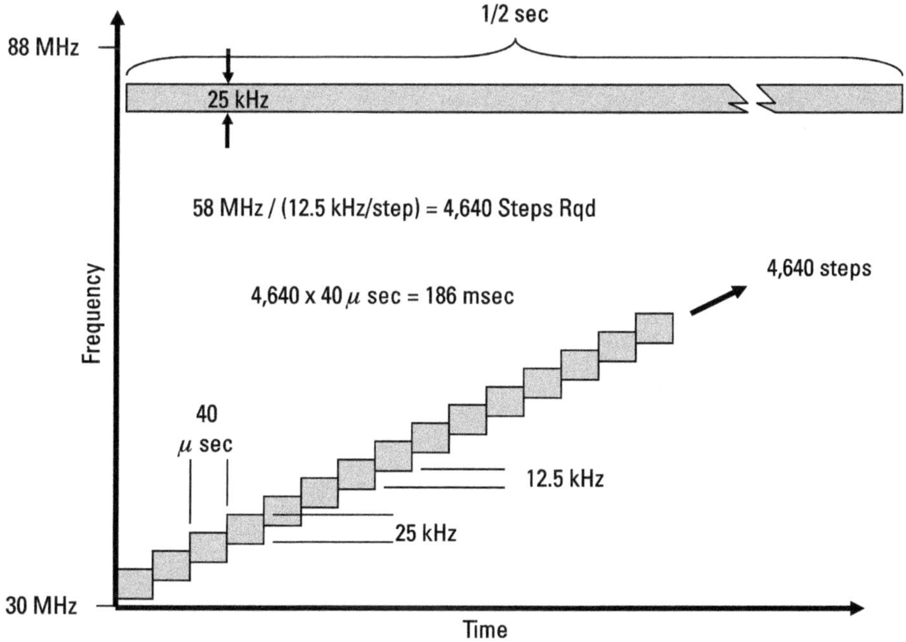

Figure 6.31 A 25 kHz search bandwidth with 50% overlap and 40 μsec dwell per step will cover 58 MHz in 186 msec.

However, this assumes that we have an optimum search and that the signal will be instantly recognized as our signal of interest. To make the problem more interesting, let us assume that we have a processor which can recognize the modulation of the signal in 200 μs. This means that we must hold at each frequency for that dwell time, so it takes 928 ms to cover the 58-MHz search range.

$$200\mu s \times 4{,}640 = 928 \text{ ms}$$

The search does not find the signal within the specified 0.5 second, as shown in Figure 6.32.

6.5.13 Increase the Receiver Bandwidth

If the search receiver bandwidth is increased to 150 kHz (covering six target signal channels) and we assume that the 200-μs processing time also allows determination of the signal frequency within the bandwidth, the search is enhanced (see Figure 6.33). Now it takes only 773 steps to cover the frequency range of interest.

$$4{,}640/6 = 773$$

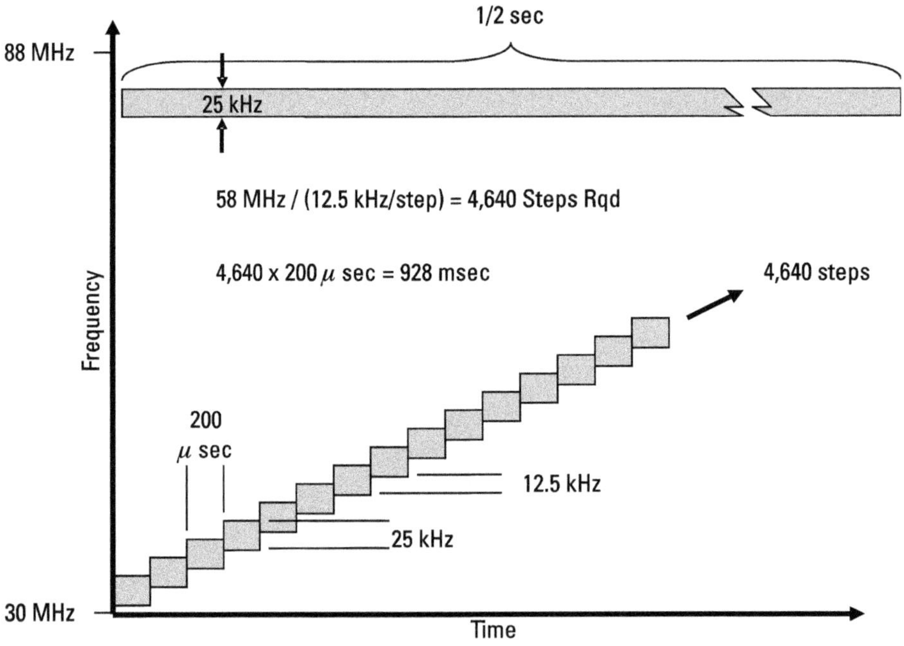

Figure 6.32 A 25 kHz search bandwidth with 50% overlap and 200 μsec dwell per step will cover 58 MHz in 928 msec.

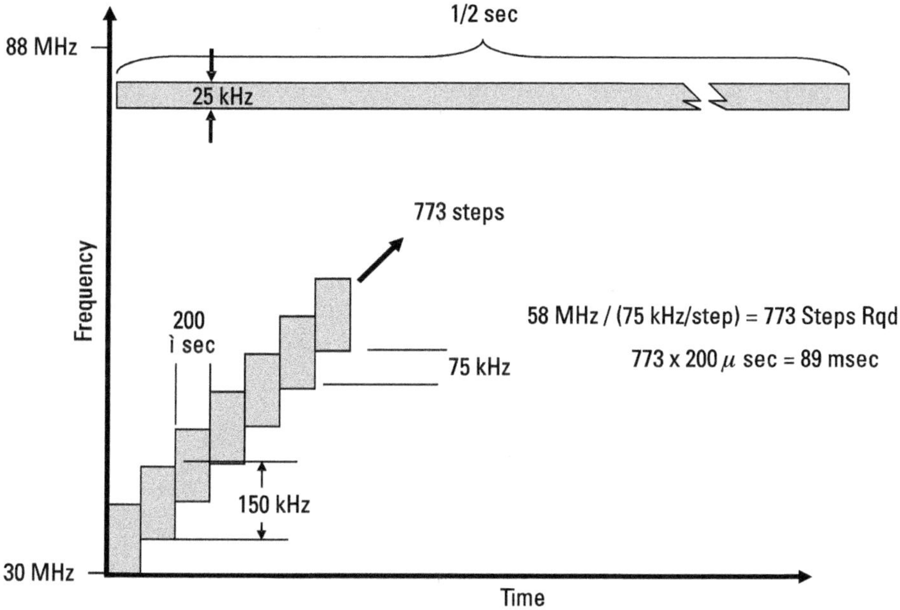

Figure 6.33 A 150 kHz search bandwidth with 50% overlap and 200 μsec dwell per step will cover 58 MHz in 89 msec.

At 200 μs per receiver tuning step, it takes only 155 ms to cover the 2,320 channels (with 50% overlap). See Figure 6.33.

$$773 \times 200 \mu s = 155 \text{ ms}$$

Note that this increase in bandwidth would reduce the receiver sensitivity by almost 8 dB. If the receiver gets much wider than this, the increasing probability of multiple signals in the bandwidth becomes problematic.

6.5.14 Add a Direction Finder

To make the problem even more interesting, let us assume that our receiver is part of a direction finding system and that we must determine the direction of arrival (DOA) of the signal of interest. Our direction finder requires 1 ms to determine the DOA. This adds only 1 ms to our search time if there are no other signals present.

We have discussed the density of the tactical communications environment, and considered that a number often used in the evaluation of tactical systems is 10% channel occupancy. This means that our 58-MHz range of interest would be expected to contain:

$$2{,}320 \times 0.1 = 232 \text{ signals}$$

With 150-kHz search bandwidth and 50% overlap, the receiver can cover the 2,088 empty channels in 139 ms.

$$\left(2{,}088/6\right) \times 2 \times 200 \mu s = 139 \text{ ms}$$

However the 232 occupied channels will require an additional 232 ms. Thus, our search requires 371 ms (139 + 232), so our search strategy yields 100% probability of finding and determining DOA for the target signal.

Note that we will return to this problem when we consider searching for frequency hopping signals in Chapter 7.

6.5.15 Search with a Digital Receiver

Here, we will consider only finding our signal of interest (between 30 and 88 MHz) using an FFT channelizer. We will take the limitation that the receiver use a standard vitrual machine environment (VME) bus format, which limits our data rate to 40 MBps. To provide the Nyquist sampling rate, we thus limit our input frequency band to 20 MHz.

Figure 6.34 shows the block diagram of the digital receiver. Using the FFT, we will cover the full frequency search range of interest in three steps as shown in Figure 6.35.

6.6 Location of Communications Emitters

One of the most important requirements placed on EW systems is the location of threat emitters. Communication emitters pose particular challenges because

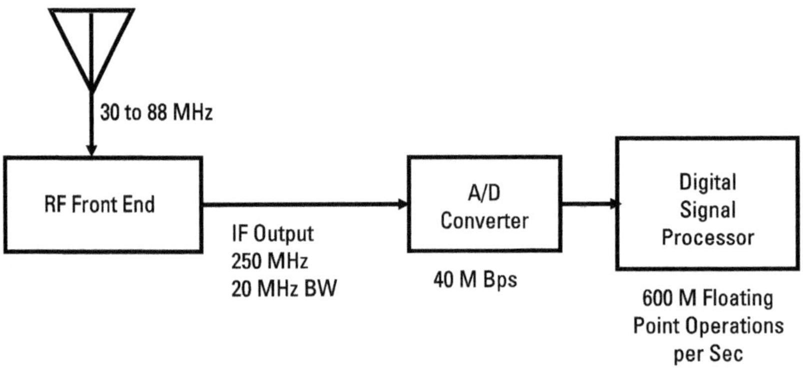

Figure 6.34 A digitally tuned receiver can be quickly tuned to any part of the band at any time.

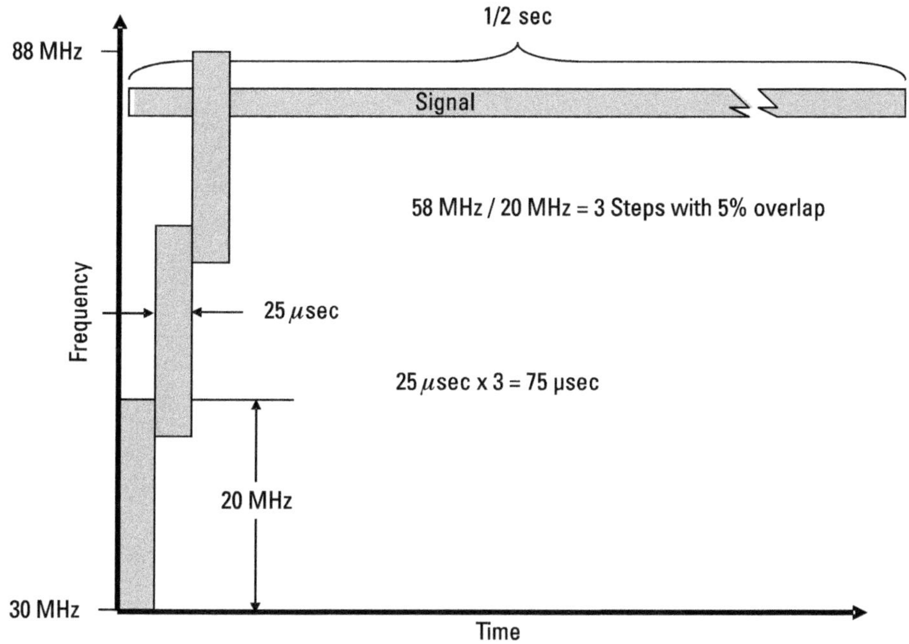

Figure 6.35 A digitally tuned receiver moves to discrete frequency assignments.

of their relatively low frequencies. Lower frequency implies larger wavelength and hence larger antenna apertures. In general, communications electronic support (ES) systems are required to provide instantaneous 360° angular coverage and adequate sensitivity to locate distant emitters. They must typically be able to accept all communications modulations, including those associated with low probability of intercept (LPI) transmissions (which we will discuss in Chapter 7). In all cases, communications ES systems deal with noncooperative (i.e., hostile) emitters. Thus, the techniques available for location of cooperative systems are by definition unavailable.

In this section, we will discuss the common approaches, and most important techniques. We will first discuss the location of normal (i.e., non-LPI) emitters here, and then in Chapter 7, we will cover location of LPI emitters. In the discussions of all system applications, the high signal density expected in the modern military environment will be an important consideration.

6.6.1 Triangulation

Triangulation is the most common approach to the location of noncooperative communications emitters. As shown in Figure 6.36, this involves the use of two or more receiving systems at different locations. Each such system must be able to determine the DOA of the target signal. It must also have some way to

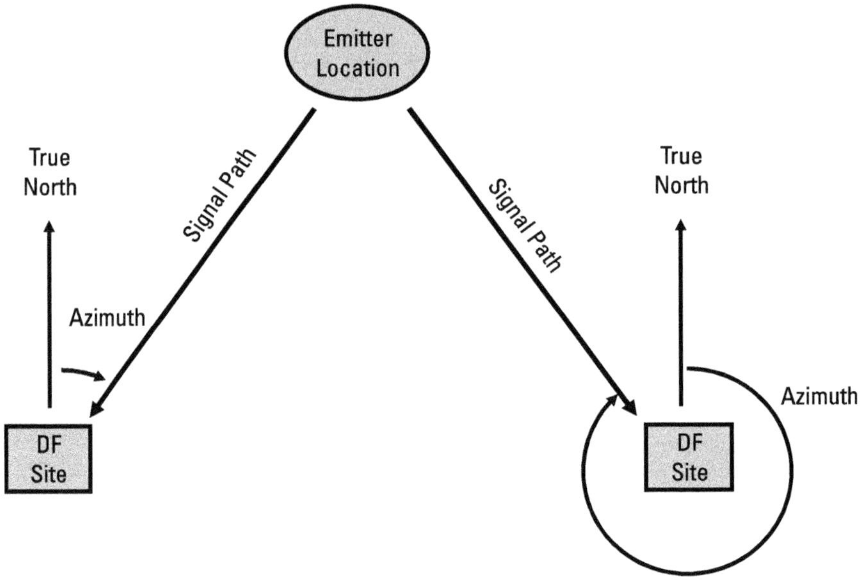

Figure 6.36 Triangulation is the location of an emitter by determining the azimuth of arrival of a signal at multiple known site locations.

establish an angular reference, typically true North. For convenience we will call these the direction finding (DF) systems in the following discussion.

Because terrain obstruction or some other condition might cause two DF systems to see different signals (in the typical dense signal environment), it is common practice to perform triangulation with three or more DF systems. As shown in Figure 6.37, the DOA vectors from three DF systems will form a triangle. Ideally, all three would cross at the emitter location, and if the triangle is small enough, the three line intersections can be averaged to calculate the reported emitter location.

These DF sites are normally quite distant from each other, so the DOA information must be communicated to a single analysis location before the emitter location can be calculated. This also implies that the location of each DF site is known.

It is important that each of the DF sites be able to receive the target signal. If the DF systems are mounted on flying platforms, they will normally be expected to have LOS to the target emitter. Ground-based systems can be expected to provide more accurate location if the terrain allows LOS, but should be able to determine the location of over-the-horizon emitters with some acceptable accuracy.

Note that the optimum geometry for triangulation provides 90° of angle between the two DF sites as seen from the emitter location.

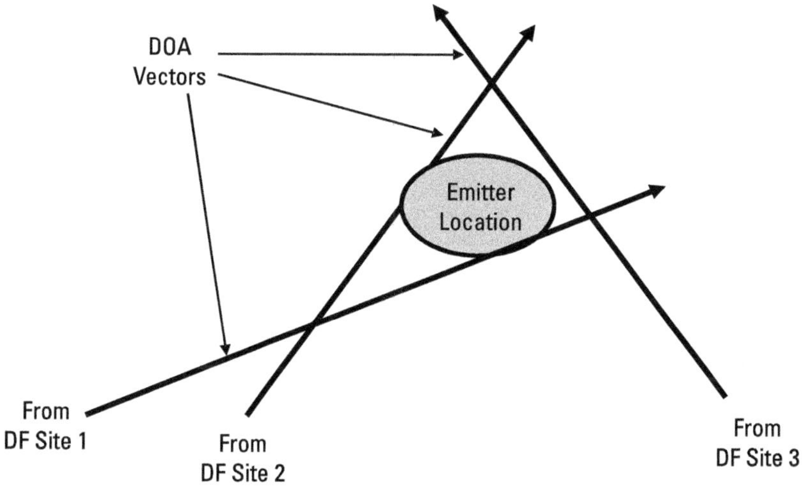

Figure 6.37 Triangulation is normally accomplished from three sites so that the three direction of arrival vectors will form a triangle. The smaller the triangle, the higher quality the emitter location.

Triangulation can also be performed from a single, moving DF system, as shown in Figure 6.38. This normally only applies to airborne platforms. The lines of bearing should still cross at 90° at the target. Therefore, the speed of the platform on which the DF system is mounted and the distance between the

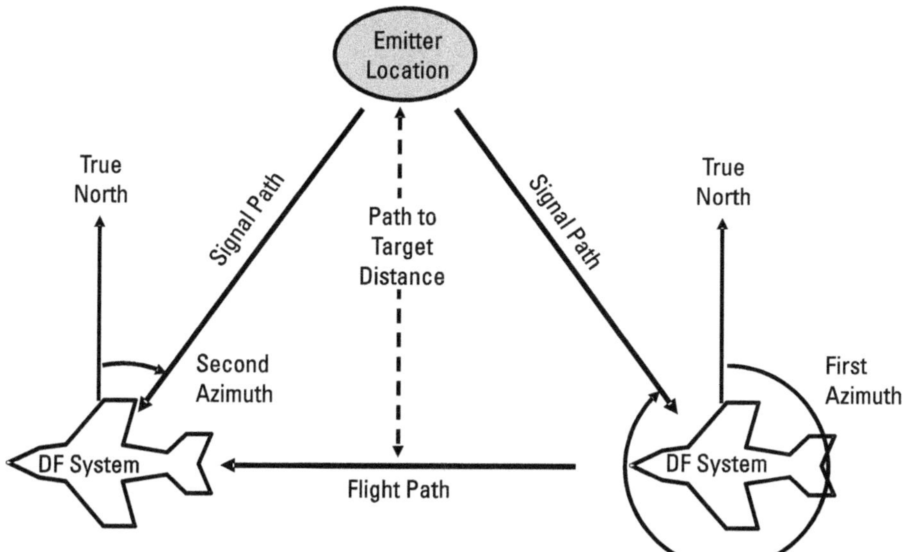

Figure 6.38 A moving DF system can perform triangulation with azimuth angles taken at different times along its flight path.

flight path and the target will dictate the time required for an accurate emitter location.

For example, if the DF platform is flying at 100 knots and passes about 30 km from the target emitter, it will take almost 10 minutes to achieve the optimum location geometry. This may be quite practical for stationary emitters, but may be too slow to track moving emitters. For this approach to yield acceptable accuracy, the movement of target emitters must not be greater than the required location accuracy over the time when data is being collected. Note that acceptance of less than optimum geometry (hence location accuracy) may provide the best operational performance.

6.6.2 Single Site Location

There are two cases in which the location of a hostile transmitter can be determined from the azimuth and range from a single emitter location site. One applies to ground-based systems dealing with signals below about 30 MHz and the second applies to airborne systems.

Signals below approximately 30 MHz can be located by a single site locator (SSL) as shown in Figure 6.39. These signals are refracted by the ionosphere. They are said to be reflected by the Ionosphere because they return with the reciprocal angle as shown in Figure 6.40. If both the azimuth and elevation angle of the signal arriving at the emitter location site are measured, the transmitter can be located. The range is calculated from the elevation angle and the height of the ionosphere at the reflection point because the angle of reflection from the ionosphere is the same as the angle of incidence. The most difficult part of this process is the accurate characterization of the ionosphere at the point of

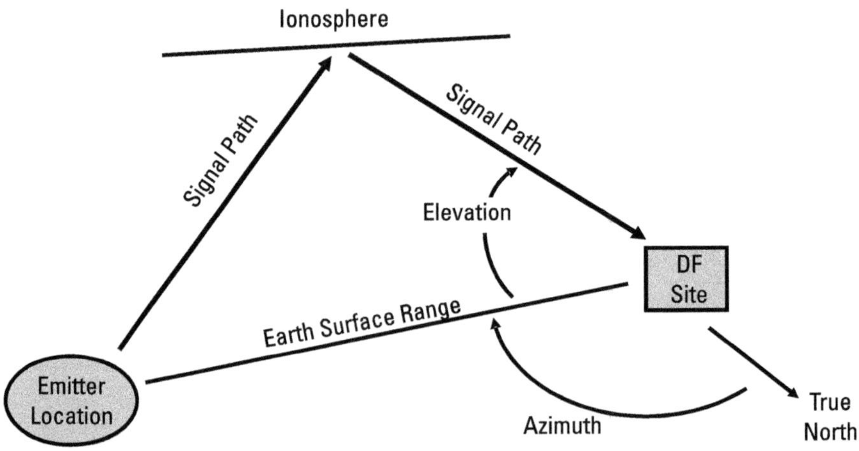

Figure 6.39 The location of an emitter below about 30MHz can be determined by measuring the azimuth and elevation from a single DF site.

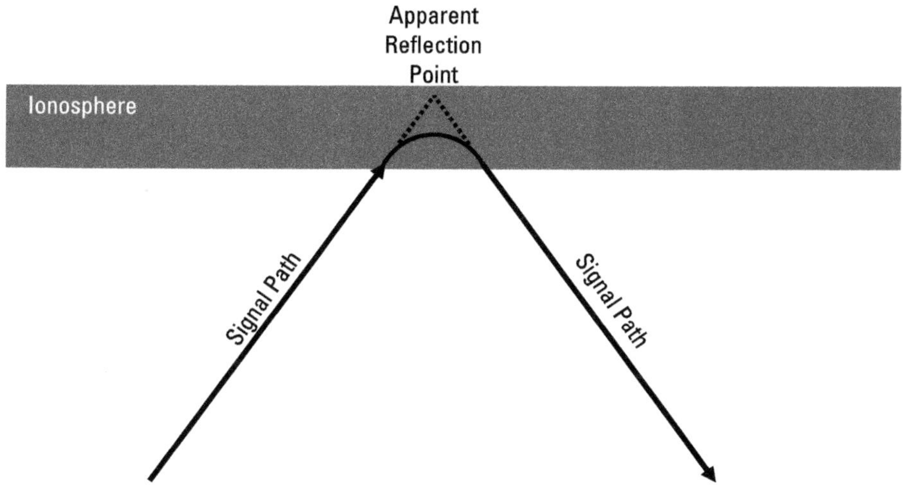

Figure 6.40 Signals below about 30 MHz appear to be reflected by the ionosphere.

reflection. Normally, the range calculation is significantly less accurate than the azimuth measurement, causing an elongated zone of location probability.

If an airborne emitter location system measures both the azimuth and elevation to a noncooperative emitter on the ground, the emitter location can be calculated as shown in Figure 6.41. The range determination requires that the aircraft know its location over the ground and its elevation. It must also have a digital map of the local terrain. The Earth surface range to the emitter is the distance from the subvehicle point to the intersection of the signal path vector with the ground.

6.6.3 Other Location Approaches

Precision emitter location approaches, to be described later, use comparison of parameters of a target signal as received at two distant sites to calculate a mathematically derived locus of possible emitter locations as shown in Figure 6.42. The techniques used can place the emitter very close to this locus, but the locus is typically many kilometers long. By adding a third site, a second and a third locus curve can be calculated. These three locus curves cross at the emitter location.

6.6.4 RMS Error

The accuracy of DOA measurement systems is typically stated in terms of the root mean square (RMS) error. This is considered the effective accuracy of a DF system. This does not define the peak errors that might be present. The system could conceivably have a relatively small RMS error even though there are a

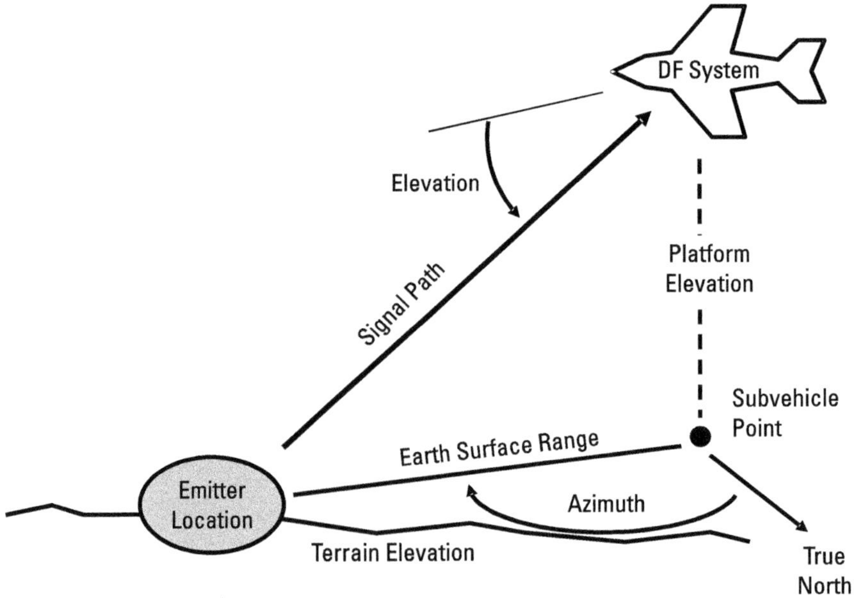

Figure 6.41 An emitter on the earth's surface can be located from an airborne DF system by measurement of azimuth and elevation.

very few large peak errors. It is assumed, when defining the RMS error of a DF system that the errors are caused by randomly varying conditions, such as noise. There have been systems in which there were known large systematic errors caused by the way the system was implemented. When these few large errors were averaged with many lower errors, an acceptable RMS error was achieved. However, there were predictable conditions in which errors several times the RMS error values were experienced, reducing the operational dependability of emitter location. Where this kind of known peak errors are corrected in processing, a proper RMS error specification is achieved.

To determine the RMS error, a large number of DF measurements are made at fairly evenly distributed frequencies and angles of arrival. For each data collection point, the true angle of arrival must be known. In ground systems, this is accomplished by use of a calibrated turntable on which the DF system is mounted or by use of an independent tracker that measures the true angle to the test transmitter at a significantly higher accuracy that that specified for the DF system (ideally a full order of magnitude). In airborne DF systems, the true angle of arrival is calculated from the known location of the test transmitter and the location and orientation of the airborne platform from its inertial navigation system (INS).

Each time a DOA is measured by the DF system, it is subtracted from the true angle of arrival. This error measurement is then squared. The squared

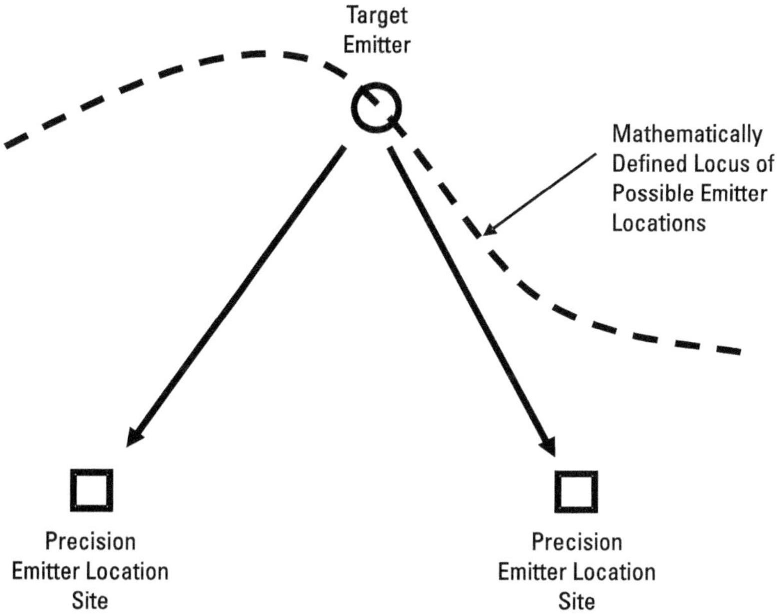

Figure 6.42 Two precision emitter location sites determine a mathematically defined locus of possible emitter locations from analysis of the same signal as received at two emitter location sites that are distant from one another.

errors are then averaged and the square root is taken. This is the RMS error of the system. The RMS error can be broken into two components as follows:

$$(\text{RMS Error})^2 = (\text{Standard Deviation})^2 + (\text{Mean Error})^2$$

Thus, if the mean error is mathematically removed, the RMS error equals the standard deviation from the true angle of arrival. If the causes of errors can be considered normally distributed, the standard deviation is 34%. Thus, as shown in Figure 6.43, the RMS error lines describe an area around the true line of bearing that have a 68% chance of containing any measured angle of arrival. Looking at this a different way, it means that if the system measures a specific angle, there is a 68% chance that the true emitter location is within the wedge-shaped area shown. This assumes that the measured mean error has been removed during data processing.

6.6.5 Calibration

Calibration involves the collection of error data as described above. However, this error data is used to generate calibration tables. These tables, in computer memory, hold the angular correction for many values of measured DOA and frequency. When a direction of arrival is measured at a particular frequency, it

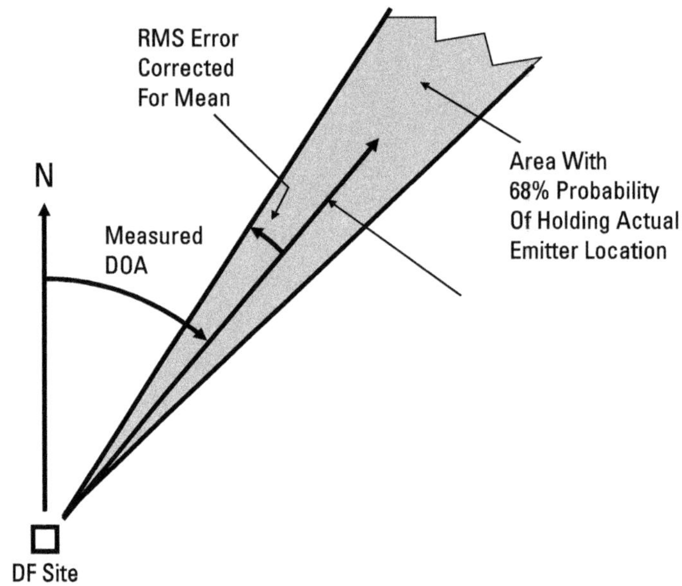

Figure 6.43 The wedge of area between the ±RMS error – mean error values from the measured DOA has a 68% probability of containing the actual emitter location.

is adjusted by the calculated angular error and the corrected angle of arrival is reported out. If a measured DOA falls between two calibration points (in angle and or frequency), the correction factor is determined by interpolation between the two closest stored calibration points. Note that slightly different calibration schemes yield better results for some specific DF techniques. These will be discussed along with those techniques.

6.6.6 CEP

Circular error probable (CEP) is a bombing and artillery term that refers to the radius of a circle around an aiming stake in which half of a number of dropped bombs or fired artillery shells fall. We use this term in emitter location system evaluation to indicate the radius of a circle around a measured emitter location which has a 50% probability of containing the true emitter location as shown in Figure 6.44. The smaller the CEP, the more accurate the system. The term 90% CEP is also used to describe the circle around the measure location with a 90% chance of containing the true emitter location. Figure 6.45 shows the CEP and the RMS errors for two DF systems from which the location of an emitter has been measured. Note that these two systems have ideal geometry to the target (i.e., 90° as seen from the target).

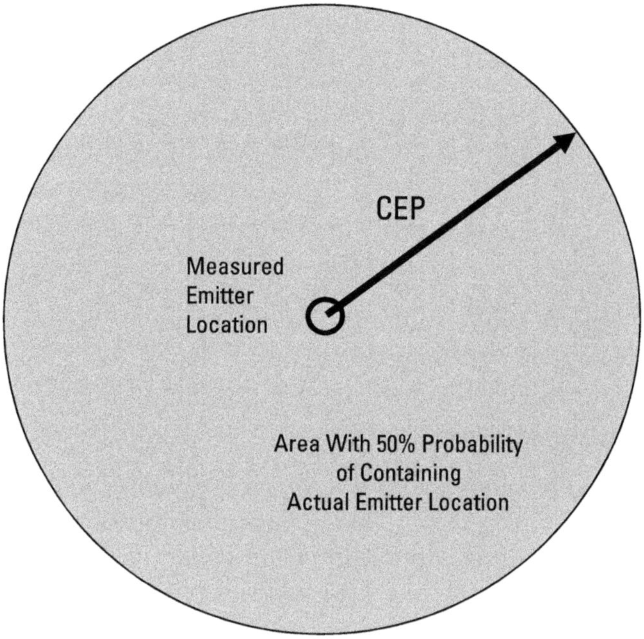

Figure 6.44 The CEP is the radius of a circle around the measured emitter location which has a 50% probability of containing the actual emitter location.

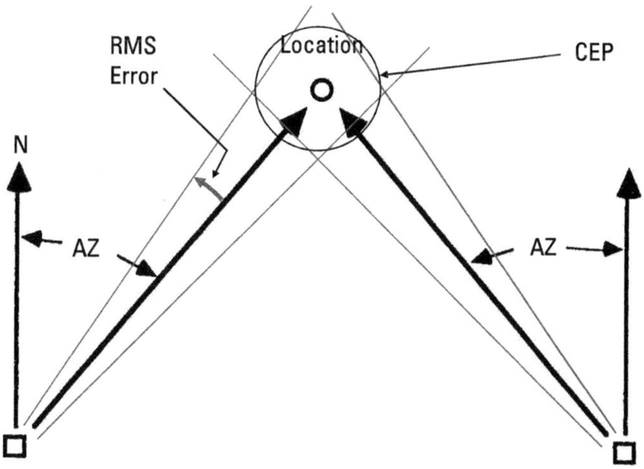

Figure 6.45 The CEP is related to the RMS error of two DF sites which triangulate to calculate the measured location of a target emitter.

6.6.7 EEP

The elliptical error probable (EEP) is the ellipse that has a 50% probability of containing the actual emitter when a location has been measured by two sites that do not have ideal geometry to the target. The 90% EEP is also often considered. The EEP may be drawn on a map as shown in Figure 6.46 to indicate not only the measured location of the emitter, but also the confidence a commander can place in the location measurement.

The CEP can also be determined from the EEP by the following formula:

$$CEP = 0.75 \times \mathrm{SQRT}\left(a^2 + b^2\right)$$

where a and b are the semi-major and semi-minor axes of the EEP ellipse.

The CEP and EEP are also defined for precision emitter location techniques, and these will be described later.

6.6.8 Site Location and North Reference

For triangulation or single site emitter location to be performed, the location of each DF site must be known and input to the process. For angle of arrival (AOA) systems, there must also be a directional reference (often to true North). Site location is also required for the precision emitter location techniques mentioned earlier. As shown in Figure 6.47, errors in site location and reference direction will cause errors in the AOA determined for target emitters. This figure is deliberate exaggerated to show the effects of errors. Typically, site location and reference direction errors are of the order of magnitude of the measurement accuracy errors. As you will see, in later examples these errors are typically only a few degrees.

Figure 6.48 (also deliberately exaggerated) shows the location errors caused by measurement, site location and directional reference errors. If an error contribution is fixed, it must be directly added to the location accuracy. Site location errors are typically considered fixed. However, when sources of errors are random and independent of each other, they are "RMSed" together. That

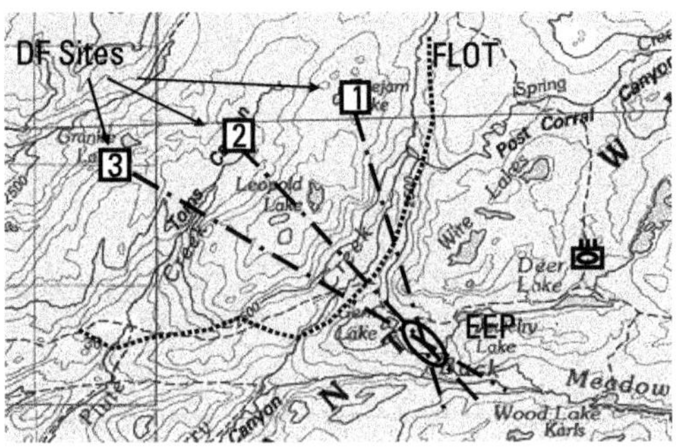

Figure 6.46 The EEP for a measured emitter location can be superimposed on a tactical map to give a commander the appropriate level of confidence in a measured emitter location accuracy.

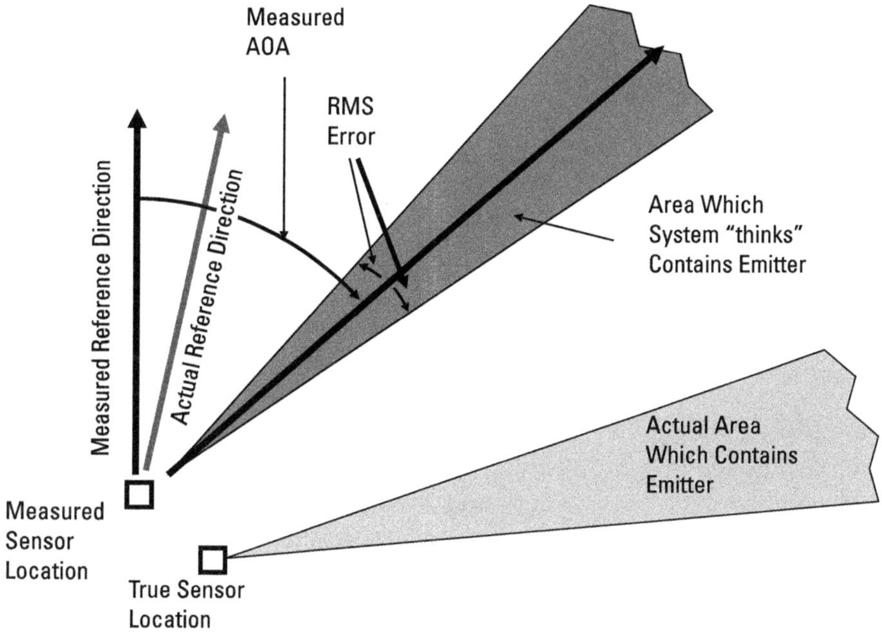

Figure 6.47 In an AOA system, the sensor location error and reference direction error cause inacuracy in the reported emitter location.

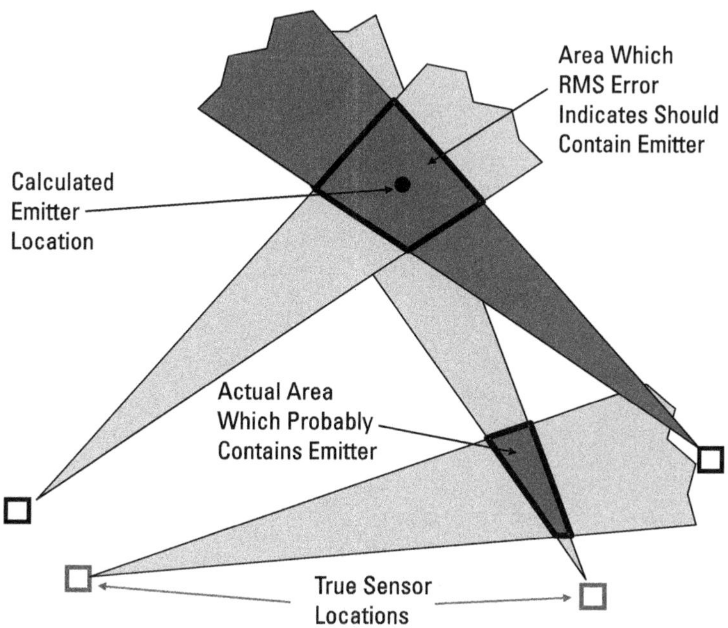

Figure 6.48 The accuracy of the location of a hostile emitter by AOA systems is a function of the measurement error and also the error in sensor locations and reference directions.

is, the resulting RMS error is the square root of the average of the squares of the various error contributions.

Before the mid-1980s, the location of DF sites was quite challenging. Ground-based DF systems required that the DF site location be determined by survey techniques and entered into the system manually. The North reference required either that the DF antenna array be oriented and stabilized to a specified orientation, or that the antenna array orientation be automatically measured and input. Automatic North sensing was particularly important for mobile sites.

A magnetometer is an instrument which senses the local magnetic field and provides an electronic output. It is functionally a digital reading magnetic compass. When a magnetometer was integrated into the antenna array of a ground-based system, its (magnetic) North reference could be automatically entered into the computer in which the triangulation was being performed. The local declination (i.e., variation of magnetic North from true North) had to be manually input to the system to calculate the azimuth reference from each site. The magnetometer accuracy was typically about 1.5°. As shown in Figure 6.49, the magnetometer was often integrated into the DF array of an AOA system. This avoided the difficult process of orienting the antenna array to magnetic North, significantly reducing the system deployment time.

Shipboard DF systems on large platforms could get their location and orientation references from the ships' navigation systems, which have been quite

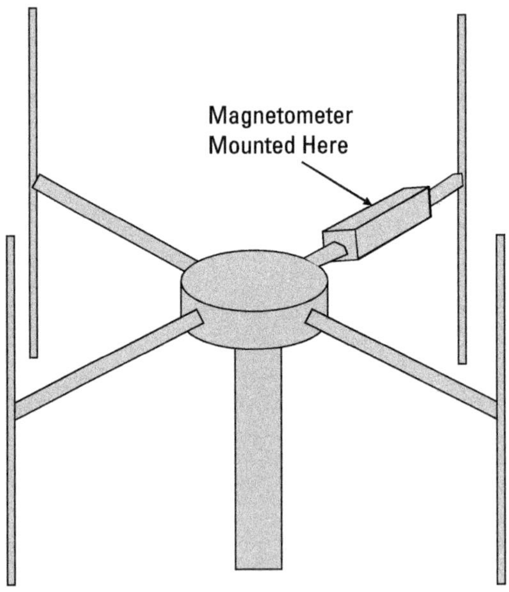

Figure 6.49 A magnetometer mounted in a direction finding array measures the orientation of the array relative to magnetic north.

accurate for many years. The ship's inertial navigation system (INS) can be manually corrected by a highly trained navigator to provide long term location and directional accuracy.

Airborne DF systems also required that the location and orientation of each DF system be known and entered into the triangulation calculation. This was provided from the aircraft's INS, which required extensive initialization procedures before each aircraft mission. An INS derived its North reference from two mechanically spinning gyroscopes (oriented 90° apart) and its lateral location reference from three orthogonally oriented accelerometers, as shown in Figure 6.50. Each gyroscope can only measure angular motion perpendicular to its axis of rotation, hence the requirement for two gyroscopes to provide three-dimensional orientation. Each of the accelerometer outputs is integrated once to provide lateral velocity and a second time to provide location change (each in one dimension). The gyroscopes and accelerometers were mounted on a mechanically controlled platform within the INS, which remained in a stable orientation as the aircraft maneuvered. After the aircraft left the compass rose on the airfield or the aircraft was launched from an aircraft carrier, the location and orientation accuracy decreased linearly with time because of the drift of gyroscopes and the accumulated error of accelerometers. Hence, the accuracy of emitter location from airborne platforms was a function of the mission duration.

Also, effective airborne DF systems were constrained to deployment in large enough platforms to support INS installations (which were about 2 cubic feet in volume).

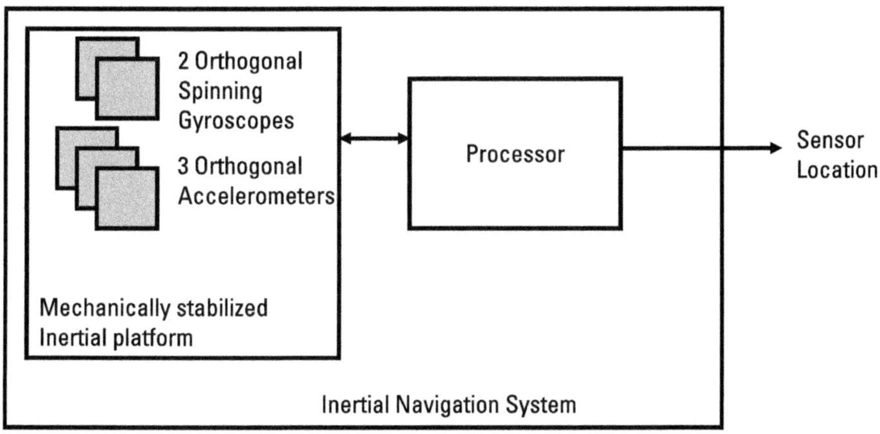

Figure 6.50 An older inertial navigation system required a mechanically stabilized inertial platform which kept itself oriented with two gyroscopes 90° apart and measured lateral motion with three orthogonal accelerometers. The location and orientation accuracy degrade linearly with time since system calibration.

In the late 1980s, the Global Positioning System (GPS) satellites were placed in orbit and small, inexpensive, rugged GPS receivers became available. GPS has had a significant impact on the way we locate mobile assets. Now the location of small aircraft, ground vehicles, and even dismounted individuals can be automatically measured (electronically) with adequate accuracy to support emitter location. This allowed the many low-cost DF systems to provide significantly better location accuracy.

GPS has also had a significant impact on the way INS devices work. Because the absolute location can be directly measured at any time, INS location accuracy is no longer a function of mission duration. As shown in Figure 6.51, inputs from the inertial platform are updated with data from the GPS receiver. Location is measured directly by GPS, and angular updates can be derived from multiple location measurements.

Because of the development of new types of accelerometers and gyroscopes, and significant electronics miniaturization, the INS system can now be implemented in significantly less size and weight, and with no moving parts. Ring laser gyroscopes bounce a laser pulse around closed path (three precise mirrors). By measuring the time to get around the circular path, it determines angular velocity. The velocity is integrated to determine orientation. Three ring laser gyroscopes are required to determine the three-axis orientation. Piezoelectric accelerometers have now replaced the old weight-on-a-spring type. There are also very small piezoelectric gyroscopes that measure angular velocity.

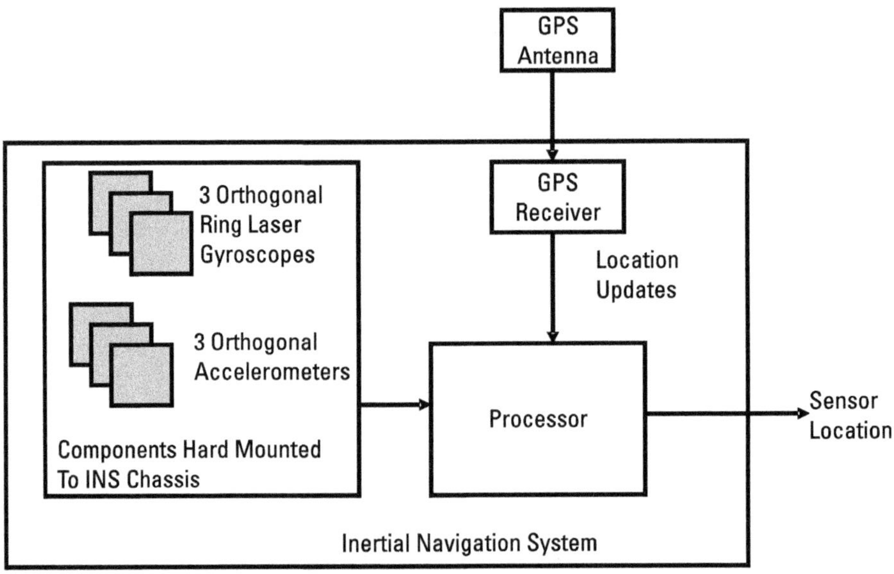

Figure 6.51 A GPS enhanced inertial navigation system uses location inputs from the GPS receiver to provide long term location accuracy.

An additional value of GPS is to provide a very accurate clock at fixed or mobile emitter location sites. This clock function is required for the precision location techniques that we will be discussing. The GPS receiver/processor synchronizes itself to atomic clocks in the GPS satellites. This has the effect of creating a virtual atomic clock in one printed circuit board and an antenna. (Note that an actual atomic clock is bigger than a bread box.) GPS has, therefore enabled the use of precision emitter location techniques in small platforms.

6.6.9 Moderate Accuracy Techniques

Since moderate accuracy systems are direction finders, their accuracy is most conveniently defined in terms of their RMS angular accuracy. A fairly good number for moderate accuracy is 2.5° RMS. This is the accuracy achievable in most DF approaches without calibration. We will be talking more about calibration later, but for now, calibration means systematically measuring and correcting errors in the measurement of AOA of transmitted signals

There are many moderate accuracy systems in use, and they are considered adequate for the development of electronic order of battle information. That is, they can locate enemy transmitters with enough precision to allow analysis of the types of military organizations present, their physical proximity, and their movements. This information is used by expert analysts to determine the enemy's order of battle and to predict the enemy's tactical intentions.

These systems are also relatively small, light, and inexpensive. In general, the higher the system accuracy, the more accurate site location and reference must be. This has been a significant problem in smaller scale (lower cost) systems. However, this has become much easier with the increasing availability of small, low cost inertial measurement units (IMUs). Combined with GPS location reference, IMUs can provide adequate location and angle reference for moderate accuracy DF systems.

Two typical moderate accuracy techniques used for communications emitter location are Watson-Watt and Doppler.

6.6.10 Watson-Watt Direction Finding Technique

As shown in Figure 6.52, a Watson-Watt DF system has three receivers connected to a circularly disposed antenna array with an even number (four or greater) of antennas plus a reference antenna in the center of the array. The circular array has a diameter of about one-quarter wavelength.

Two of the outside antennas (opposite each other in the array) are switched to two of the receivers, and the center reference antenna is connected to the third receiver. In processing, the amplitude difference between the signals at the two outside antennas is referenced to (i.e., divided by) the amplitude of the signal at the center reference antenna. This combination of signals produces the

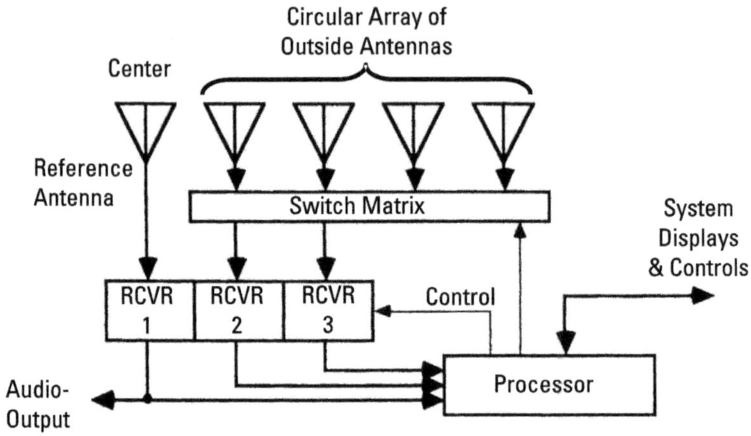

Figure 6.52 The Watson-Watt DF system uses an array with multiple outside antennas and a center reference antenna. Opposite outside antennas are switched into receivers 2 and 3

cardioid gain pattern (gain versus direction of arrival) around the three antennas as shown in Figure 6.53. By switching another pair of opposite antennas into receivers 2 and 3, a second cardoid pattern is formed. At the moment of switching, we therefore have two points on the cardiod. After sequentially switching all of the opposite pairs a few times, the DOA of the signal can be calculated.

The Watson-Watt technique works against all types of signal modulations and, without calibration, achieves about 2.5° RMS error.

6.6.11 Doppler Direction Finding Technique

If one antenna is rotated around another antenna as shown in Figure 6.54, the moving antenna (A) will receive a transmitted signal at a different frequency from that received at the fixed antenna (B). As the moving antenna moves toward the transmitter, the receiving frequency will be increased by the Doppler shift. As it moves away, the frequency will be reduced. This frequency variation is sinusoidal, and can be used to determine the DOA of the transmitted signal. Note that the emitter is in the direction at which the negative going zero crossing of the sine wave in this figure occurs.

In practice, multiple, circularly disposed antennas are sequentially switched into one receiver (A), while another receiver (B) is connected to a central antenna in the array as shown in Figure 6.55. Each time the system switches one of the outside antennas into receiver A, the phase change in the received signal is measured. After a few revolutions, the system can construct the sinusoidal variation of frequency (in antenna A versus antenna B) from the phase change data and thus determine the AOA of the transmitted signal.

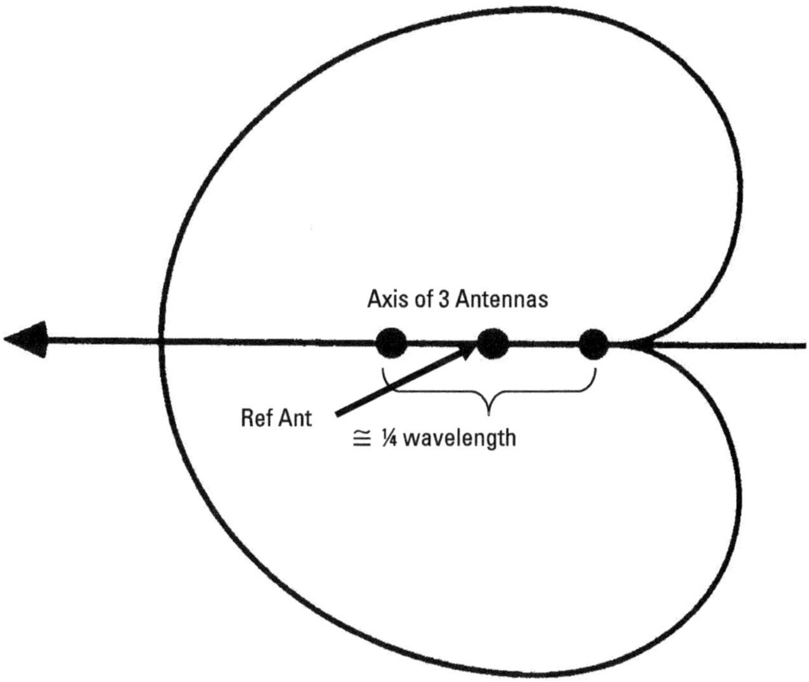

Figure 6.53 When the difference between two opposite outside antennas is normallied to the central reference antenna in a Watson-watt array, the result is a cardioid pattern of antenna array gain vs. angle of arrival.

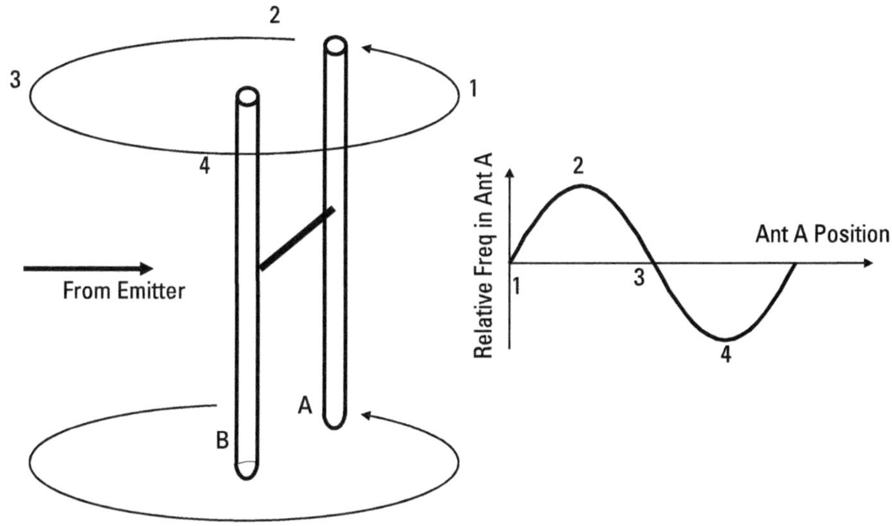

Figure 6.54 If antenna A is rotated around fixed antenna B, the frequency of a received broadcast signal varies sinusoidally with the rotation angle relative to the direction to the emitter.

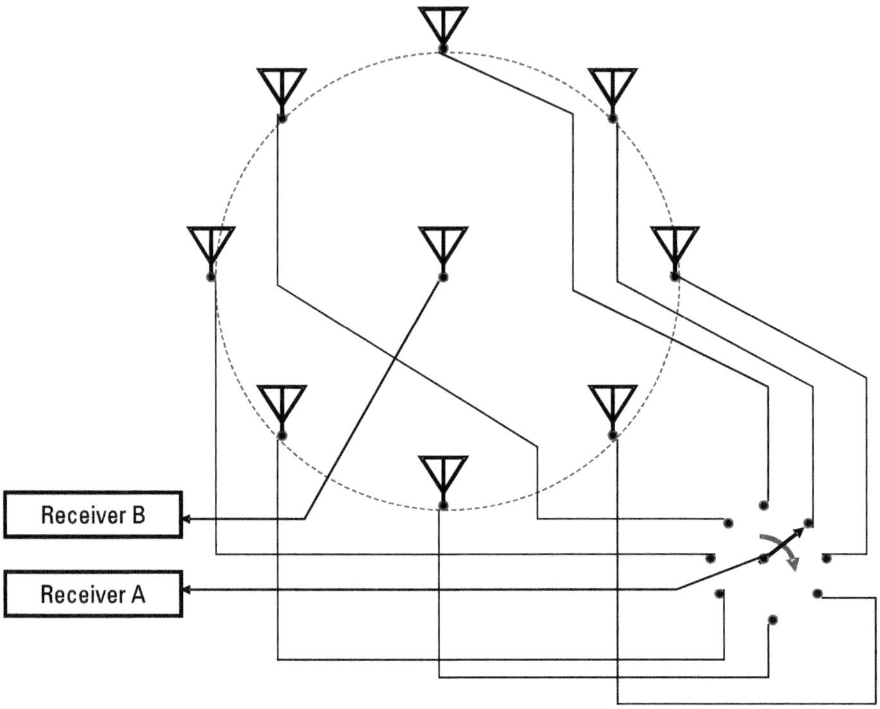

Figure 6.55 In a Doppler DF system, outside antennas are sequentially switched to one receiver (A) and a central antenna is connected to another receiver (B).

The Doppler technique is widely used in commercial applications and can have as few as three outside antennas plus the central reference antenna. It typically achieves about 2.5° RMS accuracy. However, this technique has difficulty with frequency modulated signals unless their modulation can be clearly separated from the apparent Doppler shifts of the sequentially switched outside antennas.

6.6.12 Location Accuracy

As shown in Figure 6.56, the linear error (Δ) in the location of a hostile emitter is a function of the angular error and the distance to the emitter. The formula is:

$$\text{Linear error} = \text{Tan (angular error)} \times \text{distance}$$

At 20 km, an angular error of 2.5° from the indicated line of bearing causes a linear error (Δ) of 873m.

The way that we determine the tactical usefulness of an emitter location system is by the CEP that it can provide. To evaluate the effective location

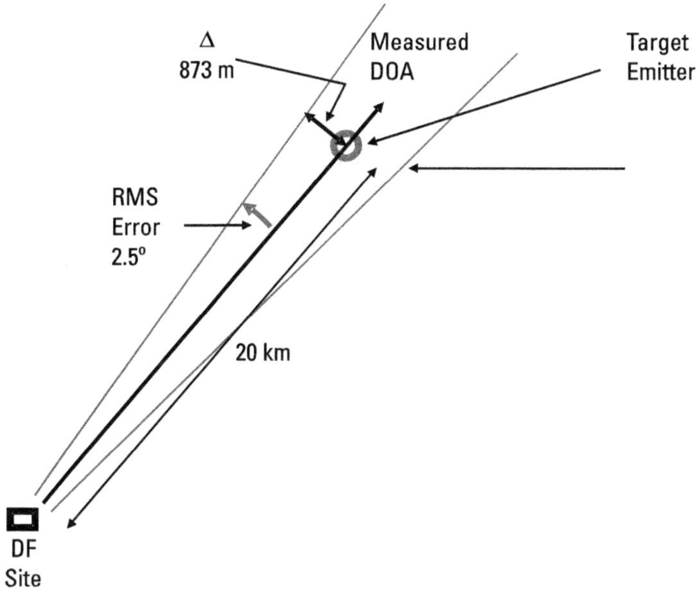

Figure 6.56 The linear error caused by a 2.5° angular error at 20 km distance is 873 meters.

accuracy of moderate accuracy DF systems, we will calculate the CEP provided by two 2.5° DF systems, each 20 km from a target emitter. We will take the case of ideal tactical geometry, that is, the two sites are 90° apart as seen from the emitter location.

To calculate the CEP for this situation, we will first determine the area included in the area within the RMS error angle limits from the two DF sites as shown in Figure 6.57. All but the most rigorous mathematicians will forgive us for approximating this area as a square with 2Δ on a side. You will recall from Section 6.6.4 that if the mean error of a DF system is removed from the RMS error, the remainder is the standard deviation (σ). We will assume for this problem that this has been done. The area of the angular wedge between the indicated direction of arrival and the 1 standard deviation (1σ) line has a 34.13% chance of containing the true AOA.

The square area between the two 1σ lines has edges 2Δ long. By the math presented above, the probability that the square contains the actual emitter location is 46.6%. The CEP for the hostile emitter location is the radius of a circle with a 50% chance of containing the location. It can be calculated from the formula:

$$CEP = \text{sqrt}\left[4\Delta^2 \times 1.074/\pi\right]$$

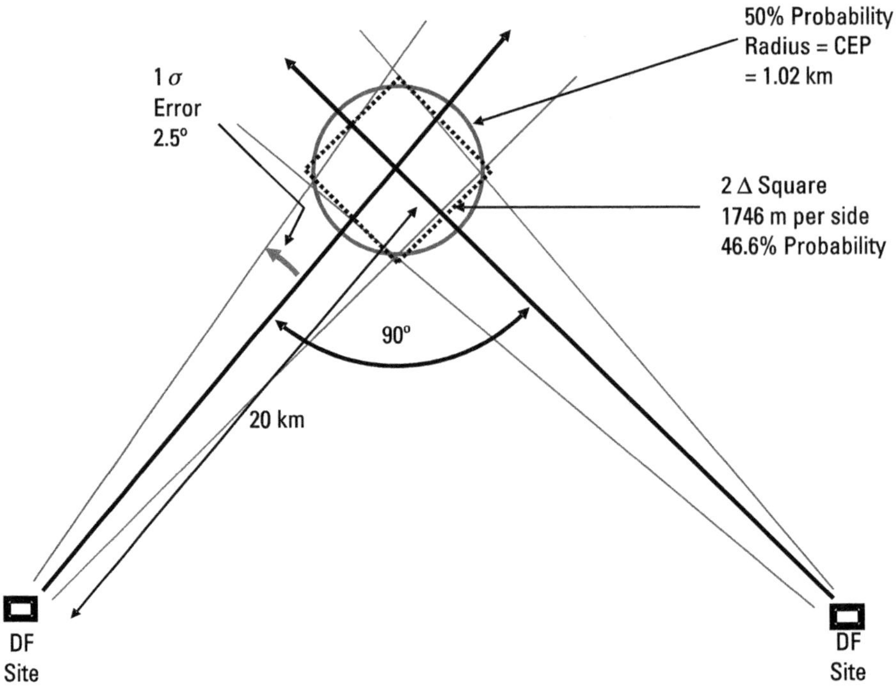

Figure 6.57 The area enclosed by the ± 1 σ lines from two ideally placed DF sites has a 46.6% probability of containing the actual emitter location.

Note that the 1.073 term is to increase the 46.6% probability that the square contains the emitter to the required 50% for the circle of radius CEP.

Now we plug the linear error value into the formula to determine the CEP to be 1.02 km.

6.6.13 High-Accuracy Techniques

When we speak of high-accuracy emitter location techniques, we are generally talking about interferometer direction finding. Interferometers can generally be calibrated to provide on the order of 1° RMS error. Some configurations provide better than that and some have less accuracy. The interferometer is a direction finder, determining only the AOA of the signal. Emitter location is determined from one of the techniques (such as triangulation) discussed earlier.

We will begin by discussing single baseline interferometers and then will cover correlative and multiple baseline interferometers.

6.6.14 Single Baseline Interferometer

Although virtually all interferometer systems employ multiple baselines, the single baseline interferometer uses one baseline at a time. The presence of multiple baselines allows for the resolution of ambiguities. It also allows multiple, independent measurements to be averaged to reduce the impact of multipath and other equipment-based sources of error.

Figure 6.58 is a basic block diagram of an interferometric DF system. Signals from two antennas are compared in phase, and the DOA of the signal is determined from the measured phase difference. Remember that we characterize the transmitted signal as a sine wave traveling at the speed of light. One cycle (360 phase degrees) of the traveling sine wave is called the wavelength. The relation between the frequency of the transmitted signal and its wavelength is defined by the formula:

$$c = \lambda f$$

where c is the speed of light (3×10^8 m/s), λ is the wavelength (in meters), and f is the frequency in cycles per second (units are 1/sec).

The interferometric principle is best explained by consideration of the interferometric triangle as shown in Figure 6.59. The two antennas from Figure 6.58 form a baseline. It is assumed that the distance between the two antennas and their precise location are known precisely. The wavefront is a line perpendicular to the direction from which the signal is arriving at the direction finding station. This is a line of constant phase for the arriving signal. The signal expands spherically from the transmitting antenna, so the wavefront is actually a circular segment. However, since the baseline can be assumed to be much shorter than the distance from the transmitter, it is very reasonable to show the wavefront as a straight line in this drawing. The precise location of the station

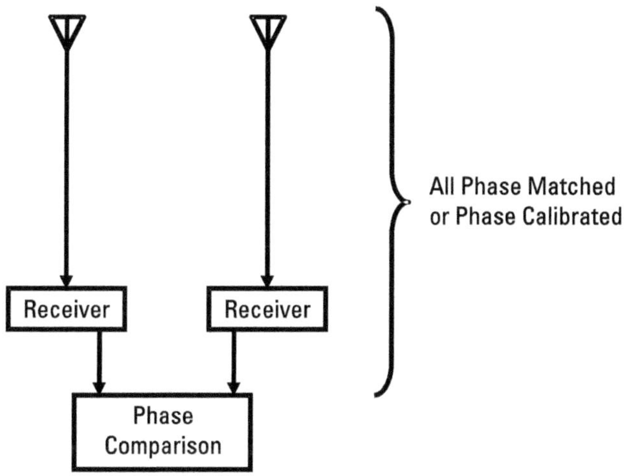

Figure 6.58 The interferometer compares the phase of a signal at two antennas and uses the phase difference to calculate the angle of arrival.

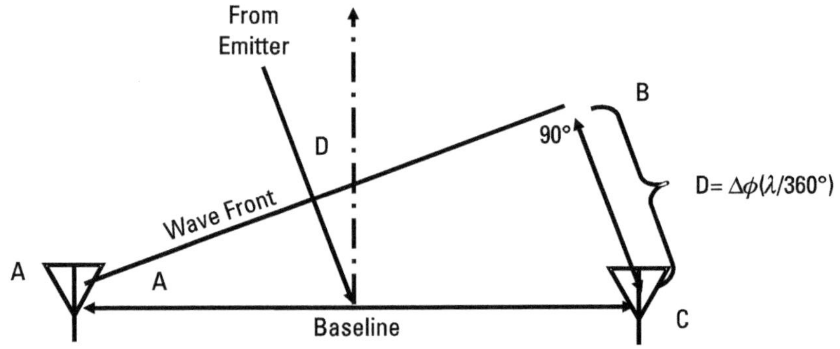

Figure 6.59 The operation of an interferometer is best understood through consideration of the interferometric triangle.

is taken to be the center of the baseline. Because the signal has the same phase along the wavefront, the phases at point A and point B are equal. Hence, the phase difference between the signals at the two antennas (i.e., points A and C) is equal to the phase difference between the signal at points B and C.

The length of line BC is known from the formula:

$$BC = \Delta\Phi\left(\lambda/360^\circ\right)$$

where $\Delta\Phi$ is the phase difference and λ is the signal wavelength.

The angle at point B in the diagram is 90° by definition, so the angle at point A (call it angle A) is defined by:

$$A = \arcsin\left(BC/AC\right)$$

where AC is the length of the baseline.

The AOA of the signal is reported out relative to the perpendicular to the baseline at its center point, because the interferometer provides maximum accuracy at that angle. Note that the ratio of phase degrees to angular degrees is maximum here. By construction, you can see that angle D is equal to angle A.

Interferometers can use almost any type of antenna. Figure 6.60 shows a typical interferometer array which might be mounted on a metal surface, such as the skin of an aircraft or the hull of a ship. A horizontal array as shown would measure azimuth of arrival, while a vertical array would measure elevation AOA. These antennas are cavity-backed spirals, which have a large front to back ratio, and thus provide only 180° of angular coverage. The spacing of the antennas in this array determine accuracy and ambiguity. The end antennas have a very large spacing, and thus provide excellent accuracy. However, their phase response is as shown in Figure 6.61. Note that the same phase difference (between the signals at the two antennas) can represent several different angles

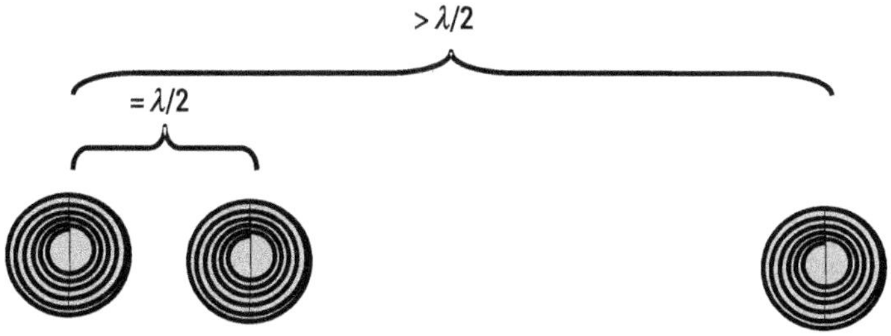

Figure 6.60 Three cavity backed spiral antennas are often used for interferometer direction finding on aircraft or ships.

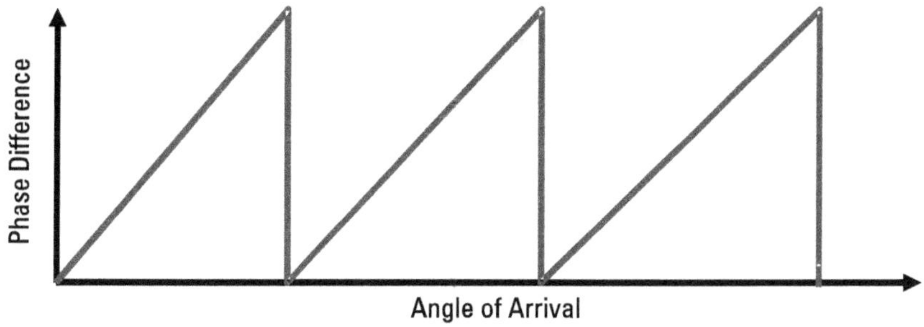

Figure 6.61 Phase difference vs. angle of arrival in two antennas spaced much farther than a half wavelength is highly ambiguous.

of arrival. This ambiguity is resolved by the two left antennas, which are spaced not more than half a wavelength apart, and thus have no ambiguity.

Ground-based systems often use arrays of vertical dipoles as shown in Figure 6.62. To avoid the ambiguities shown in Figure 6.61, the antennas must be less than half a wavelength apart. However, if the antennas are less than one-tenth of a wavelength apart, the interferometer is considered inadequately accurate. Thus a single array can provide direction finding only over a 5-to-1 frequency range. Some systems have multiple dipole arrays stacked vertically. Each array has different length dipoles with different spacing (smaller and closer dipoles used over higher frequency ranges). Note that the four antennas make six baselines as shown in Figure 6.63.

Because these dipole arrays cover 360° of azimuth, the interferometer has a front-back ambiguity as shown in Figure 6.64, because signals arriving from either of the two angles shown would create the same phase difference. This problem is resolved as shown in Figure 6.65 by making a second measurement with a different pair of antennas. The correct AOA is correlated in the two measurements, while the ambiguous AOAs do not correlate.

Figure 6.66 shows a typical interferometer DF system. The antennas are switched into the phase comparison two at a time, and the DOA are measured.

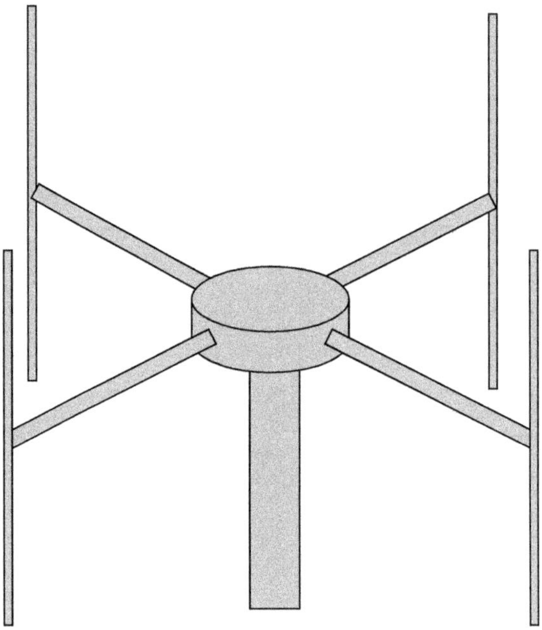

Figure 6.62 Ground based interferometers often use arrays of vertical dipole antennas to provide 360° coverage.

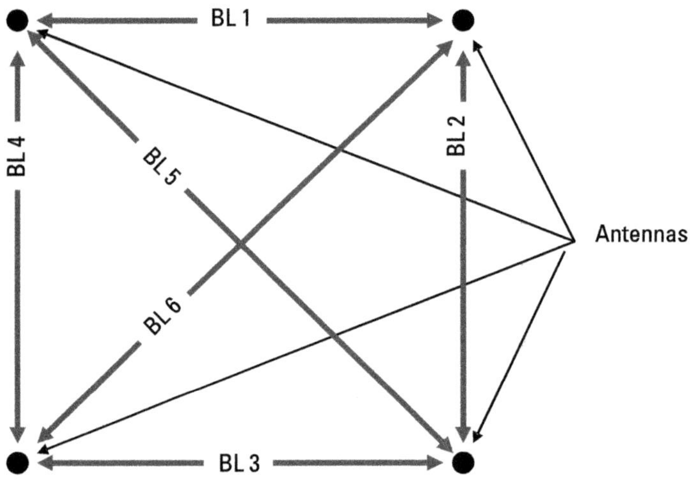

Figure 6.63 An array of four antennas has six interferometric baselines.

If there are four antennas, the six baselines are used sequentially. Often, each baseline is measured twice with the two antenna inputs switched to balance out any tiny differences in signal path length. The twelve AOA results are then averaged and the DOA are reported.

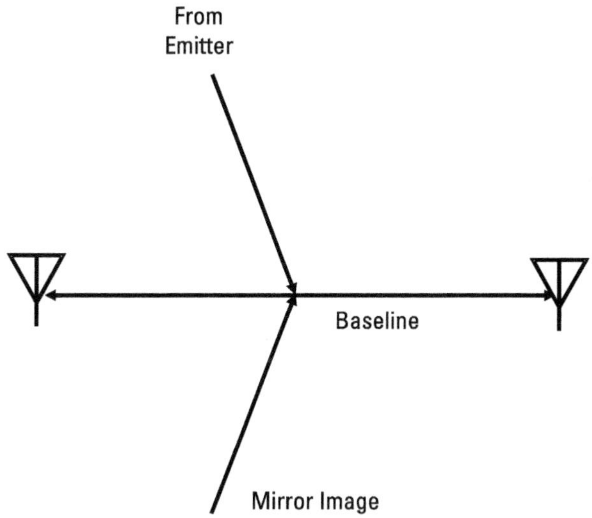

Figure 6.64 The phase difference between signals at two 360° antennas is the same for the signal from the emitter direction and a signal from the mirror image direction.

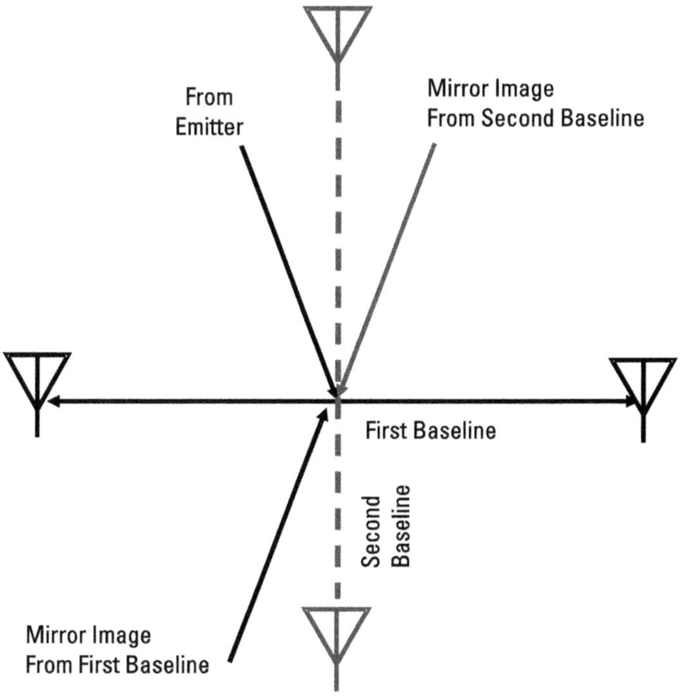

Figure 6.65 A second baseline will have its front-back ambiguity at a different angle of arrival from that of the first baseline.

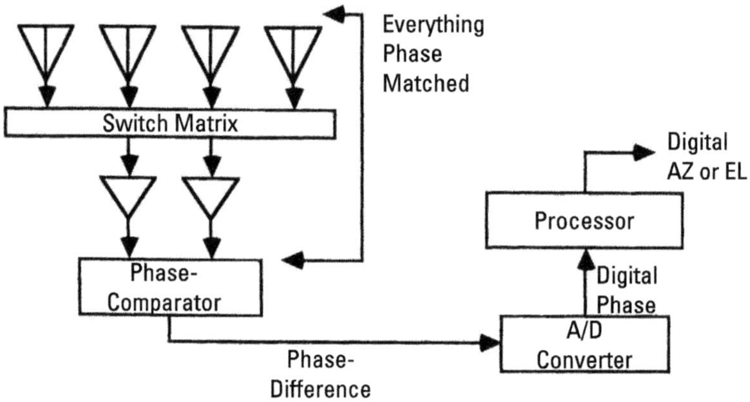

Figure 6.66 An interferometer system sequentially switches two of its antennas into a phase measurement receiver and the angle of arrival is calculated for each baseline in turn.

6.6.15 Multiple Baseline Precision Interferometer

Although it is typically applied only at microwave frequencies, the multiple baseline interferometer can be used in any frequency range as long as the length of the antenna array can be accommodated. As shown in Figure 6.67, There are multiple baselines, all greater than a half-wavelength. In the figure, the baselines are 5, 14, and 15 half-wavelengths.

The phase measurements from all three baselines are used in a single calculation, using modulo arithmetic, to determine the AOA and resolve all ambiguities. The advantage of this type of interferometer is that it can produce up to 10 times the accuracy of the single baseline interferometer. The disadvantage at lower frequencies is that the arrays become extremely long.

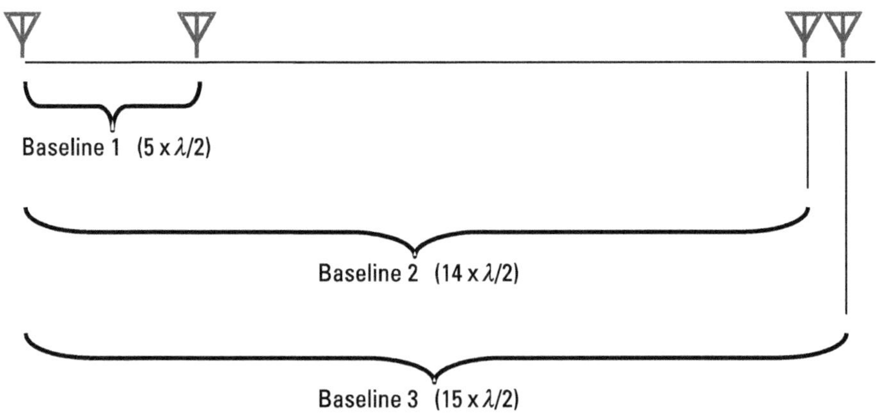

Figure 6.67 The multiple baseline precision interferometer calculates angle of arrival to high precision from phase differences in multiple very long baselines.

6.6.16 Correlative Interferometer

The correlative interferometer system uses a large number of antennas, typically five to nine. Each pair of antennas creates a baseline, so there are many baselines. The antennas are spaced more than a half-wavelength apart, typically one to two wavelengths as shown in Figure 6.68. There are ambiguities in the calculations from all baselines. However, the large number of DOA measurements allows a robust mathematical analysis of the correlation data. The correct AOA will have a greater correlation value and will be reported.

6.6.17 Precision Emitter Location Techniques

In general, these techniques provide emitter location with sufficient accuracy to support targeting. This means that the location accuracy is expected to be equal to the burst radius of a weapon (tens of meters) However, there are other applications which may profit from extremely accurate location, for example, determining if two emitters are co-located.

We will discuss two precision techniques, time difference of arrival (TDOA) and frequency difference of arrival (FDOA), then the combination of the two techniques. Both TDOA and FDOA require the presence of a highly accurate reference oscillator at each receiver site. Earlier, this required an atomic clock at each location, but now GPS provides the equivalent at significantly lower size and weight.

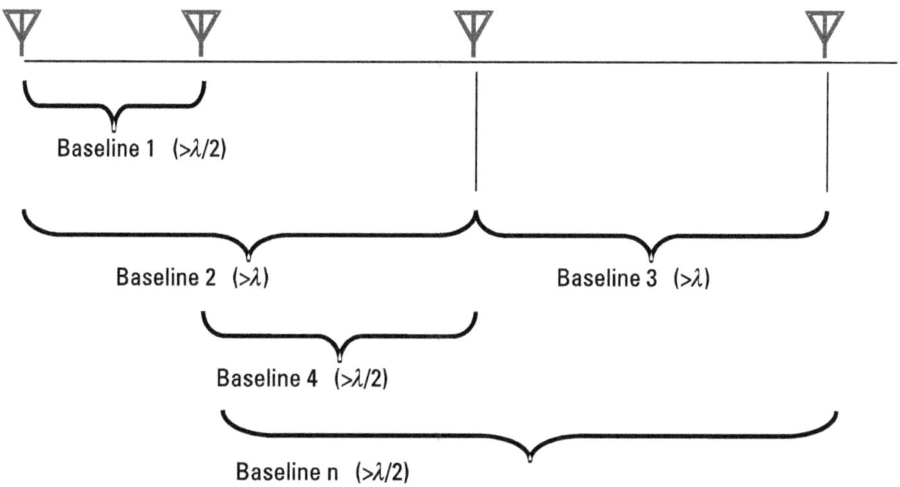

Figure 6.68 The correlative interferometer uses many baselines, all of which are greater than a half wavelength.

6.6.18 TDOA

TDOA depends on the fact that signals travel at the speed of light; thus, a single signal will arrive at two receiving sites at a time difference that is proportional to the difference in distance, as shown in Figure 6.69. If we knew the precise time at which the signal left the transmitter and the time at which it reached each receiver, we could calculate the distance from each receiver site to the transmitter and would thus know the precise emitter location. This is done in cooperative systems such as GPS in which the transmitted signal carries information about the time the signal is transmitted.

However, when dealing with hostile signals, we have no way to know the time that the signal leaves the transmitter. We thus can only determine the difference between the two times of arrival. Because communication signals are continuous, the only way to determine this time of arrival difference is to delay the received signal in the receiver closest to the emitter until the modulation from the two signals is correlated (see Figure 6.70). This requires that each receiver have a variable delay capability. (Either might be closest to the emitter.) The whole range of relative delays, in effect, searches through the area of possible emitter locations.

In practice, the received modulation is digitized at each receiver each time the relative delay is changed, and the resulting digital signals are correlated at a single location. The accuracy of the correlation (tens of nanoseconds) requires that the received signals be sampled and digitized at a very high rate.

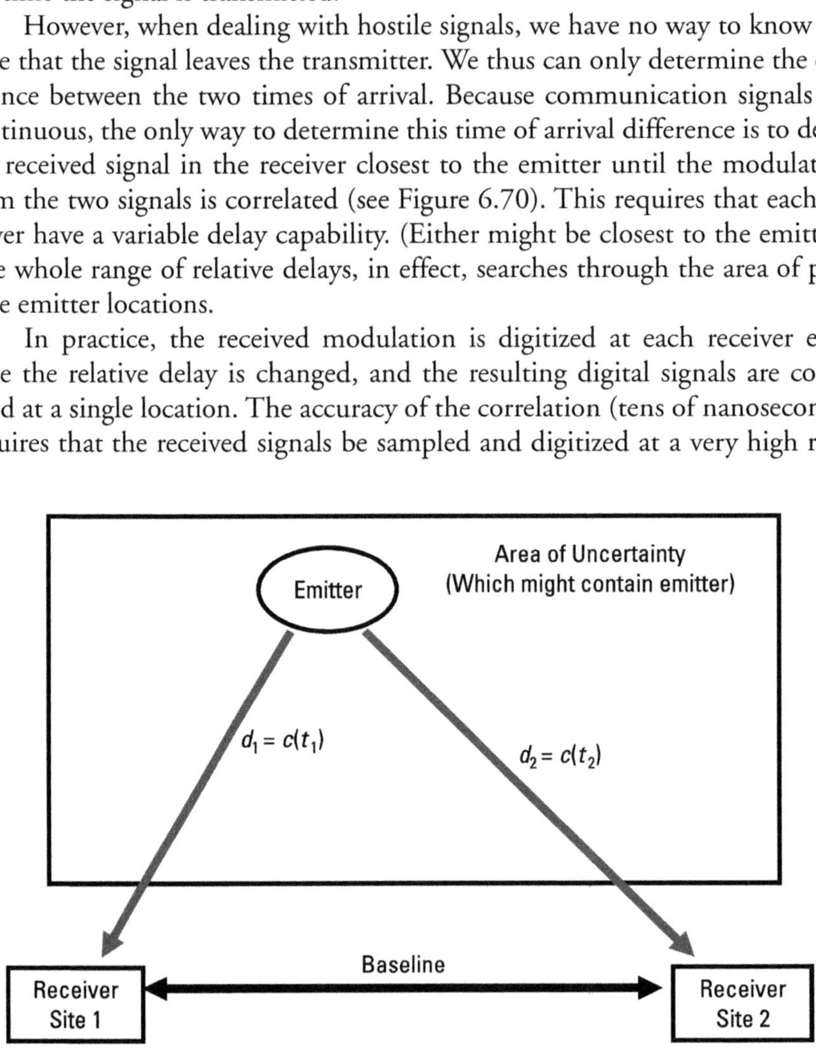

Figure 6.69 Since the signal travels at the speed of light, the time difference of arrival is proportional to the difference in distance to the two receiving sites.

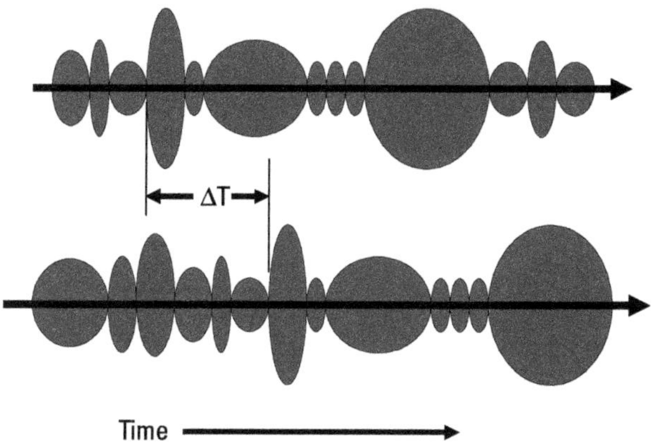

Figure 6.70 A single analog signal received at two distance stations will have the same modulation, but offset in time by the difference in distance.

This requires significant link bandwidth between the two receiving sites and the location at which correlation is performed.

As shown in Figure 6.71, the correlation of these two digitized signals will form a soft correlation peak when the delay is equal to the difference in time of arrival of the signal at the two receivers.

If the target emitter transmits a digital signal, and the two receiving sites can demodulation the received signals to recover the digital data, the two receivers output the same digital signal (offset in time by the relative propagation delays) so the correlation can possibly be more precise. The auto-correlation of a digital signal forms what is called a thumb-tack correlation as the relative delay changes. When the two digital signals are not synchronized, the correlation is

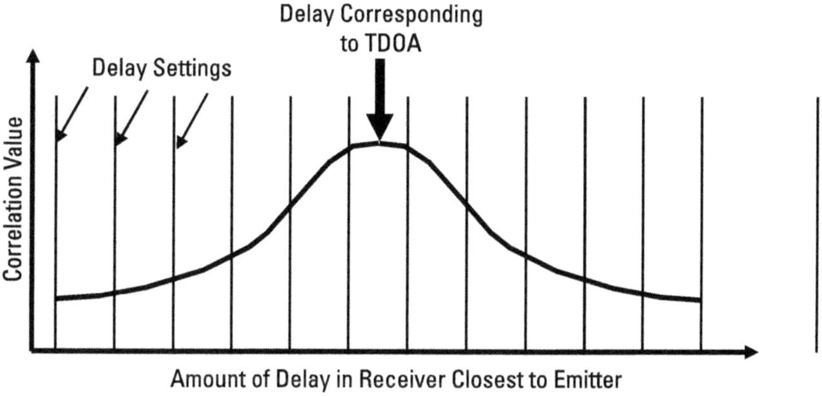

Figure 6.71 Delaying one of the two received analog signals will produce a soft correlation peak when the delay is equal to the time difference of arrival.

about 50%. When the signal from the closest receiver is delayed by the difference in time of arrival (within 1 bit) the correlation rises above 50%. When the delay is appropriate to bring the data from the two signals into synchronization, the correlation rises to about 100%. This is called thumb-tack correlation and is shown in Figure 6.72. It is important to note that this may not be practical because it requires delay increments smaller than the transmitted bit period. If the area of uncertainty (as in Figure 6.69) is large, the time to perform the correlation can become extremely long and/or the link bandwidths can become impractical.

6.6.19 Isochrones

Once the time difference is known, the difference in distance is known. A fixed difference in distance defines a hyperbolic surface in space. This surface intersects the Earth (assuming flat Earth) in a hyperbolic location contour, which is called an isochrone. The emitter is now known to lie along this hyperbola. If the time difference is measured extremely accurately, the emitter will be very close to this line (tens of meters), but the line is infinite in length. Figure 6.73 shows a family of isochrones, each for a different TDOA.

The actual location of the signal is determined by use of a third receiving station as shown in Figure 6.74. Each pair of receiving stations forms a baseline. Each baseline defines an isochrone. The isochrones from the two baselines shown in the figure cross at the emitter location. Actually, there is a third baseline (formed by receivers 1 and 3) that will define a third isochrone to cross the other two at the emitter location.

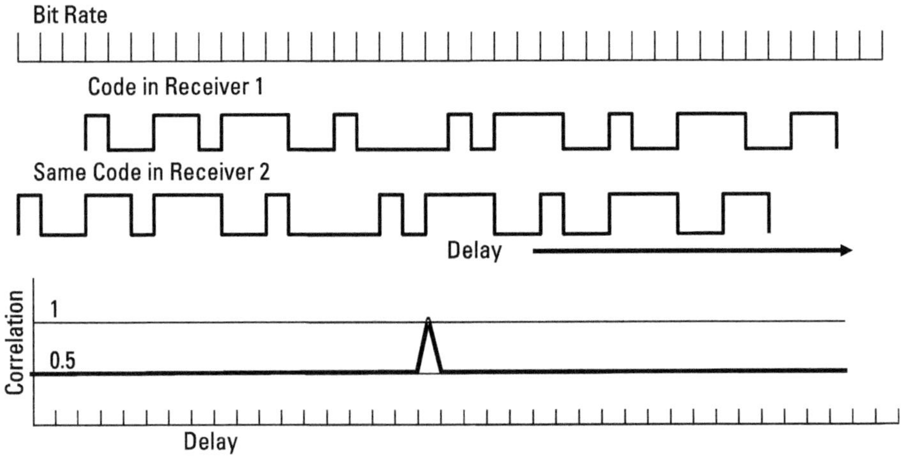

Figure 6.72 If two digital signals are the same, sliding one through the other in time will produce a sharp correlation peak when the two signals are synchronized.

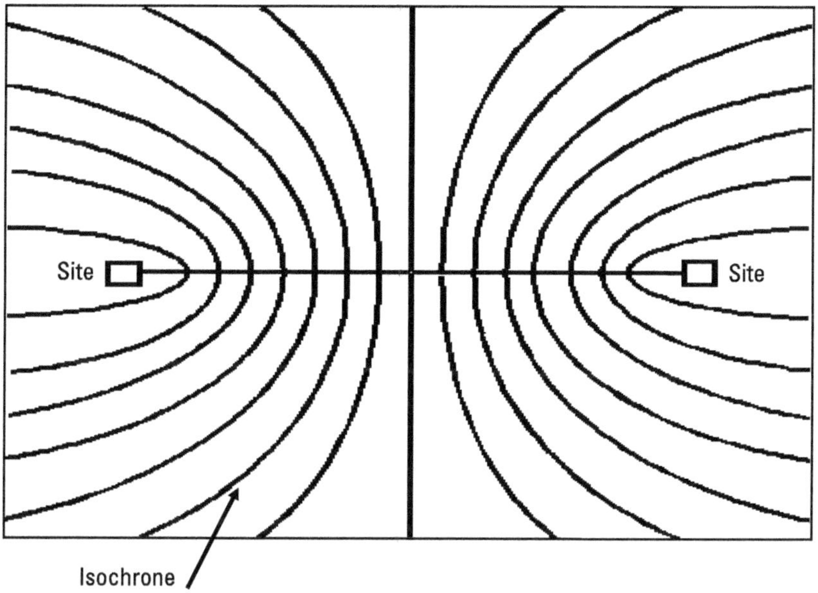

Figure 6.73 Each value of time difference produces a hyperbolic locus of possible locations called an isochrone.

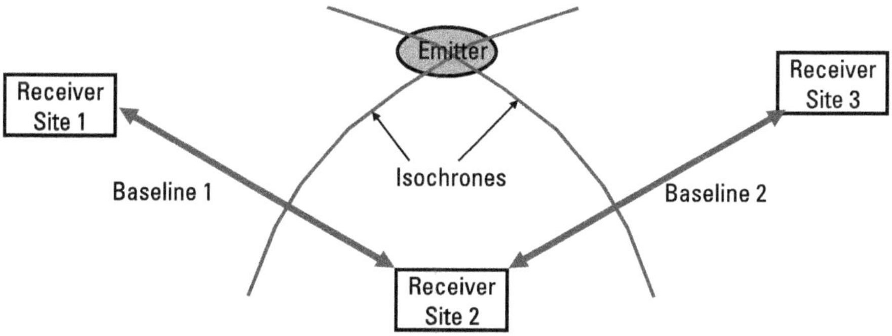

Figure 6.74 The target emitter is located at the intersection of isochrones from two baselines.

6.6.20 FDOA

This technique requires moving platforms and is primarily useful against fixed emitters on the Earth's surface.

When either the transmitter or the receiver is moving, a received signal will be received at a frequency different from the transmitted frequency. Here we have a fixed transmitter and a moving receiver. The frequency difference, caused by the Doppler shift, is determined by the formula:

$$\Delta F = F \times V \times \cos(\theta)/c$$

where ΔF is the change in the received signal relative to the transmitted frequency minus the Doppler shift, F is the transmitted frequency, V is the magnitude of the velocity of the moving receiver, θ is the true spherical angle between the velocity vector of the receiver and the DOA of the signal, and c is the speed of light.

Figure 6.75 shows two moving receivers, each receiving the same signal. Each receiver receives the signal at a frequency determined by its velocity vector and the direction of arrival of the target signal. The two moving receivers form a baseline. The received frequency at each receiver is the transmitted frequency plus the applicable Doppler shift ($F + \Delta F$). The FDOA is the difference between the two received frequencies.

For any frequency difference of arrival, there is a complex curved surface that is a locus of all of the emitter locations that would produce the measured frequency difference. If the target emitter is on the Earth's surface, the curved locus surface defines an Earth surface curve that is the locus of possible emitter locations.

Because the two receivers can have any velocity vectors (i.e., any speed in any direction), the shape of this curve can have much variety in its shape. To make it easy on our human eyes, Figure 6.76 is drawn for the special case in which the two receivers are traveling in the same direction at the same speed,

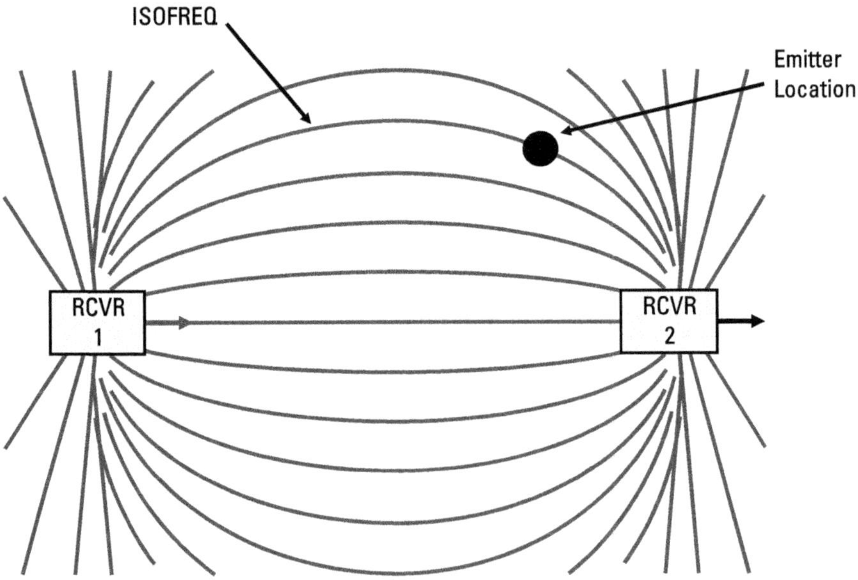

Figure 6.75 Each frequency difference of arrival defines a contour that will contain the location of the emitter.

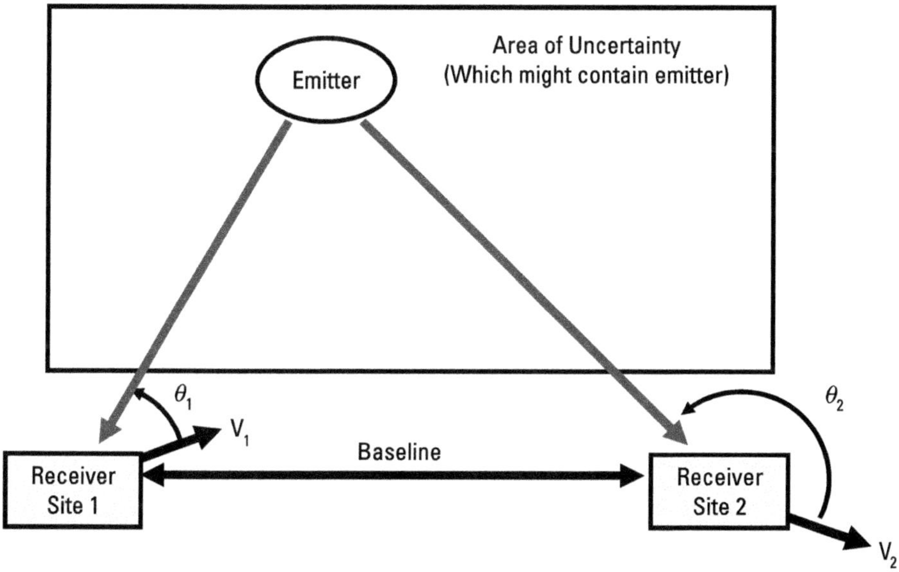

Figure 6.76 Receivers on two moving platforms will receive signal from an emitter at different frequencies, depending on the velocity vectors of the platforms.

although not necessarily in a tail chase. This figure shows a family of frequency difference value curves called isofreqs. They are also sometimes called isoDopps. Each isofreq is the locus of possible emitter locations for a specific FDOA. If the emitter location is as shown in this figure, the system only knows that it is somewhere along the indicated isofreq line, from the FDOA over the baseline formed by receivers 1 and 2.

To determine the actual emitter location, a third moving receiver must be added as shown in Figure 6.77. Now a second baseline is formed by receivers 2 and 3, so a second isofreq can be calculated. This second isofreq will cross the first at the emitter location. Like the TDOA approach, there is actually a third baseline formed by receivers 1 and 3, which creates a third isofreq curve passing through the emitter location.

6.6.21 Frequency Difference Measurement

An FDOA system just measures the frequency of the signal as received at each receiver location. This requires an extremely accurate frequency reference that, in the past, required a Cesium beam clock, but can now use the frequency reference output from a GPS receiver. Unlike TDOA, it is not necessary to perform a time-consuming correlation procedure; the frequency is simply measured at each location and the values subtracted. This can be accomplished with much

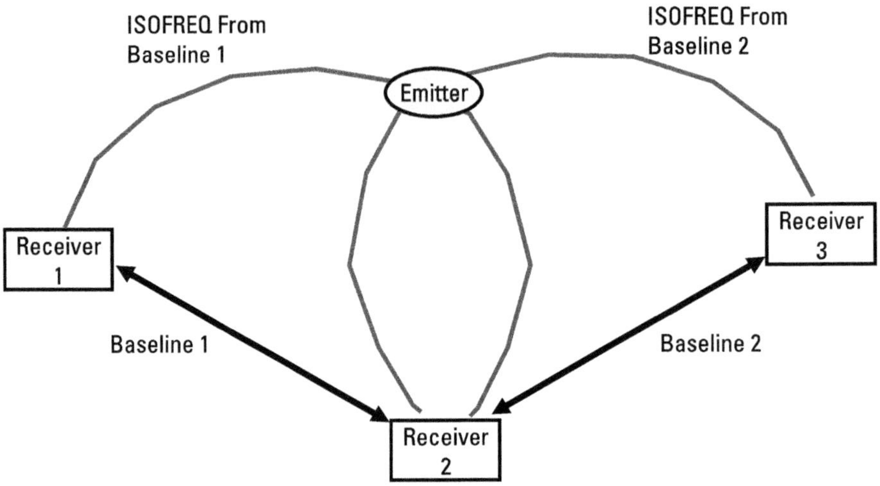

Figure 6.77 The emitter location is determined by the intersection of isofrequs from two baselines.

narrower data links connecting the three receiver platforms to the location at which the FDOA calculations are performed.

However, if the emitter is moving, its movement will create a Doppler shift of similar magnitude to that caused by the movement of the three receivers. Thus, it will be difficult to determine the proper isofreq contours. Unless there are many moving receivers (each measuring received frequency) and very powerful processing capability, it may be impractical to perform FDOA on moving target emitters.

6.6.22 TDOA and FDOA

The critical element in an FDOA receiver, like the TDOA receiver, is the presence of an extremely accurate time/frequency reference. With the wide availability of GPS, this can be implemented in small moving platforms. This means that both TDOA and FDOA are typically performed when the receivers are mounted on helicopters or fixed wing aircraft. As shown in Figure 6.78, each baseline allows calculation of both isochrone and isofreq contours. This means that each baseline can determine the emitter location at the intersection of an isocrone and an isofreq.

Because three receiver platforms are normally present, there will be three baselines and thus six defined contours through the emitter location (three isochrones and three isofreqs). The additional measurement parameter allows better location accuracy than would be provided by TDOA or FDOA processing alone.

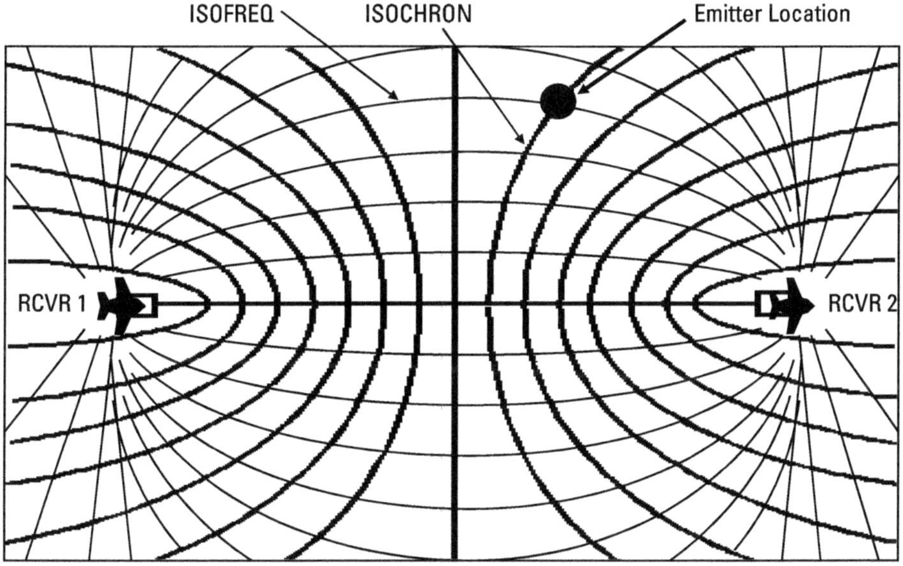

Figure 6.78 If both the time and frequency difference of arrival are determined for two moving platforms, both isochrones and isofreqs are defined.

6.6.23 Calculation of CEP for TDOA and FDOA Emitter Location Systems

The elliptical error probable (EEP) for precision emitter location systems is plotted on a map centered on the calculated emitter location. This describes not only the calculated emitter location, but also the confidence in the accuracy of the location. As in all emitter location approaches, the EEP is an ellipse that has a 50% probability of containing the actual emitter location. The 90% EEP has a 90% probability. However, in comparing different emitter location approaches, the important parameter is CEP or 90% CEP. As stated earlier, the CEP is related to the EEP by the formula:

$$CEP = 0.75\text{sqrt}\left(a^2 + b^2\right)$$

where a and b are the semi-major and semi-minor axes of the EEP ellipse.

6.6.24 References That Give Closed Form Formulas for TDOA and FDOA Accuracy

Reference [2] gives closed form formulas for the 1 standard deviation (1σ) width of the isochrones generated by a TDOA emitter location system and the isofreqs generated by an FDOA emitter location systems in terms of the various sources of error.

The $\pm 1\sigma$ lines shown in Figure 6.79 define the width of the isochron or isofreq, that is, the uncertainty in the actual course of the line. In a normally

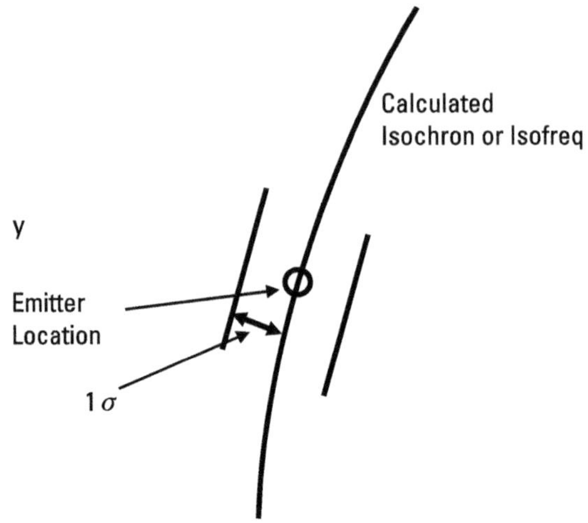

Figure 6.79 The "width" of the Isochron or Isofreq is often defined as the seperation of the $\pm 1\sigma$ contours from the calculated curve.

distributed function (i.e., the amount of error), 1σ is the point at which there is a 34.13% probability that the answer is closer to the correct value. Thus, there is a 68.26% probability that the actual emitter location lies between the $\pm 1\sigma$ lines. In Figure 6.80, the isochrons or isofreqs from two baselines cross at the calculated emitter location. The $\pm 1\sigma$ lines from the two baselines form a parallelogram that has a 46.59% chance of containing the actual emitter location (assuming the error function is Gaussian).

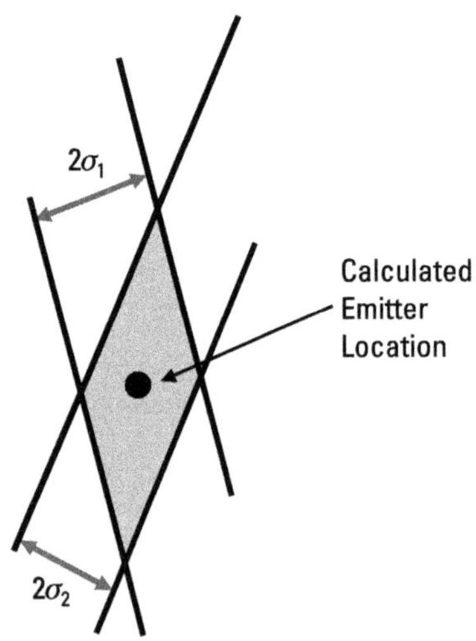

Figure 6.80 The $\pm 1\sigma$ error lines from the two baselines form a parallelogram.

If you draw an ellipse oriented with the parallelogram, but defining an area with a 50% probability of containing the actual emitter location as shown in Figure 6.81, this will be the EEP.

Formulas for CEP using only geometric error sources are also discussed in [3], which also provides approaches to defining the parallelogram from the intercept geometry.

The relationship between the dimensions of the EEP ellipse and the CEP come from [4].

6.6.25 Scatter Plots

A more accurate way to determine the EEP and CEP for a TDOA or FDOA emitter location system is to run location calculations from an intercept geometry many times on a computer (perhaps 1,000 times). During each calculation, the value of each variable is randomly selected according to its probability distribution (e.g., Gaussian with some stated standard deviation). For each calculation, plot the calculated location relative to the correct emitter location. Then draw an ellipse centered on the actual emitter location and sized to contain 50% (or 90%) of the plotted emitter locations. Figure 6.82 shows such an EEP.

6.6.26 Precision Location of LPI Emitters

There are significant issues associated with the precision location of low probability of intercept (LPI) emitters. This will be covered in Chapter 7.

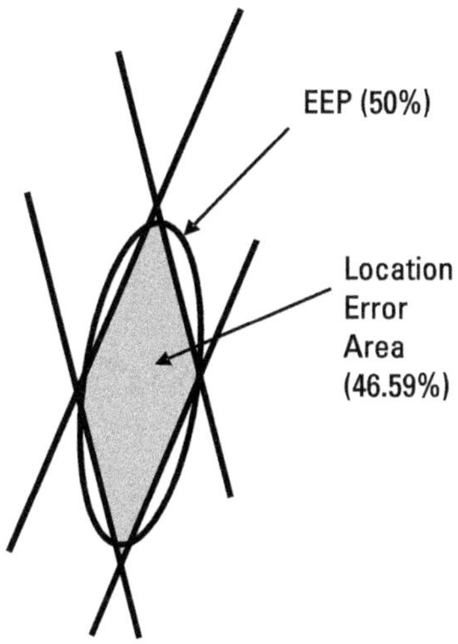

Figure 6.81 An ellipse with area 1.073 times the area of the parallelogram and matching its orientation is the EEP.

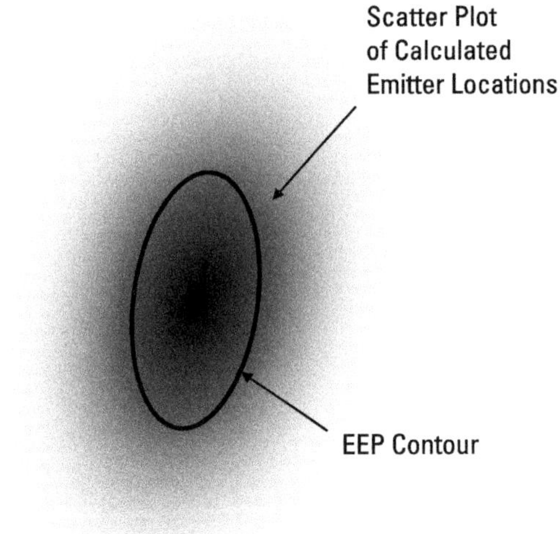

Figure 6.82 The plotted locations from many simulated TDOA or FDOA measurements with normally distributed error values form an elliptical pattern. The ellipse containing 50% of these solutions is the EEP.

6.7 Communication Jamming

The purpose of communication is to take information from one location to another. It follows then that the purpose of communication jamming is to prevent an enemy's information from reaching the intended location. Figure 6.83 shows a communication jamming situation. There is a desired signal link from a transmitter to a receiver, and a jamming link from a jammer to the receiver. The desired signal transmitter power (P_S) combines with the desired signal antenna gain (G_S) in the direction of the receiver to form the desired signal effective radiated power (ERP_S). The distance from the desired signal transmitter to the receiver (d_S) is used in calculation of propagation losses. P_J, G_J, ERP_J, and d_J

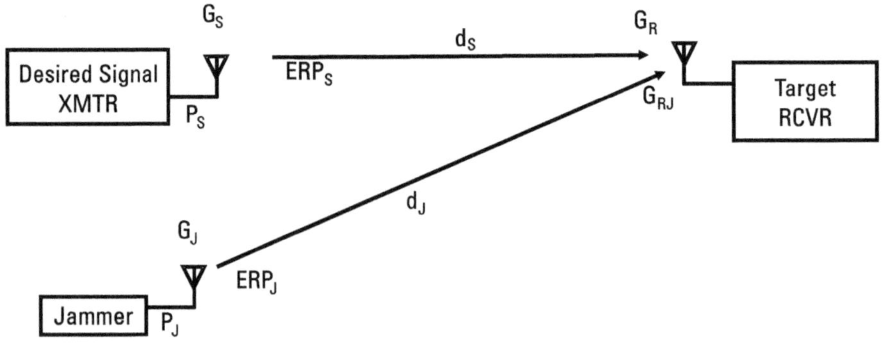

Figure 6.83 The communication jamming situation includes a desired signal link from the desired signal transmitter to the receiver and a jamming link from the jammer to the receiver.

are the equivalent values for the jamming link. As in any jamming, communication jamming involves causing an undesired signal to be received by a receiver in such a way that it cannot properly receive the desired signal. Each of the links in the figure is a communication link as described earlier. The received power from the desired signal link is called S and is determined from the formula:

$$S = ERP_S - L_S + G_R$$

where S is the desired signal received power in the receiver (in dBm), ERP_S is the effective radiated power of the desired signal transmitter in the direction of the receiver (in dBm), L_S is the link loss between the desired signal transmitter and the receiver (in decibels), and G_R is the receiving antenna gain in the direction of the desired signal transmitter (in decibels).

The received power from the jammer is called J and is determined from the formula:

$$J = ERP_J - L_J + G_{RJ}$$

where J is the jamming signal received power in the receiver (in dBm), ERP_J is the effective radiated power of the desired signal transmitter in the direction of the receiver (in dBm), L_J is the link loss between the jammer and the receiver (in decibels), and G_{RJ} is the receiving antenna gain in the direction of the jammer (in decibels).

The losses in each of these links include all of the elements discussed in section 6.4 and in Chapter 5:

- LOS or two-ray propagation loss;
- Atmospheric loss;
- Rain loss;
- KED.

These losses apply as appropriate to each link. The two links do not have to have the same propagation models.

6.7.1 Jam the Receiver

You always jam the receiver, not the transmitter. This seems obvious, but it is easy to get confused in complex situations. A particular source of confusion in this matter comes from radar jamming. A radar usually has its transmitter and receiver in the same location (and usually using the same antenna), so it is desirable to retro-directively jam a radar, that is, to transmit the jamming signal to the location from which the transmitted signal originates. Because communication signals must have the transmitter and receiver in different locations, you need to remember to jam the receiver (not the transmitter). For example, if you are jamming a UAV data link as shown in Figure 6.84, the jamming signal must be directed at the ground station, because the data link carries information from the UAV to the ground station. Jamming directed at the UAV will have no impact on the data link, because it carries information from the UAV to the ground station.

6.7.2 Jamming a Net

If you are jamming an enemy communication net as shown in Figure 6.85, all of the enemy communication stations are most likely transceivers, each having both transmit and receive functions. In a push-to-talk net, one station will be transmitting (because the operator has pushed the transmit switch) and the others in the net will be receiving. When a jamming signal is transmitted into the area of the net, it will be received by all stations which are in the receive mode. The signal flow from the jammer to each receiver is a one way link. Each of these links can be defined; however, it is usually practical to define an average jamming link to all receivers in the net. When we discuss specific techniques, we will use drawings that show one transmitter, one jammer, and one receiver, but in reality, that receiver could be any one of the members of this enemy net who are in a receiving mode. Thus, to jam this typical receiver is to jam the whole net. However, as shown in Figure 6.86, there can be significant difference in the distance from the jammer to the receiving stations in the net. This must be considered when calculating the appropriate communication jamming parameters for the net.

One other point to be made from this diagram is that the transmitting station can also be received by a receiver associated with a jammer. This is an

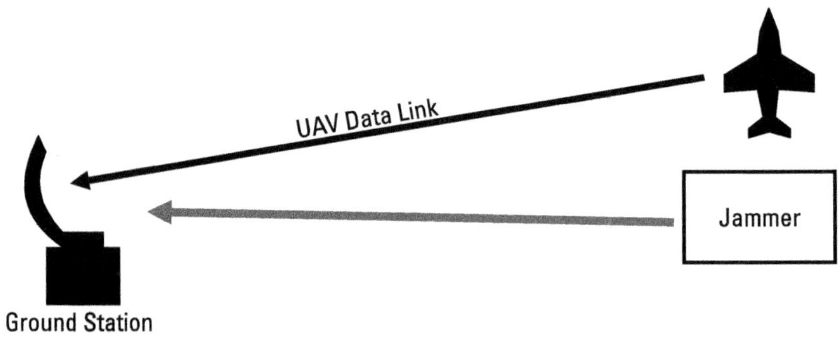

Figure 6.84 A UAV data link carries information from the UAV to the ground station, so a jammer must broadcast to the ground station to jam this link.

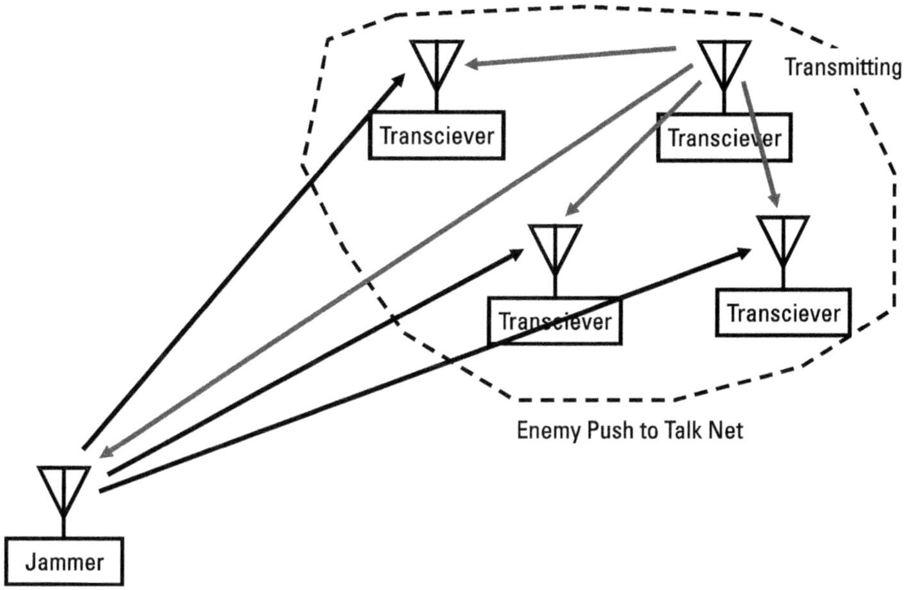

Figure 6.85 When a jammer jams an enemy push to talk net, it broadcasts to each transceiver in the net which is currently in receive mode. A receiver at the jammer location would also be able to receive the transmitting station signal.

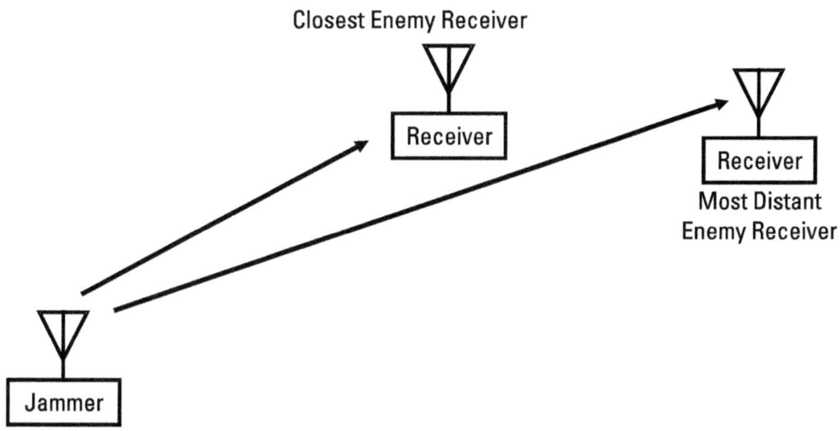

Figure 6.86 When jamming an enemy net, it is important to consider the link distance to the most distant receiving member of the net.

intercept link and will be an important consideration in some kinds of complex jamming techniques to be discussed later in Chapter 7.

6.7.3 Jamming-to-Signal Ratio

The ratio of the received jamming signal power to the received desired signal power in the receiver is called the jamming-to-signal ratio (*J*/*S*). It is expressed in decibels. Because both of the received power values are in dBm (i.e., logarithmic), *J*/*S* can be found by subtracting *S* from *J*. *J* and *S* are as defined by the above formulas, so *J*/*S* can be further defined by the formula:

$$J/S = J - S = ERP_J - ERP_S - L_J + L_S + G_{RJ} - G_R$$

with all of the terms as defined above.

Because communication transceivers often have whip antennas, they transmit and receive more or less equally over 360° of azimuth. This means that the gain of the receiving antenna will be the same in the direction of the desired signal and jamming transmitters. With the two antenna gains equal, the *J*/*S* formula simplifies to:

$$J/S = ERP_J - ERP_S - L_J + L_S$$

6.7.4 Propagation Models

In Section 6.4, we discussed the three propagation models which most commonly characterize tactical communication link performance. In Section 6.7.3, we talked about desired signal, intercept, and jamming links, which are all tactical communication links. It is important to realize that each can have any propagation model. That is the reason that we left the loss terms in the *J*/*S* equations as just losses, rather than simplifying the equations to remove some of the common terms.

Because any link might have any loss model, it is necessary when approaching a communication jamming problem to first determine the appropriate loss model for each of the links involved. This involves consideration of the geometry and often the calculation of the Fresnel zone distance for each link. For air-to-air situations where the desired signal transmitter, receiver, and jammer are all far from the ground, both the desired signal and jamming links have LOS propagation. This is also normally true when the jammed communication takes place at microwave frequencies and narrow directional antennas are used. However, when the problem involves ground-to-ground or air-to-ground jamming at VHF and UHF, the only way to determine the required propagation model is to calculate the Fresnel zone distance for each link.

6.7.5 Ground-Based Communication Jamming

Let us jump right into the most complex situation: the target communication link and the jammer are all ground based as shown in Figure 6.87. In this problem, the target link is operating at 250 MHz over 5 km with a 1-W transmitter. Both the transmit and receive antennas are 2-dBi whip antennas that are mounted 2m above the ground. The jammer has a 500-W transmitter and a 12-dBi log periodic antenna mounted on a 30-m mast. It is 50 km from the target receiver. All three stations are within LOS of each other. We want to determine the *J*/*S* which is achieved.

To solve the problem, the first step is calculation of the Fresnel zone distance for the desired and jamming links. The formula for Fresnel zone distance (from Section 6.4.5) is:

$$FZ(km) = \left[h_T(m) \times h_R(m) \times F(MHz)\right]/24{,}000$$

For the desired signal link, FZ is:

$$[2 \times 2 \times 250]/24{,}000 = 0.0417 \text{ km} = 41.7\text{m}$$

For the jamming link, FZ is:

$$[30 \times 2 \times 250]/24{,}000 = 0.625 \text{ km} = 625\text{m}$$

In each case, the link distance is far greater than the Fresnel zone distance, so two-ray propagation applies as shown in Figure 6.88.

Because the receiving antenna is a whip, it has the same gain toward the jammer and the desired signal transmitter, thus the formula for J/S is:

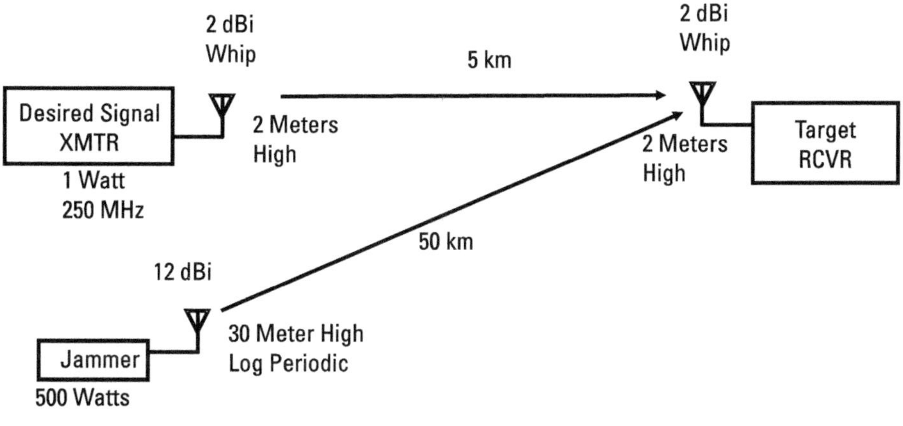

Figure 6.87 The J/S achieved from a ground based jammer depends on jamming geometry.

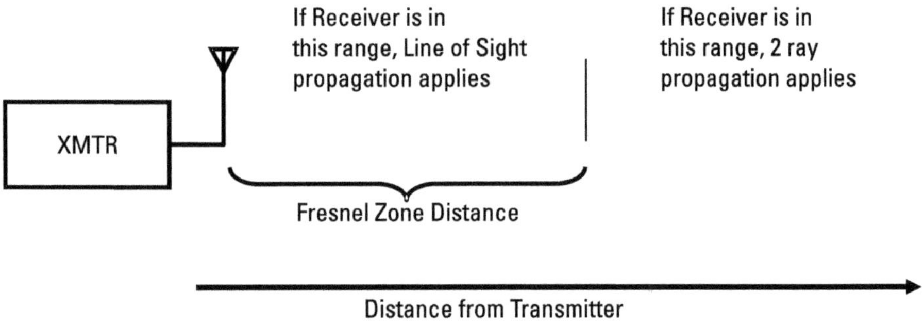

Figure 6.88 The applicable propagation model depends on the relationship between the link distance and the Fresnel zone distance.

$$J/S(\text{dB}) = ERP_J(\text{dBm}) - ERP_S(\text{dBm}) - Loss_J(\text{dB}) + Loss_S(\text{dB})$$

The ERP of the jammer is:

$$\begin{aligned} ERP(\text{dBm}) &= P_T(\text{dBm}) + G_T(\text{dB}) \\ &= 10\log(500{,}000 \text{ mw}) + 12 \text{ dB} = 57 + 12 = 69 \text{ dBm} \end{aligned}$$

The ERP of the desired signal transmitter is:

$$ERP(\text{dBm}) = 10\log(1{,}000 \text{ mw}) + 2 \text{ dB} = 32 \text{ dBm}$$

The two-ray loss for either link is (from Section 6.4.2):

$$Loss(\text{dB}) = 120 + 40\log d(\text{km}) - 20\log h_T(\text{m}) - 20\log h_R(\text{m})$$

For the jamming link, the loss is:

$$[120 + 68 - 29.5 - 6] = 152.5 \text{ dB}$$

For the desired signal link, the loss is:

$$[120 + 28 - 6 - 6] = 136 \text{ dB}$$

So the J/S is:

$$J/S(\text{dB}) = 69 \text{ dBm} - 32 \text{ dBm} - 152.5 \text{ dB} + 136 \text{ dB} = 20.5 \text{ dB}$$

6.7.6 Formula Simplification

If you are working a series of problems in which you know the propagation for both the desired and jamming links will be two-ray, you could use a simplified formula for J/S:

$$J/S(\text{dB}) = ERP_J(\text{dBm}) - ERP_S(\text{dBm}) - Loss_J(\text{dB}) + Loss_S(\text{dB})$$
$$= ERP_J(\text{dBm}) - ERP_S(\text{dBm}) - \left(120 + 40\log d_J - 20\log h_J - 20\log h_R\right)$$
$$+120 + 40\log d_S - 20\log h_T - 20\log h_R$$

where d_J is the distance from the jammer to the target receiver in kilometers, d_S is the distance from the desired signal transmitter to the target receiver in kilometers, h_J is the height of the jammer antenna in meters, h_S is the height of the desired signal transmitter antenna in meters, and h_R is the height of the target receiver antenna in meters.

Because the receiving antenna is the same for both links, this formula simplifies to:

$$J/S = ERP_J - ERP_S - 40\log d_J + 20\log h_J + 40\log d_S - 20\log h_T$$

6.7.7 Airborne Communications Jamming

Now consider the case shown in Figure 6.89. We are jamming the same communications net, but now our jammer is mounted on a helicopter that is hover-

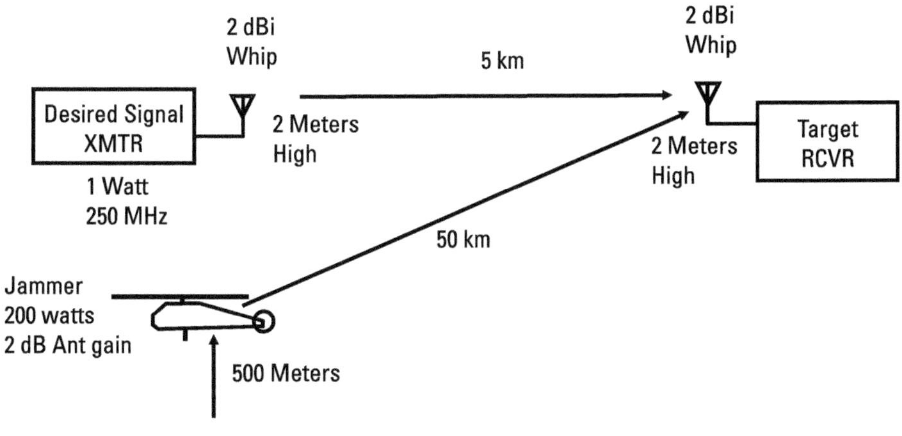

Figure 6.89 The J/S achieved from an airborne jammer is generally increased significantly by the jammer elevation.

ing at 500m. The jammer is still 50 km from the target receiver. The jamming transmitter outputs 200W and the jamming antenna is now a 2-dB folded dipole antenna on the belly of the helicopter. What is the J/S?

First, we need to determine the Fresnel zone distance for the jamming link.

$$FZ(km) = [h_T \times h_R \times F]/24{,}000 = [1{,}000 \times 2 \times 250]/24{,}000 = 20.8\,\text{km}$$

Because the jammer is more than 20.8 km from the receiver, the jamming link propagation is two-ray.

The jamming link loss is thus:

$$Loss_J = 120 + 40\log d - 20\log h_T - 20\log h_R = 120 + 68 - 6 - 54 = 128 \text{ dB}$$

The jamming ERP is now:

$$ERP_J - 10\log(200{,}000 \text{ mw}) + 2 \text{ dBi} = 53 \text{ dBi} + 2 \text{ dB} = 55 \text{ dBm}$$

The J/S is then:

$$\begin{aligned} J/S(\text{dB}) &= ERP_J - ERP_S - Loss_J + Loss_S \\ &= 55 \text{ dBm} - 32 \text{ dBm} - 128 \text{ dB} + 136 \text{ dB} = 31 \text{ dB} \end{aligned}$$

Because the jammer is elevated, it creates almost 10 dB more J/S even though the jammer ERP is reduced by 14 dB.

6.7.8 High Altitude Communication Jammer

Consider the jamming situation shown in Figure 6.90. A fixed wing aircraft flying at 3,000-m altitude jams a 250-MHz communication net with stations 5 km apart. All stations in the target net are transceivers with 2-m-high whip antennas (2-dB gain). The output power from each transceiver's transmitter is 1W. The jamming aircraft is 50 km from the area over which the target net is operating. The jammer outputs 100W into a 3-dBi antenna. What J/S is achieved?

First, we must determine the appropriate propagation models for each link. The target link Fresnel zone distance is:

$$FZ = (2 \times 2 \times 250)/24{,}000 = 0.0417 \text{ km} = 47.7\text{m}$$

This is far less than the 5-km transmission path, so two-ray propagation is appropriate for the target link. The target link loss is thus:

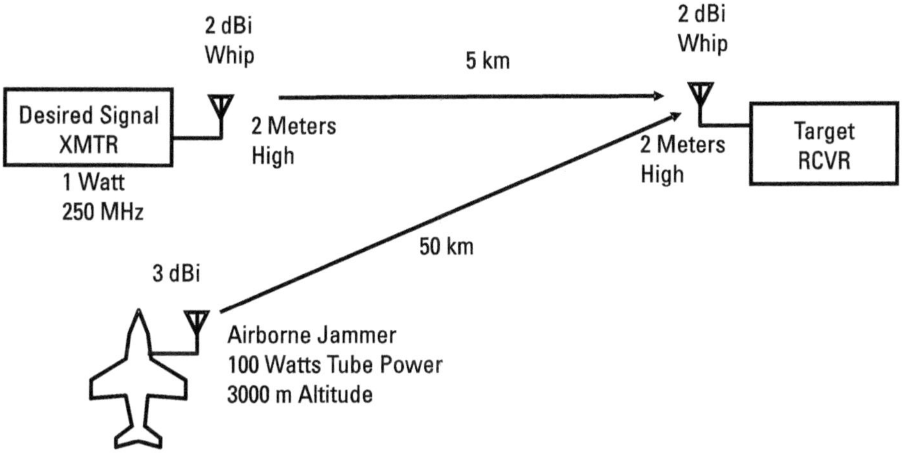

Figure 6.90 Significant J/S can be achieved from a high altitude airborne jammer.

$$LOSS_S = 120_S = 120 + 40\log(\text{dist}) - 20\log(h_T) - 20\log(h_R)$$
$$= 120 + 40\log(5) - 20\log(2) - 20\log(2) = 120 + 28 - 6 - 6 = 136 \text{ dB}$$

The jamming link Fresnel zone distance is:

$$FZ = (3{,}000 \times 2 \times 240)/24{,}000 = 62.5 \text{ km}$$

Because the Fresnel zone distance is greater than the jamming link transmission distance, line of sight propagation applies. The loss in the jamming link is:

$$\begin{aligned} LOSS_J &= 32.4 + 20\log(\text{dist}) + 20\log(\text{frequency}) \\ &= 32.4 + 20\log(50) + 20\log(250) \\ &= 32.4 + 34 + 48 = 114.4 \text{ dB} \end{aligned}$$

The ERP of the target link transmitters is 30 dBm (i.e., 1W) + 2 dBi = 32 dBm

The ERP if the jammer is 50 dBm (i.e., 100W) + 3 dBi = 53 dBm

The J/S is:

$$J/S = ERP_J - ERP_S - LOSS_J + LOSS_S = 53 - 32 - 114.4 + 136 = 42.6 \text{ dB}$$

The fact that the airborne jammer link has LOS loss allows it to generate very high J/S against the target net which has two-ray loss.

6.7.9 Stand-In Jamming

Now we consider a stand-in jammer operating against the same target net described in the above problem. This is a low-power jammer that is very close to the receiver. In this case, there might be a number of low power jammers emplaced throughout the area in which the target net is operating. Each jammer has 5-W ERP from a 0.5-m-high whip antenna. Figure 6.91 shows one such jammer located 500m from a receiver. We will consider this the typical jamming case (i.e., there are assumed to be stand-in jammers about 500m from each transceiver in the target net). What J/S is achieved?

The desired signal link as described above operates with two-ray propagation. Its ERP is 32 dBm and its link loss is 136 dB.

Now calculate the FZ for the jamming link:

$$FZ = \left(h_T \times h_R \times freq\right)/24{,}000 = (0.5 \times 2 \times 250)/24{,}000 = 0.01 \text{ km} = 10 \text{ m}$$

This is less than the 500-m jamming link distance, so two-ray propagation applies.

The jammer ERP is 37 dBm (5W).

The jamming link loss is:

$$LOSS_J = 120 + 40\log(\text{dist}) - 20\log(h_T) - 20\log(h_R)$$
$$= 120 + 40\log(0.5) - 20\log(0.5) - 20\log(2) = 120 - 12 + 6 - 6 = 108 \text{ dB}$$

The J/S is:

$$J/S = ERP_J - ERP_S - LOSS_J + LOSS_S = 37 - 32 - 108 + 136 = 33 \text{ dB}$$

A high J/S is achieved with a low power jammer because the jammer is very close to the target receiver.

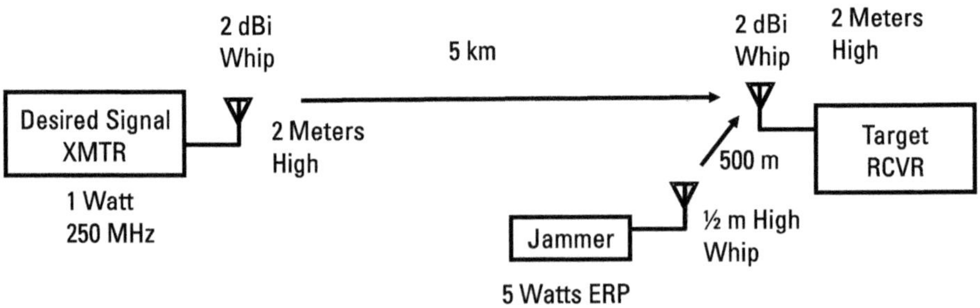

Figure 6.91 Stand-in jamming can provide high J/S with low jammer power.

6.7.10 Jam Microwave UAV Link

Next, we consider jamming UAV links from the ground. A UAV must have a command link (uplink) from the control station and a data link (downlink) back to the control station. We will cover the jamming of each link. Both links operate at approximately 5 GHz.

Figure 6.92 shows the UAV command link. The control station has a 20-dBi dish antenna which has 20-dBi gain and 15-dB side-lobe isolation. That is, the average side lobe is 15 dB below the main beam boresight gain (which is the gain toward the UAV). The uplink transmitter has 1-W transmitter power. The UAV is 20 km from the ground station and has a 3-dBi whip antenna. The downlink transmitter (on the UAV) outputs 1W into its antenna. The jammer has a 10-dBi log periodic antenna and has 100-W jamming power into its antenna.

Because both links operate at microwave frequencies, LOS propagation applies.

6.7.10.1 Command Link

First, consider jamming the command link, with the jammer antenna directed toward the UAV as shown in Figure 6.92. What J/S is achieved?

The desired signal ERP is 30 dBm (1W) + 20 dB = 50 dBm. The jammer ERP is 50 dBm (100W) + 10 dB = 60 dBm

Because the command station is 20 km from the UAV, the command link loss is:

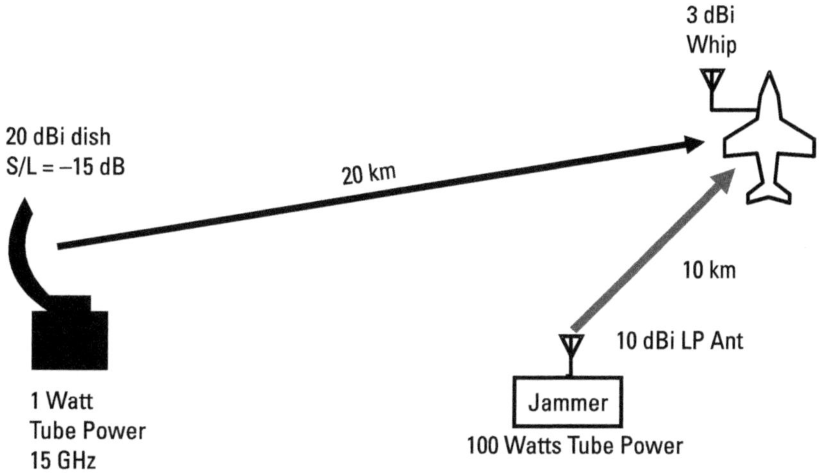

Figure 6.92 Jamming a UAV up link requires transmission to the UAV.

$$\begin{aligned} LOSS_S &= 32.4 + 20\log(\text{dist}) + 20\log(\text{frequency}) \\ &= 32.4 + 20\log(20) + 20\log(5{,}000) \\ &= 32.4 + 26 + 74 = 132.4 \text{ dB} \end{aligned}$$

The jammer is 10 km from the UAV, so the jamming link loss is:

$$\begin{aligned} LOSS_J &= 32.4 + 20\log(\text{dist}) + 20\log(\text{frequency}) \\ &= 32.4 + 20\log(10) + 20\log(5{,}000) \\ &= 32.4 + 20 + 74 = 126.4 \text{ dB} \end{aligned}$$

As the receiving antenna on the UAV is a whip, it will have equal gain toward the ground station and the jammer. Thus, the J/S is given by:

$$J/S = ERP_J - ERP_S - LOSS_J + LOSS_S = 60 - 50 - 126.4 + 132.4 = 16 \text{ dB}$$

6.7.10.2 Data Link

Now consider jamming the data link as shown in Figure 6.93. The jammer is 20 km from the control station and its antenna is directed into a side lobe of the control station antenna. What J/S is achieved?

The data link transmitter has 1-W transmitter power and a 3-dBi antenna. The desired signal link ERP is 30 dBm (1W) + 3 dBi = 33 dBm. The desired signal link loss is the same as calculated for the command link, 132.4 dB.

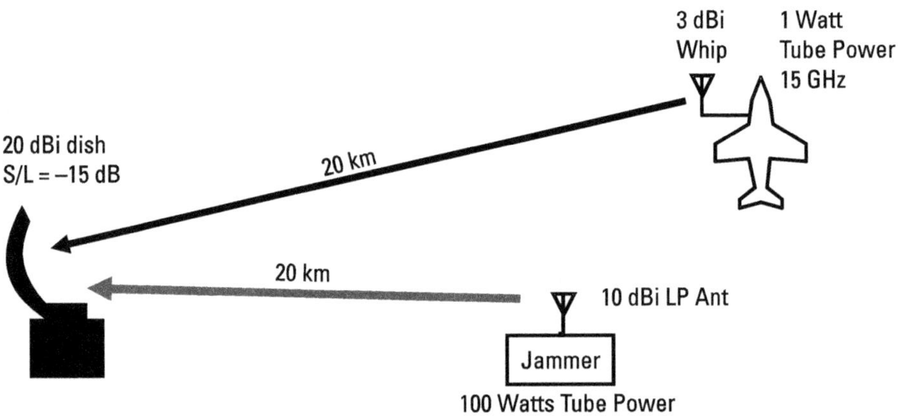

Figure 6.93 Jamming a UAV down link requires transmission to the ground station location.

The jammer ERP as calculated above is 60 dBm. Because the jammer is 20 km from the control station, the jamming link loss is the same as the desired signal loss (132.4 dB).

The control station antenna is directional. Its gain toward the UAV (G_R) is 20 dBi, but its gain in the direction of the jammer (G_{RJ}) (which is in a side lobe) is 15 dB less, or 5 dBi. Thus, the J/S is given by the formula:

$$J/S = ERP_J - ERP_S - LOSS_J + LOSS_S + G_{RJ} - G_R$$
$$= 60 - 33 - 132.4 + 132.4 + 5 - 20 = 12 \text{ dB}$$

References

[1] Gibson, J. D., (ed.), *Communications Handbook*, Ch. 84: Boca Raton, FL: CRC Press, 1997.

[2] Chestnut, P., "Emitter Location Accuracy Using TDOA and Differential Doppler," *IEEE Transactions on Aerospace and Electronic Systems*, Vol. 18, March 1982.

[3] Adamy, D., *EW 102: A Second Course in Electronic Warfare*, Norwood, MA: Artech House, 2004.

[4] Wegner, L. H., "On the Accuracy Analysis of Airborne Techniques for Passively Locating Electromagnetic Emitters," RAND Report, R-722-PR, June 1971.

7

Modern Communications Threats

7.1 Introduction

Communication threats are changing in significant and challenging ways. The increasing use of low probability of intercept (LPI) communication has become a significant challenge to electronic warfare (EW) communication links. Also, there are now air defense missiles and associated radars that make significant use of interconnecting data links. Unmanned aerial vehicles (UAVs) are in widespread and growing use for reconnaissance, EW, and weapon delivery. They are extremely dependent on interconnection with ground stations by command and data links. Finally, cell phones are widely used not only for command and control function in nonsymmetrical warfare situations, but are used to set off improvised explosive devices (IEDs).

As in Chapter 4, where modern radar threats were described, these modern communication threats are described generically. This allows the description of EW techniques without dealing with classified information. Later, when you are applying EW in real-world situations, you can plug in the parameters you learn from classified sources.

7.2 LPI Communication Signals

Signals associated with LPI communications have special modulations designed to make them difficult for normal types of receivers to detect. Ideally, a hostile receiver will not even be able to determine that such a signal is present. This is accomplished by spreading the frequency range over which the LPI signal is broadcast. Thus, they are also called spread spectrum signals. As shown in Figure 7.1, a special second modulation is applied to this type of signal to spread its spectrum. Three types of spreading modulations are used:

- *Frequency hopping:* The transmitter periodically hops to a pseudo-randomly selected frequency. The hopping range is much greater than the bandwidth of the signal carrying the information being communicated (i.e., the information bandwidth).
- *Chirp:* The transmitter is rapidly tuned across a frequency range that is significantly wider than the information bandwidth.
- *Direct sequence spread spectrum:* The signal is digitized at a rate much higher than required to carry the information, thereby spreading the energy of the signal across a wide bandwidth.

There are also LPI signals in which more than one of the above spreading techniques is employed.

The spreading demodulator (in the receiver) shown in Figure 7.1 must be synchronized with the spreading modulator (in the transmitter) to reverse the spreading modulation (see Figure 7.2). This returns the signal to the same bandwidth that it had before spreading. We call this the information bandwidth. The synchronization requires that both modulator and demodulator be controlled by the same pseudo-random function, based on a digital code sequence. In addition, the code in the receiver must be in phase with the transmitter code. This requires a synchronization procedure at system start-up and at any time the receiver or transmitter has been out of communication for an extended time. Except for the synchronization requirements, the spreading/dispreading process is transparent to the people or computers passing information

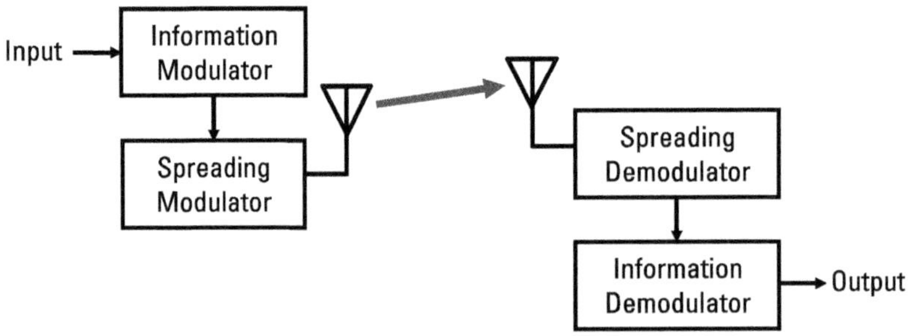

Figure 7.1 LPI communication systems add special frequency spreading modulations for transmission security.

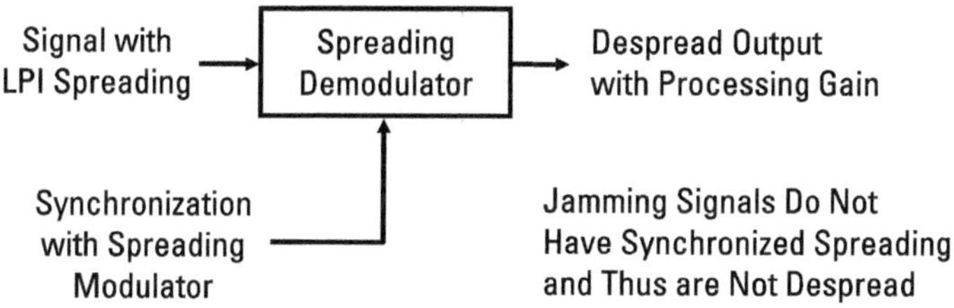

Figure 7.2 Synchronization with the spreading modulator allows the spreading modulation to be removed from the intended received signal but not from jamming signals.

from the transmitting location to the receiving location. In some circumstances, synchronization requires a delay before transmission can begin.

We will discuss each of the frequency spreading techniques in sections that also describe the techniques used to jam them. Note that there is a section in Chapter 2 discussing the generation and uses of codes.

7.2.1 Processing Gain

Removing the spreading modulation from an LPI signal is said to create a processing gain. This refers to the fact that the spread signal has a very low signal-to-noise ratio (SNR) when received by a normal receiver. After dispreading, the received signal has a significantly higher SNR. However, signals that do not have precisely the correct spreading modulation will not be despread and thus are not subject to the processing gain. Further, the spreading demodulator will actually spread a narrowband signal, reducing its signal strength in the output channel as shown in Figure 7.3.

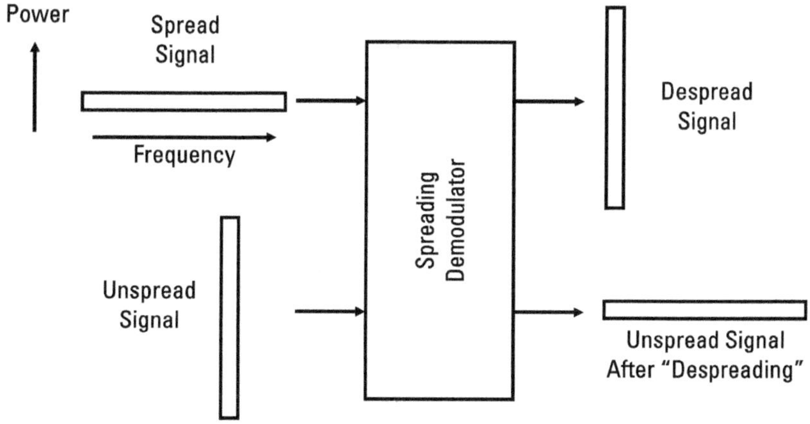

Figure 7.3 The spreading demodulator compresses matching LPI signal into its information bandwidth. It also spreads a narrowband signal.

7.2.2 Antijam Advantage

Figure 7.4 shows the antijam advantage for an LPI communication system. The antijam advantage is the amount of signal power that must be received at an LPI system receiver location to provide the same jamming-to-signal ratio (J/S) that would be achieved if the entire jamming signal power were within the bandwidth of a nonspread system receiver. This is the ratio between the information bandwidth and the transmission bandwidth of the LPI signal. This assumes that the jamming signal is spread across the whole spread spectrum frequency range of the LPI signal. As we will see, there are sophisticated jamming techniques that partially overcome this advantage in some cases.

7.2.3 LPI Signals Must Be Digital

As you will see in this chapter, each of the spectrum-spreading techniques requires that the input signals be in digital form. Digitizing allows the signal to be time-compressed and broadcast between transmission gaps required in some spreading schemes. It can also be required by the nature of the modulation approach. Because the requirement is specific to the spreading technique used, this matter will be discussed in the applicable following sections. Note that Chapter 5 covers digital communication in more detail than it is treated in this chapter.

The implication is that successful jamming of a spread spectrum signal requires only 0-dB J/S and that it may require significantly less than 100% duty cycle. Jamming of digital signals is effective in that it causes bit errors. The bit error rate is the number of incorrectly received bits divided by the total number of bits received. As shown in Figure 7.5, the bit error rate can never be greater than 50%, regardless of the J/S. At 0-dB J/S, the bit error rate is almost 50%. Increasing the jamming power above this point causes very few additional errors. A widely honored, experience-based assumption is that when the bit error

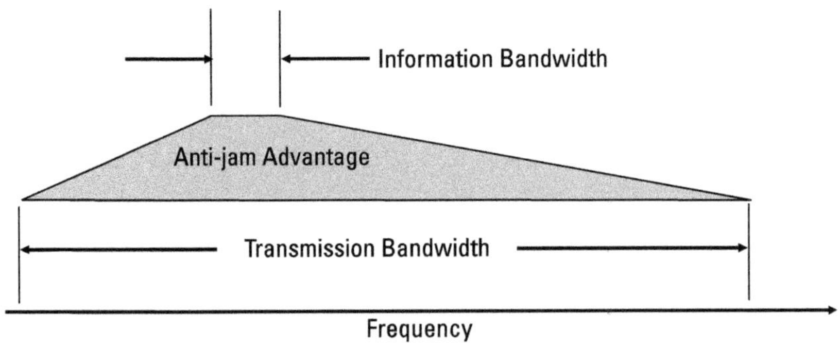

Figure 7.4 The anti-jam advantage of LPI communication is the ratio of the transmission bandwidth to the information bandwidth.

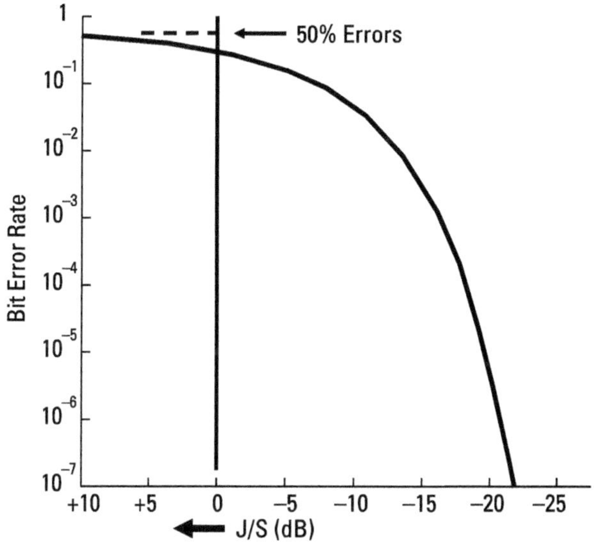

Figure 7.5 The bit error rate in a digital signal receiver cannot exceed 50%. 0 dB J/S causes close to that level of errors.

rate is at least 33% over a few milliseconds, no information can be recovered from the jammed signal. (Some authors place this as low as 20%.)

As you will see in the following sections, the digital nature of LPI signals allows some clever jamming techniques to be employed.

7.3 Frequency-Hopping Signals

Frequency hoppers are arguably the most important of the LPI signals both because they are widely used and because they can provide very wide frequency spreading.

Figure 7.6 shows the frequency versus time plot for a frequency-hopping signal. The signal pauses at one frequency for a short period of time and then moves to another, randomly selected frequency. The dwell time at one frequency is called the hop duration. The hopping rate is the number of hops per second. The hopping range is the frequency range over which the transmission frequencies can be selected. The whole signal bandwidth is moved to the assigned frequency for each hop. A typical example is the Jaguar VHF Frequency-Hopping Radio. It has a signal bandwidth of 25 kHz and its hop range can be as wide as 30 to 88 MHz (i.e., 58-MHz hopping range).

The block diagram of a frequency hopping transmitter is shown in Figure 7.7. A digitally modulated signal is converted to hop frequencies using a synthesizer that is tuned to pseudo-randomly selected frequencies. The front end

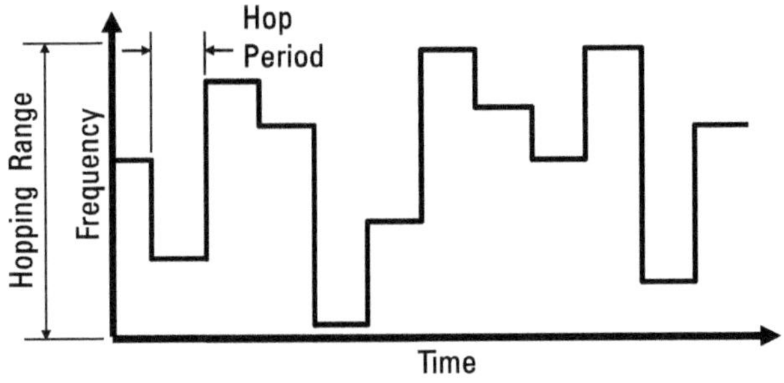

Figure 7.6 A frequency hopping signal changes its transmit frequency many times during a message.

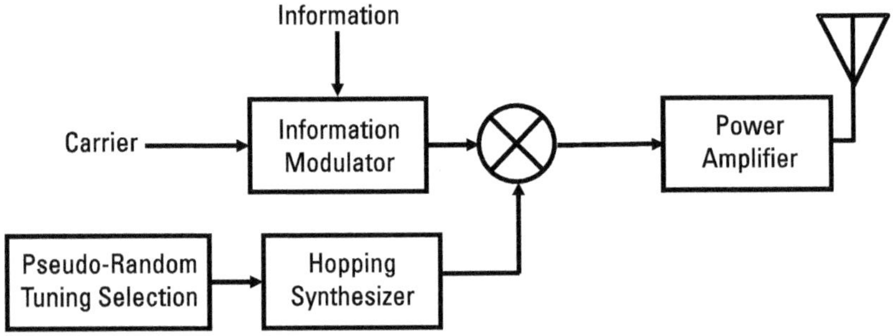

Figure 7.7 The frequency hopping transmitter includes a pseudo-randomly tuned synthesizer to rapidly hop the transmitted signal over a wide frequency range.

of the frequency-hopper receiver has a synthesizer that is tuned to the same frequency as the transmitter synthesizer. This requires a synchronization scheme common to the transmitter and receiver. When a receiver is first turned on, it is necessary to go through a lengthy synchronization procedure. Each time a new signal is received, the receiver must go through a limited resynchronization procedure. To allow for this synchronization period, a short tone may be inserted into the earpiece of a frequency-hopping transceiver when the transmit key is depressed to delay the start of a voice transmission. When digital data is transmitted, this delay can be automatic.

7.3.1 Slow and Fast Hoppers

Frequency-hopping systems can be either slow hoppers or fast hoppers. A slow hopper (such as the Jaguar mentioned above) carries multiple bits during each hop. A fast hopper changes to multiple hop frequencies during each bit of data. These two waveforms are shown in Figure 7.8.

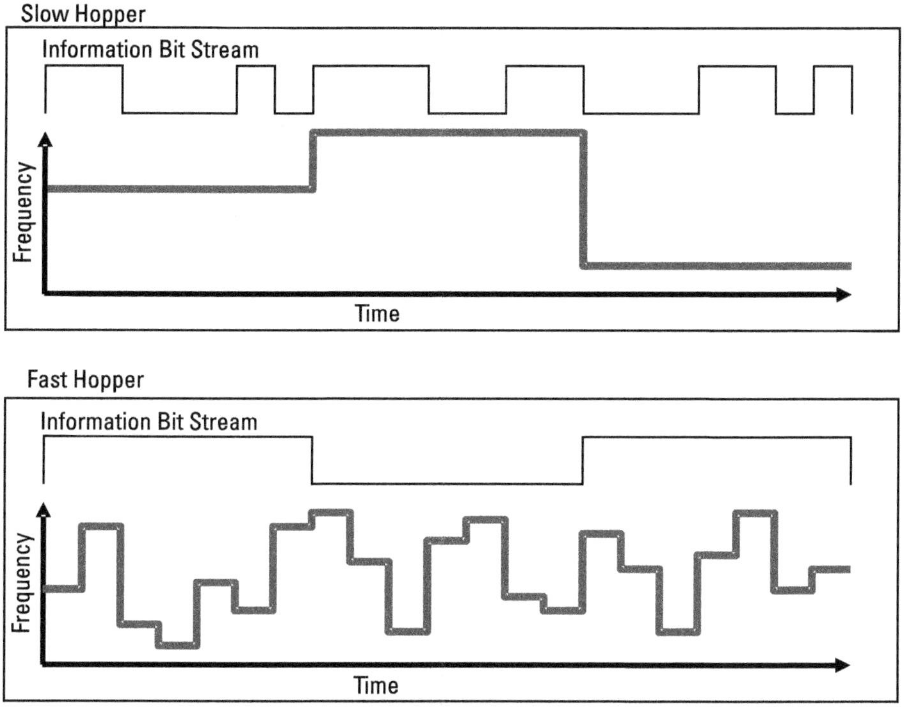

Figure 7.8 A slow hopper transmits multiple bits per hop; a fast hopper has multiple hops per bit.

7.3.2 Slow Hopper

The slow hopper uses a phase-lock-loop synthesizer as shown in Figure 7.9. This synthesizer can be designed to cover a very wide frequency range and to support many hop frequencies. For example, the Jaguar has a 25-kHz bandwidth and can hop over 58 MHz. This provides 2,320 maximum hop frequencies. Note that this system also has smaller hopping ranges (256 and 512 hops selectable within the 58 MHz to avoid high occupancy frequency ranges).

Because all of the signal power remains at a single transmission frequency for long enough to transmit multiple bits, the slow hopper is relatively easy for a receiver to detect. However, the constantly changing and unpredictable frequency makes it difficult to perform important EW functions such as emitter location and jamming.

The bandwidth of the feedback loop in a phase-lock-loop synthesizer is designed to optimize its performance. The wider the bandwidth, the faster the synthesizer can come up on a new frequency; the narrower the bandwidth, the higher the signal quality. A typical synthesizer used in a frequency-hopping system will be close enough to its final hop frequency in a time equal to approximately 15% of the hop period. Thus, at 100 hops per second, the system

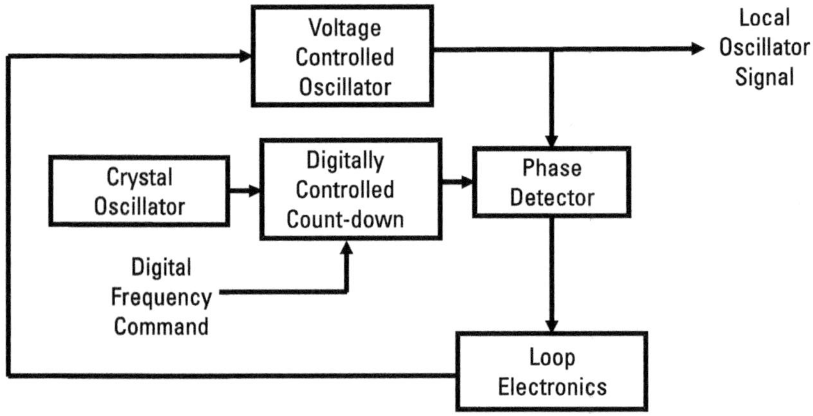

Figure 7.9 A slow hopper typically uses a phase lock loop synthesizer with the loop bandwidth optimized for settling time vs. signal quality.

would spend 1.5 ms waiting for the synthesizer to settle at the beginning of each hop. As shown in Figure 7.10, the system can transmit its information only after this settling time. This 15% drop-out of data (or voice) would make the system unusable.

To hear and understand a voice signal, we need to have continuous signal. Thus, it is necessary to digitize the input signal to the transmitter and place the digital signal into a first-in-first-out (FIFO) device. This signal might be, for example, 16 kbps. Then the signal would be clocked out of the FIFO at

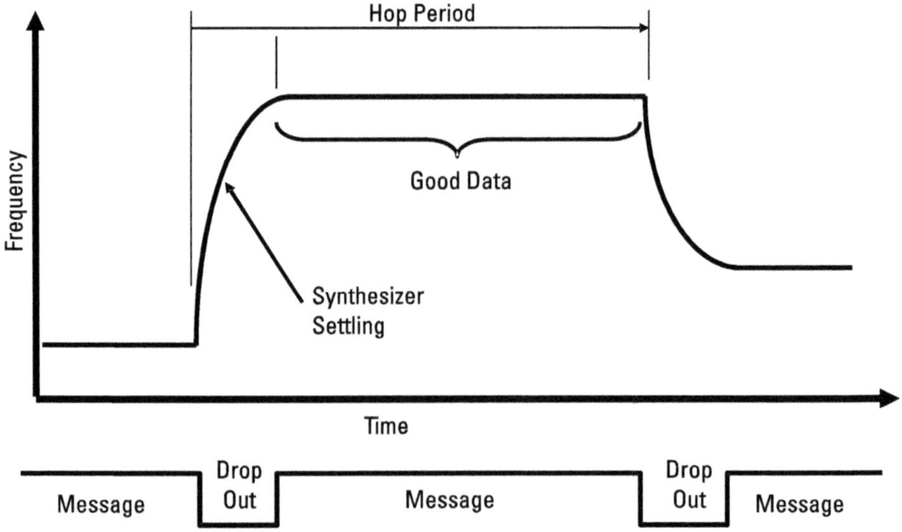

Figure 7.10 The slow hopper must delay its transmission until the synthesizer has settled on each new hop frequency.

something like 20 kbps during the time between the synthesizer settling periods. At the receiver, the process is reversed. The 20-kbps data is input to a FIFO and clocked out as a continuous signal at 16 kbps.

When the transmitter and receiver hop times and frequencies are synchronized and the settling time dropouts are removed, the frequency-hopping process is basically transparent to the user. Although the above discussion considered voice signals, the same considerations obviously apply to digital data transmissions.

7.3.3 Fast Hopper

Fast-hopping signals present significantly more challenge to hostile receivers, because they change frequencies so quickly. There is an inverse relationship between the dwell time of a signal in a receiver bandwidth and the receiver bandwidth required. An often used rule of thumb is that the dwell time must be the inverse of the bandwidth (i.e., 1-μs dwell time requires 1-MHz bandwidth). Because the bandwidth of the information carried by the system is much narrower than this, the receiver sensitivity is strongly compromised.

A synchronized receiver will remove the hopping, so the rest of the receiver can operate at the bandwidth of the information signals carried. Because a hostile receiver cannot remove the hopping, it must operate in a wider bandwidth. This makes it difficult to detect the presence of the signal, providing increased transmission security.

One problem with fast hoppers is that they require more complex synthesizers. Figure 7.11 shows a block diagram of a direct synthesizer. It has multiple oscillators and quickly switches one or more into a combining/filtering network to generate a single output frequency. Because this process is much faster than tuning a phase-lock loop, the direct synthesizer can switch frequencies multiple

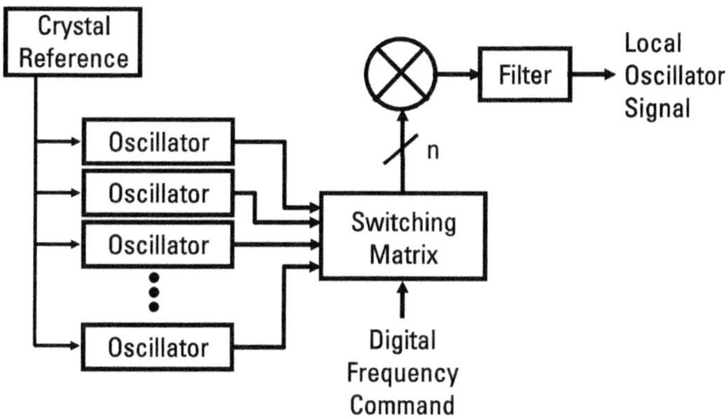

Figure 7.11 The fast hopper can be expected to use a direct synthesizer. Its increased complexity may limit the number of hopping frequencies.

times during each data bit. Because the complexity of the direct synthesizer is proportional to the number of signals it can output, a fast-hopping system can be expected to have fewer hop frequencies than a slow-hop (i.e., phase-lock-loop) system.

7.3.4 Antijam Advantage

The antijam advantage of a frequency-hopping system, either slow or fast hop, is the ratio between the hopping range and the receiver bandwidth. The total power of a received jamming signal spread over the hopping range must be increased by this factor to provide J/S equal to that achieved in a fixed frequency system. For the VHF Jaguar example, 58 MHz/25 kHz = 2,320 or 33.7 dB.

The major problem associated with the effective jamming of frequency hoppers is that the jammed system uses only one (randomly selected) channel at a time, while the jammer needs to deal with all of the channels from which the target transmitter can select.

There are three general approaches to jamming frequency hoppers: barrage jamming, partial-band jamming, and follower jamming.

7.3.5 Barrage Jamming

A barrage jammer covers the entire frequency range over which the target system hops as shown in Figure 7.12. Thus, any channel chosen by the target transmitter/receiver will be jammed. This approach has the excellent advantage that the jammer need not receive the hopping signal; thus, it eliminates the need for look-through. Because look-through is difficult to achieve in remote jammers, barrage jamming may be the ideal approach.

There are two major disadvantages to barrage jamming. One is fratricide. Barrage jamming will also jam any friendly communication (fixed frequency or hopping) that is operating in the same geographical area. The second disadvantage is that barrage jamming is notoriously inefficient. Because you need to jam all possible channels, the power per channel is determined from the formula:

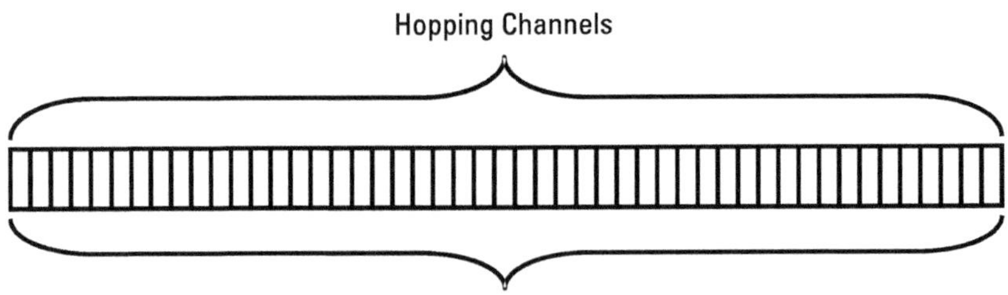

Figure 7.12 A barrage jammer divides its power among all of the hopping channels.

Power/channel = Total jammer power/number of hopping channels available

The solution to both of these problems is to place the jammer near the enemy receiver. Remember that J/S is the ratio of the received jamming signal strength to the received desired signal strength, both in the target receiver. The signal strength is reduced by the square or fourth power of the distance from the transmitter to the receiver (depending on the frequency and geometry; see Chapter 6). Therefore, as the range to the target receiver is reduced, the J/S is increased. If the range to the target receiver is significantly shorter than the range to friendly receivers, fratricide is significantly reduced.

If you know where the enemy receiver is located, this is clear. In normal tactical circumstances, an emitter location system will not tell you where the receiver is located. However, you may be able to locate the receiver from other considerations; for example, if an enemy net uses transceivers, a receiver will be located with a transmitter. A second, very important example is a radio frequency improvised explosive device (RFIED) in which the receiver is located at the weapon which is presumably near its intended target. A third example is jamming the uplink to a cell phone tower; the receiver is in the tower. In practical terms, it is much better to have a barrage jammer near the enemy, where it will cause maximum J/S and minimum fratricide to friendly communication. This consideration also applies to partial-band jamming.

An example is shown in Figure 7.13. A 1-W ERP, VHF transmitter is 10 km from its intended receiver. Both transmitter and receiver have whip antennas that are 2m above the ground. The signal hops over 1,000 channels. A barrage jammer with 1-W ERP is located 2m above the ground, 1 km from the target receiver. The propagation mode for both links will be two-ray. Using the formulas found in Section 6.2, the ratio of total jammer power to received desired signal power (which occupies only one channel at a time) is 40 dB. Dividing the jamming power over the 1,000 hopping channels reduces the power per

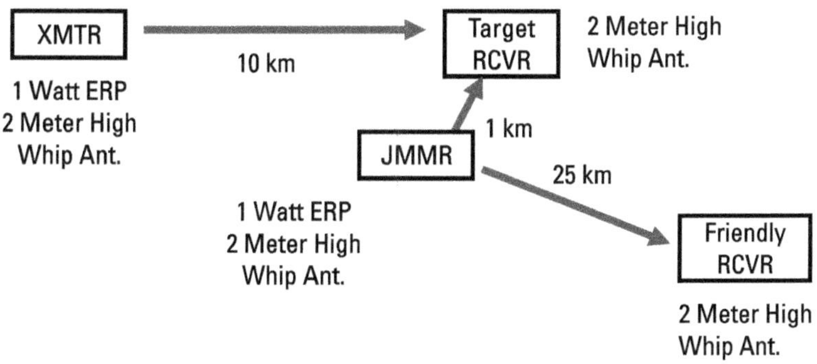

Figure 7.13 A barrage jammer 1 km from the target receiver and 25 km from a friendly receiver provides excellent J/S while avoiding fratricide.

channel by a factor of 1,000 (i.e., 30 dB). Thus, the effective J/S in the target receiver is 10 dB. (Remember from Section 7.2.3 that only 0 dB is required for effective jamming.) If a friendly receiver is 25 km from the jammer (operating over a similar 10-km link), it will be jammed with a J/S of –16 dB. If it is hopping over 1,000 channels, the effective J/S will be reduced to –46 dB.

7.3.6 Partial-Band Jamming

Partial-band jamming covers only part of the hopping range as shown in Figure 7.14. The amount of frequency range covered by the jammer is determined by these steps:

1. Determine the overall *J/S* (in decibels); the total received jamming power divided by the received desired signal power.
2. Convert the *J/S* (in decibels) into the linear form. For example, 30 dB is a ratio of 1,000.
3. Spread the jamming frequency over a band determined by:

$$\text{Nondecibel } J/S \text{ ratio} \times \text{hopping channel bandwidth}$$

In the above example, dividing the signal into 1,000 channels reduces the *J/S* by 30 dB, producing 0 dB *J/S* in each of the hopping channels covered by the jamming.

Because the target signal randomly hops over its whole hopping range, the jamming duty cycle is calculated by dividing the number of jammed channels by the total channels in the hopping range.

The required duty cycle is generally accepted as 33% for digitized voice, although some EW writers convincingly argue that 20% or even much less can be effective under many circumstances.

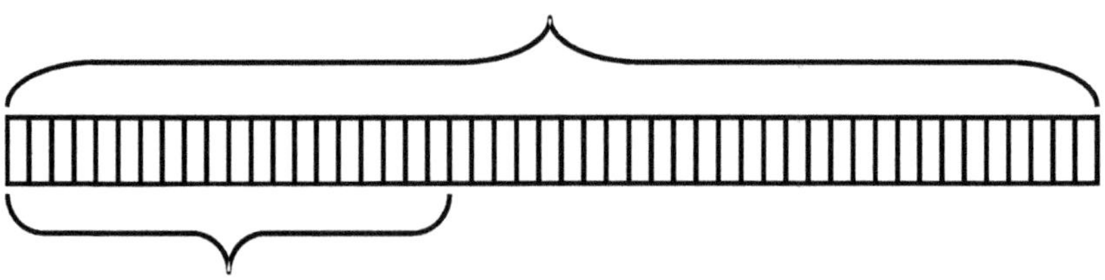

Figure 7.14 Partial band jamming distributes jamming over the number of channels which can be subjected to 0 dB J/S per channel.

A partial-band jamming example is as follows. Suppose that a frequency hopper with a 25-kHz channel bandwidth hops over 58 MHz. If a jammer can provide 29-dB total J/S, it would be spread over 794 channels (19.9 MHz) for 0 dB per channel. The total number of hop channels is:

$$58 \text{ MHz}/25 \text{ kHz} = 2{,}320$$

The jamming duty factor is:

$$794/2{,}320 = 34.2\%$$

A few important points about partial-band jamming:

- Because 0-dB jamming and 33% duty cycle produce effective jamming, this is the most efficient use of a jammer (i.e., maximum jamming effectiveness for the amount of jammer ERP available).
- The required jamming duty factor must be in every second of transmission; otherwise, useful information could get through.
- The jammed band must be moved around the hopping range. Otherwise, the target system can reduce its hopping range to avoid the jammed channels.
- If error correction codes are used by the target system, the jamming duty factor will need to be increased to provide effective jamming.

7.3.7 Swept Spot Jamming

A swept spot jammer covers part of the hopping range, but sweeps its spot over the whole range as shown in Figure 7.15. This is a special application of partial-band jamming, and can be very effective in remote jammers.

7.3.8 Follower Jammer

The follower jammer determines the frequency to which a frequency hopper is tuned in a small portion of the hop period. It then sets a jammer to that frequency to jam the rest of the hop. A wideband digital receiver can use fast Fourier transform (FFT) processing to quickly measure the signal frequency. However, the high density of the tactical signal environment gives the system an additional requirement. Figure 7.16 shows frequency versus emitter location in a very low-density environment. Each dot in the diagram represents the signal frequency and emitter location for a transmission. A frequency hopper

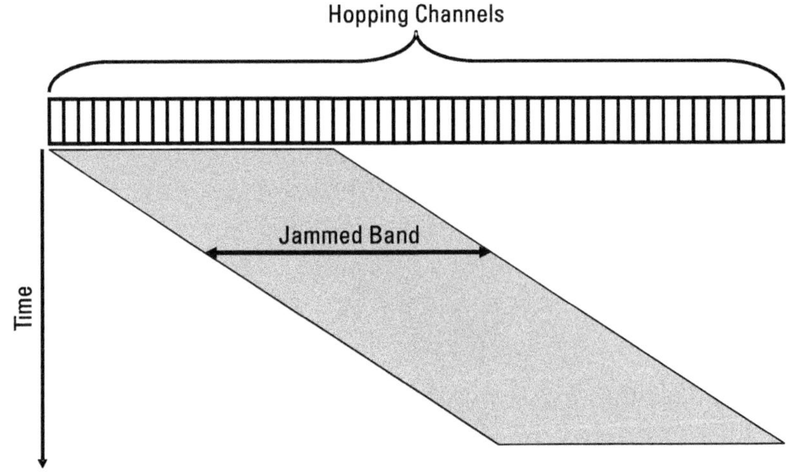

Figure 7.15 Swept spot jamming covers all of the hopping channels with a less than 100% duty cycle.

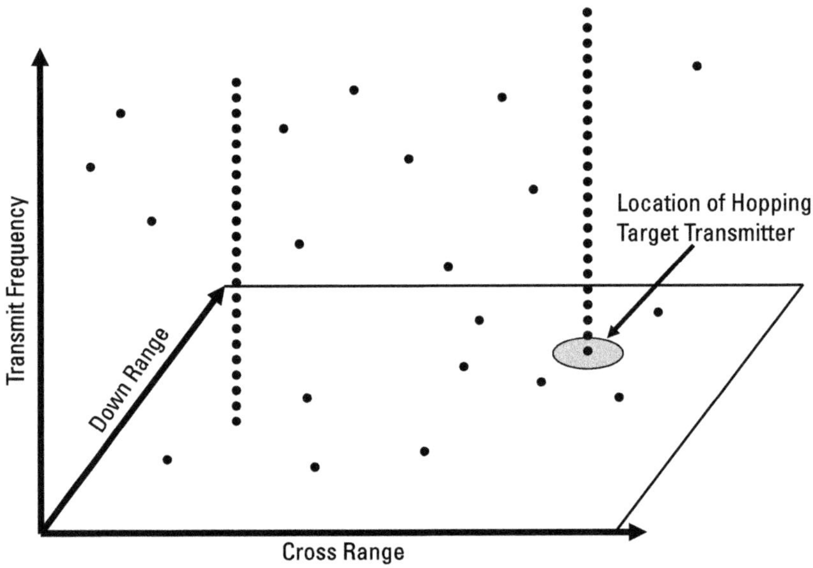

Figure 7.16 The follower jammer must apply jamming at the frequency of the emitter at the target location.

has many frequencies from a single location. In a real-world environment, up to 10% of available channels could be occupied at any instant. This means that over the 30- to 88-MHz VHF band, there would be about 232 signals (assuming 25 kHz per signal channel). A follower jammer must determine the frequency and location of each of those 232 signals and determine the frequency being emitted from the target location. The follower jammer is then set to that frequency.

An important side note: We have been saying that you jam the receiver, not the transmitter. However, by determining the frequency of a transmitter in a hostile net, we know the frequency to which the receivers in the net are tuned. Jamming at the transmitted frequency will jam all hostile receivers in the net.

Follower jamming has the great advantage that it places all of its jamming power into the channel being used by the jammed hopping system. It also has the advantage that it jams only the frequency being used (at that moment) by the enemy. There is a very low probability that friendly hopping systems will be using that frequency at that time; hence, fratricide is minimized.

Figure 7.17 shows the timing in a follower jammer. During the first part of the hop period, the hopper is settling onto its new frequency. Then the jammer must find the frequency and location of all signals present and select the frequency to jam (i.e., the frequency emitted from the target signal location). Then there is propagation delay allowance. After of all of those processes are complete, the remaining part of the hop period is available for jamming. If the jamming duration is at least one-third of the hop period, jamming will be effective.

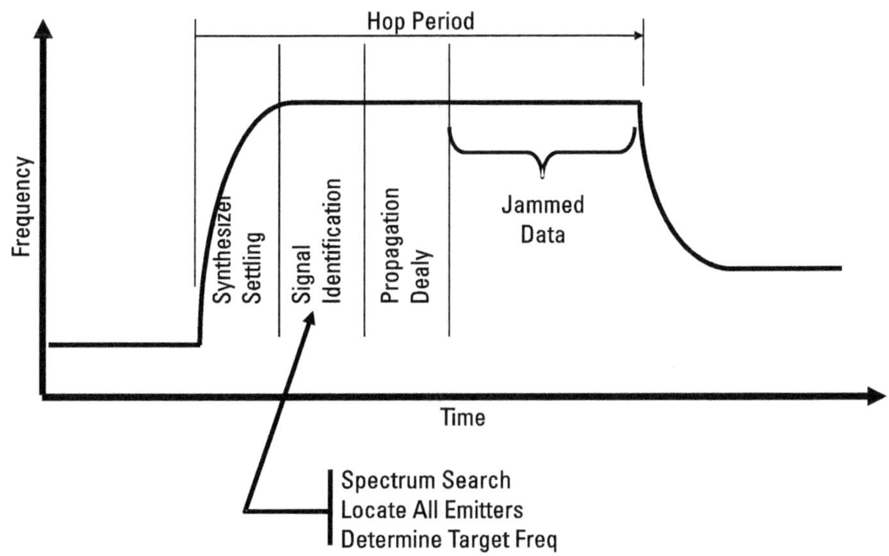

Figure 7.17 Follower jamming requires fast enough analysis to allow time for settling, propagation delay, and adequate jamming duty factor.

7.3.9 FFT Timing

The speed at which a follower jammer can determine the proper jamming frequency depends on the receiver configuration and the speed of the processor. Consider the system configuration in Figure 7.18 as an example. The jammer includes a phase-matched, two-channel interferometer to determine the direction of arrival of each received signal. The RF front end covers a portion of the frequency range of interest and outputs an intermediate frequency (IF) signal to the digitizer. The I&Q digitizer captures both the amplitude and phase of the IF signal at a very rapid sampling rate. The digital signal processor (DSP) performs an FFT to determine the phase of any signal present in any of the signal channels that it determines. The FFT will channelize the digitized IF data into a number of channels equal to half the number of samples processed. For example, if 2,000 samples are input to the FFT process, the signal will be processed into 1,000 channels. Note that I&Q samples are in effect independent, so 1,000 I&Q samples will allow analysis into 1,000 channels.

If a second digital interferometer system inputs simultaneous direction of arrival information on all signals present, the computer controlling the jammer will know the location of each received signal and can set the jammer to the instantaneous frequency of the signal at the target location (i.e., the target signal hop frequency).

A typical digital interferometric direction finder has been described in [1]. With the system restrictions defined in that column, the frequency and direction of arrival of all 232 signals assumed present in the 30- to 88-MHz range is

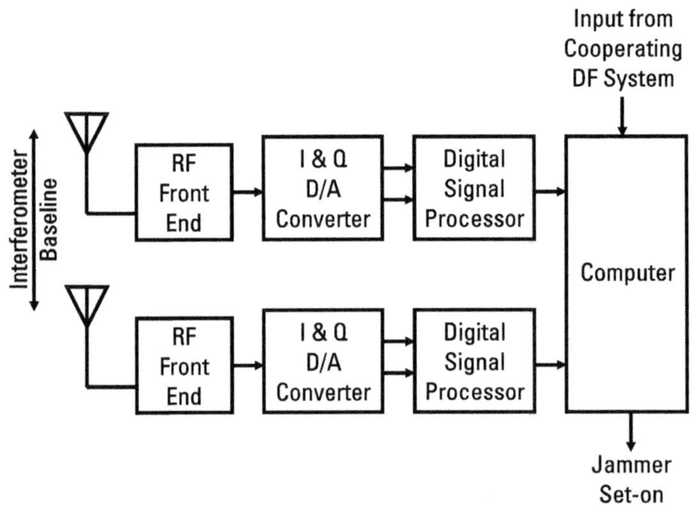

Figure 7.18 A follower jammer must determine the frequency and location of all signals present in the environment.

determined in 1.464 ms. Two such systems would cooperatively determine the emitter locations for all 232 of these signals in this amount of time.

7.3.10 Propagation Delays in Follower Jamming

Radio signals propagate at the speed of light. The signal from the transmitter must reach the jamming site. After analysis and frequency set-on, the jammer signal must reach the receiver location. Figure 7.19 shows a jamming geometry for illustration. The target system transmitter and receiver are separated by 5 km, so there must be a 16.7-μs propagation allowance built into the system. For discussion, let us place our jammer 50 km away. Now there is a 167-μs propagation delay in either direction. This means that 334 μs of the time after the transmitter has settled onto its new hop is not available for analysis or jamming.

7.3.11 Jamming Time Available

Combining the location analysis and propagation delay times in the described system and deployment geometry, 1,798 ms is unavailable for jamming. If a frequency hopper has 100 hops per second, the time available for jamming each hop is:

$$10 \text{ ms} - 15\% \text{ settling time} - 1.798 \text{ ms} = 10 - 1.5 - 1.798 \text{ ms} = 6.702 \text{ ms}$$

Compared to the time the target transmitter has available to send data (10 ms – 1.5 ms – 16.7 ms = 8.483 ms), we are jamming 80% of the transmitted bits. Thus, the jamming will be effective.

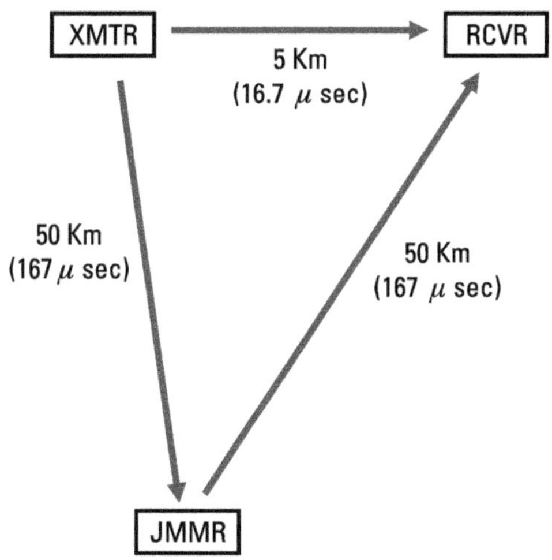

Figure 7.19 Follower jammer effectiveness can be severely impacted by propagation delays.

However, now consider a target signal at 500 hops per second. The hops are only 2 ms long, leaving 1.7 ms of data after 15% settling time. Our analysis and propagation delay time (1.798 ms) are longer than that, so this system in this deployment geometry will not effectively jam the signal.

As an added protection against jamming, signal data bits are sometimes front-loaded in the hop as shown in Figure 7.20. This reduces the amount of time available to a hostile receiver for determination of the hop frequency of a target emitter.

The point of this discussion is that it is necessary to consider the digitization parameters and the deployment to predict the effectiveness of a follower jammer. In the 500 hops per second example, a faster digitizer and/or shorter jamming range is clearly required.

7.3.12 Slow Hop Versus Fast Hop

All of the above described techniques are appropriate for slow hoppers. However, fast hoppers (with hops per bit) are not vulnerable to follower jamming. In any reasonable tactical situation, the propagation delay will make analysis and set-on impractical. Thus, fast hopping must be jammed using barrage jamming or one of the techniques described later for direct sequence spread spectrum (DSSS) signals.

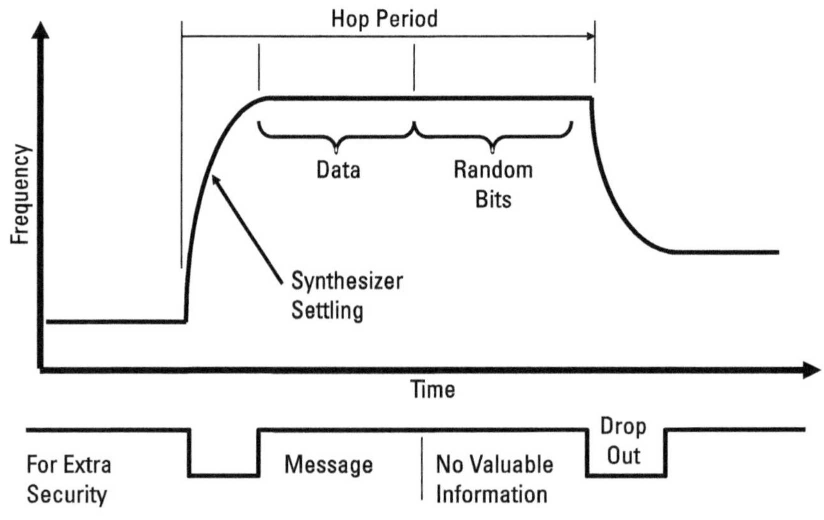

Figure 7.20 For extra anti-jam capability, signal data can be front loaded in the hop period.

7.4 Chirp Signals

Although chirp is most often associated with range resolution improvement in radars, it can also be used for antijamming protection in communication. Frequency modulation, called chirp in this case, creates a processing gain that makes the detection or jamming of signals more difficult.

There are two ways to implement chirp. One is to linearly sweep a digital signal across a frequency range much greater than its information bandwidth. The second way is to apply the chirp to every bit of a digital signal. Both have processing gain based on sweep range versus the information bandwidth of the signal. In general, the processing gain reduces the effective jamming to signal ratio (J/S) by an equivalent amount. As discussed below, there are ways to increase the effective J/S against chirp signals.

7.4.1 Wide Linear Sweep

Using the approach shown in Figure 7.21, a digitally modulated IF signal is swept across a frequency range much greater than the bandwidth of the information carried by the signal. This produces a transmitted waveform as shown in Figure 7.22. Note that the start times of the sweeps are randomly varied to prevent a hostile receiver from synchronizing with them. The intended receiver has a similar circuit with a sweeping oscillator that is synchronized to the transmitter. As noted earlier for frequency hoppers, the information must be carried in digital form so that it can be transmitted at a faster bit rate during the linear part of the sweep and returned to the constant bit rate in the receiver. Otherwise, there would be significant signal drop-outs that would interfere with communication.

Because the data is digital, optimum jamming causes about 33% bit error rate in the received signal, so partial-band jamming will provide the best practical jamming performance in unsophisticated jammers. If the chirped transmitter has a fixed sweep synchronization pattern or if the jamming signal can

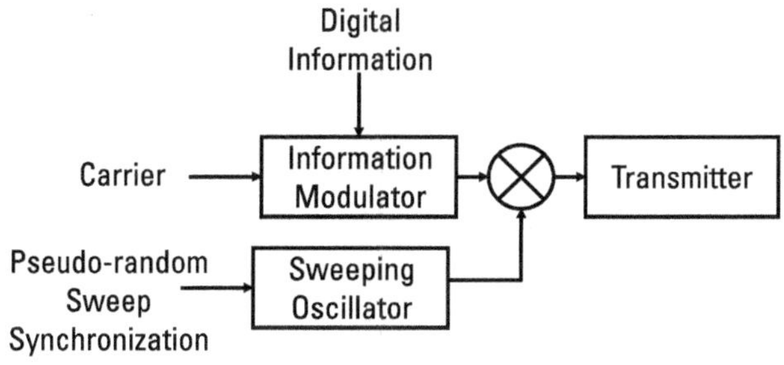

Figure 7.21 Chirp can be applied to a digital data stream to provide anti-detection and anti-jam protection.

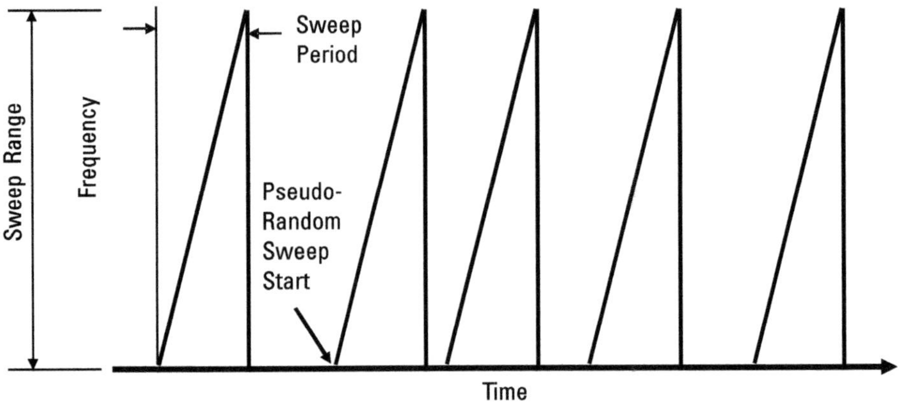

Figure 7.22 A chirped signal is swept across a large frequency range with pseudo-randomly selected start times for its sweep cycles. This precludes a hostile receiver from synchronizing to the chirp sweep.

be delayed (perhaps using a DRFM), it may be practical to analyze the chirp pattern and match it with a follower jammer. This would provide significantly better J/S by overcoming the processing gain advantage of the intended receiver. Note that the chirp may not have a constant sweep rate, but can follow any desired frequency versus time pattern.

7.4.2 Chirp on Each Bit

The chirp communication technique discussed in most literature places a chirp modulation on each data bit transmitted and recovers the digital data in the receiver as shown in Figure 7.23. The chirp can be applied either with a sweeping oscillator or using a surface acoustic wave (SAW) chirp generator. A de-chirping filter in the receiver converts signals with specific chirp characteristics into impulses because it has a linear delay versus frequency characteristic. In effect, the signal is delayed to the end of the chirp period to produce an output impulse. In this figure, an up-chirp is applied, so the de-chirp filter must have decreasing

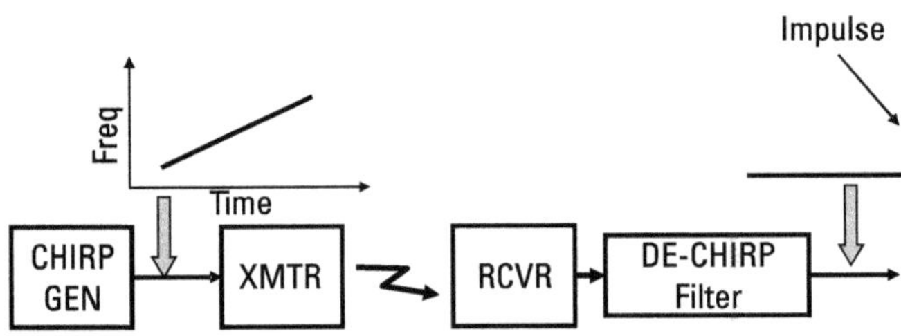

Figure 7.23 When a swept FM is placed on a bit of a digital signal, it can be processed by a matched de-chirp filter to create an impulse.

delay as the frequency increases. This chirp technique allows the digital data to be carried in two different ways: parallel binary channels or single channel with pulse position diversity.

7.4.3 Parallel Binary Channels

In some systems, logical ones cause one chirp direction (perhaps increasing frequency), while logical zeros cause the opposite chirp direction (in this case decreasing frequency). This type of system is shown in Figure 7.24. The chirp frequency slope is typically linear. In the receiver, each received bit causes an impulse output from the appropriate de-chirping filter. Note that the data stream input in the figure is 1, 0, 1, 1, 0; thus, the up-chirp filter outputs impulses for the first, third, and fourth bits while the down-chirp filter outputs impulses for the second and fifth bits. These impulses are converted into logical bits to reproduce the digital input to the transmitter.

The processing gain is the product of the chirp frequency excursion and the bit duration, which is also the ratio of the chirp excursion to the data bit rate. If analyzed in an averaging spectrum analyzer, the transmitted waveform will be as shown in Figure 7.25. This allows the end points of the chirp modulation to be determined. If noise jamming is applied across this frequency range, the J/S will be reduced by the processing gain. However, since the transmitted

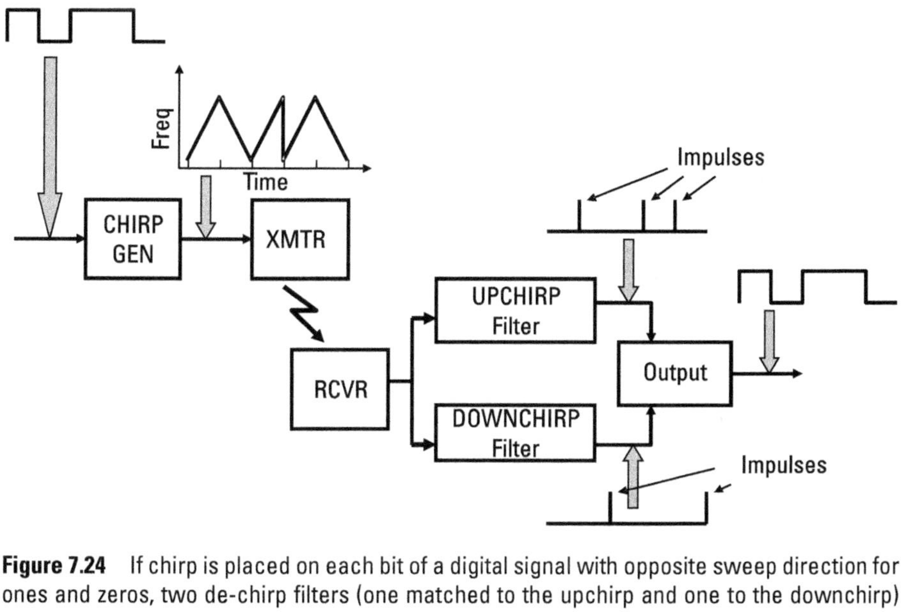

Figure 7.24 If chirp is placed on each bit of a digital signal with opposite sweep direction for ones and zeros, two de-chirp filters (one matched to the upchirp and one to the downchirp) will produce impulses for each one or zero. These impulses allow reproduction of the transmitted digital data.

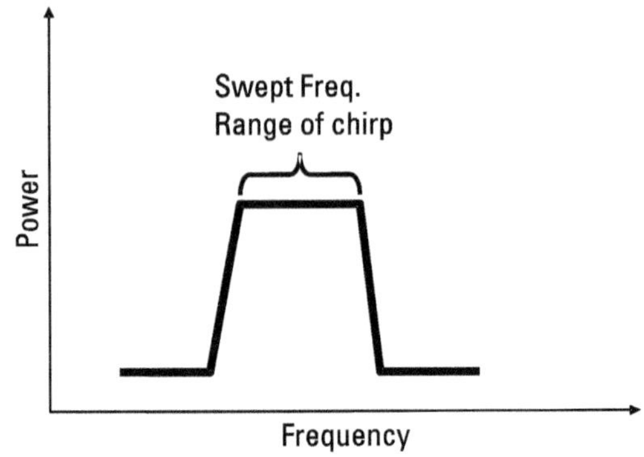

Figure 7.25 An averaging spectrum analyzer will show the frequency range covered by the chirp on a signal.

signal is digital, pulse jamming can be applied (causing bit errors while the jamming pulse is up) to increase the jamming effectiveness.

If the chirp slope and end points are determined with a spectrum analyzer, a linearly chirped signal can be used as a jamming waveform. The jamming chirp can be randomly positive or negative. Because a data signal can be expected to have roughly equal numbers of ones and zeros, half of the bits will be jammed at full J/S. A 50% bit error rate is more than enough to stop the transfer of information over the jammed channel.

7.4.4 Single Channel with Pulse Position Diversity

As shown in Figure 7.26, the timing of the impulse from the de-chirping filter in the receiver is a function of the start frequency of the chirp generator in the transmitter. Thus, if logical ones start at one frequency and logical zeros start at another frequency, the timing of impulses from the de-chirping filters allows the separation of ones and zeros by time. In this example, an up-chirp is used and the chirps on zeros begin and end at higher frequencies than for ones. This will cause the impulses from zeros to be output with less delay than the impulses for ones. Note that the output of the de-chirping filter has an impulse in the left part of the time slot when the input data is a logical zero and in the right part of the time slot when the input data is a logical one. Because the figure shows an input data stream, 1, 0, 1, 1, 0, the impulses for the first, third, and fourth bits are late and those for the second and fifth bits are early.

There is a patent for a chirp communication system that uses the time separation of ones and zeros as above, but has a pseudo-random start-frequency selection feature for security. This causes the output impulse from the de-chirping filter to have a pseudo-random time pattern. The intended receiver is synchronized with the transmitter so that this time randomness can be resolved.

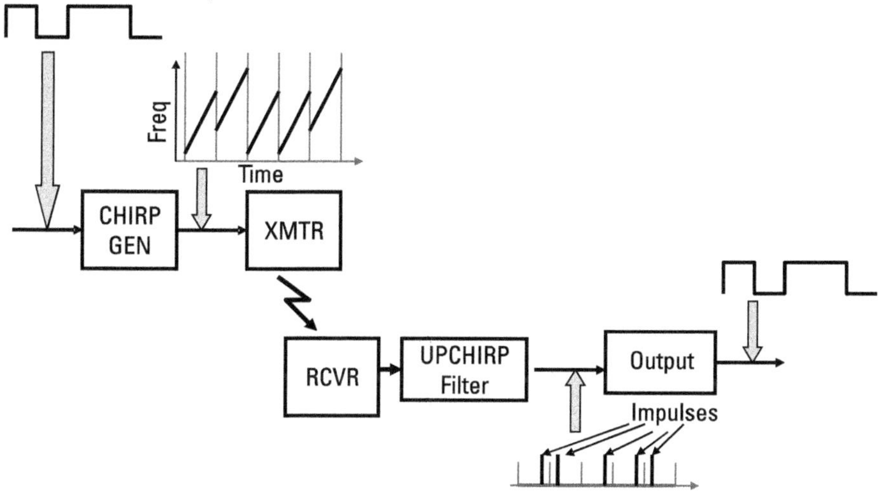

Figure 7.26 If the chirp start-frequency is different for logical ones and zeros, the impulse output of a matched de-chirp filter will have different delays allowing the reproduction of the original data stream.

Noise jamming across the chirp range will have its J/S reduced by the processing gain. Again, pulse jamming will increase jammer effectiveness and use of a chirped waveform matched to the transmitted signal (with random ones and zeros) will significantly improve the J/S.

7.5 Direct Sequence Spread Spectrum Signals

Direct sequence spread spectrum (DSSS) signals are digital signals that are spread in frequency by application of a secondary digital modulation. Digital signals have spectral characteristics as shown in Figure 7.27, with the typical

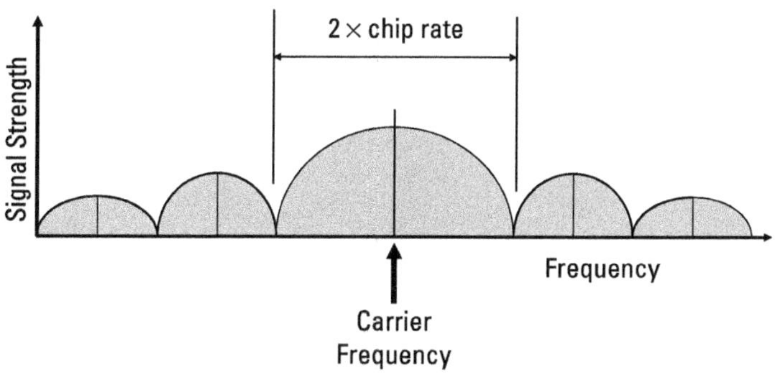

Figure 7.27 DSSS signals, like any digital signals distribute energy over a spectrum dependent on the bit rate.

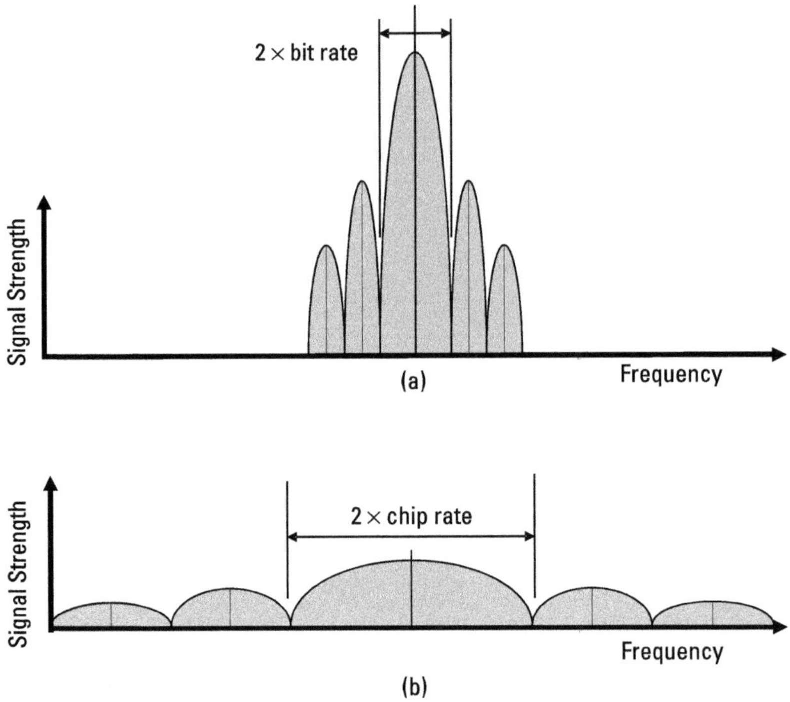

Figure 7.28 Applying a second, wider digital modulation to a digital signal spreads its spectrum and reduces its signal strength density.

null-to-null bandwidth equal to twice the bit rate of the modulation. Figure 7.28(a) shows the spectrum of the signal when only the information modulation is present. Figure 7.28(b) shows the spectrum when the higher bit rate spreading modulation has been applied. The bits in the spreading modulation are called chips. This figure is unrealistic in that the spreading modulation chip rate is only shown as five times the information modulation rate; actually, the spreading modulation is normally of the order of 100 to 1,000 times the information bit rate to provide adequate processing gain.

As shown in Figure 7.29, a de-spreading modulation is applied to the received signal to remove the spreading modulation; thus, de-spreading the signal and increasing its signal strength versus frequency by the spreading factor, for example, 30 dB if the spreading modulation chip rate is 1,000 times the information bit rate. This is a processing gain that applies only to signals the receiver is designed to receive.

The spreading modulation is a pseudo-random code. The de-spreader, shown in Figure 7.30, is the spreading demodulator in the block diagram of Figure 7.29. It applies the same modulation that was placed on the signal in

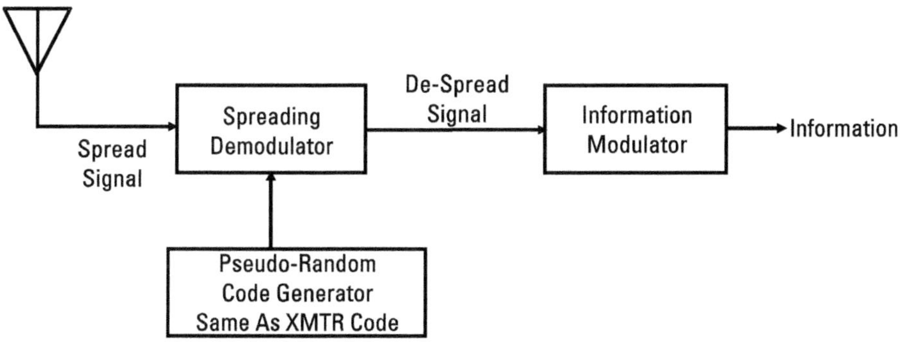

Figure 7.29 A DSSS receiver applies the same code used to spread the signal, thereby removing the spreading modulation.

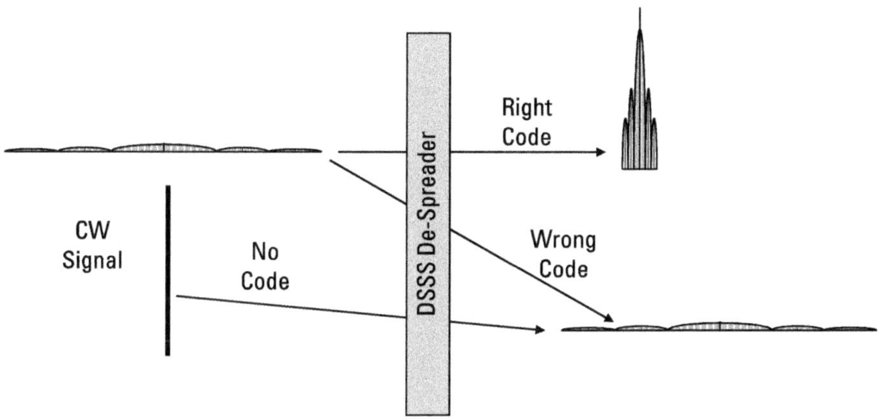

Figure 7.30 The de-spreading process also spreads and reduces signals which are not modulated with the matching code.

the transmitter. This has the effect of removing the spreading modulation from the signal, thus restoring the original information signal. If the code applied in the receiver is different than that in the transmitter, the signal is not de-spread and thus remains at its low (i.e., spread) signal strength. Note that since the de-spreading process is identical to the spreading process, a nonspread signal input to the receiver will be spread, and thus reduced by the spreading factor. This provides the antijam performance of the DSSS LPI approach.

7.5.1 Jamming DSSS Receivers

If the spreading code is known, as it might be in commercial systems, the jamming signal can be appropriately modulated and pass through the receiver enhanced by the processing gain. However, in military applications, the code will not be known, so the J/S can be expected to be reduced by the spreading factor.

As discussed in Section 7.3, a digital signal is best jammed by the creation of bit errors and J/S of 0 dB creates close to 50% bit errors (the maximum bit error rate). More jamming power has very little effect on the receiver. DSSS signals are digital, so 0-dB J/S (after the receiver processing) is adequate. Remember the processing gain for the desired signal.

Because any received jamming signal will be reduced by the same amount, it makes sense to use a simple continuous wave (CW) signal near the center frequency of the DSSS transmitter.

7.5.2 Barrage Jamming

Barrage jamming can be used against the DSSS signal, but remember that the J/S will be reduced by the receiver's processing gain, and a CW signal would be just as effective (and much easier to generate).

A barrage jammer has the advantage of simplicity of operation. No look-through is required. Thus, this type of jamming is highly compatible with simple remote jammers like those on UAVs, emplaced by artillery, or hand emplaced.

7.5.3 Pulse Jamming

Because the digital DSSS signal can be made unintelligible by a 33% bit error rate (or less under some circumstances), the jamming can have significantly less than 100% duty cycle. It is usually possible to transmit significantly higher peak power in a pulse jammer than in a continuous jammer.

Note that if the target communication system uses error correcting codes with interleaving, it may not be practical to use pulse jamming.

7.5.4 Stand-In Jamming

Going back to the basic J/S formulas in Chapter 6, you see that the J/S is strongly influenced by the range between the jammer and the target receiver. If line-of-sight propagation applies, the jammer power into the receiver (and thus the J/S) reduces by the square of the jammer to receiver range. Thus, J/S will be increased by the square of reducing range. If two-ray propagation applies, the J/S will increase by the fourth power of reducing range.

Stand-in jamming involves placing the jammer near the target receiver using remote jammers that may be turned on by command or by automatic timing. They might be barrage jammers or use some other broad spectrum jamming waveform. Ideally, stand-in jammers will be far enough from friendly communication to avoid fratricide.

7.6 DSSS and Frequency Hop

Figure 7.31 is a block diagram of a hopped DSSS transmitter. The information signal will be digital, and the direct sequence modulator will typically add a higher bit-rate digital signal to the information signal. The result will be a digital signal at the higher bit rate.

Figure 7.32 shows the spectrum of a hopped DSSS signal. Each of the humps in the spectrum is the central main lobe of a typical digital spectrum as shown in Figure 7.27. The hop frequencies will typically be picked so that the main lobes of the digital spectra overlap. For example, if the spreading chip rate were 5 Mbps, the null-to-null bandwidth of the digital spectrum would be 10 MHz. The hop frequencies might then be chosen about 6 MHz apart.

To jam this type of signal, it is necessary to place the jamming signal near the hop frequency. If, for example, pulse jamming is used, it must either be applied to each hop frequency or applied to the active hop after its frequency is detected by the jammer.

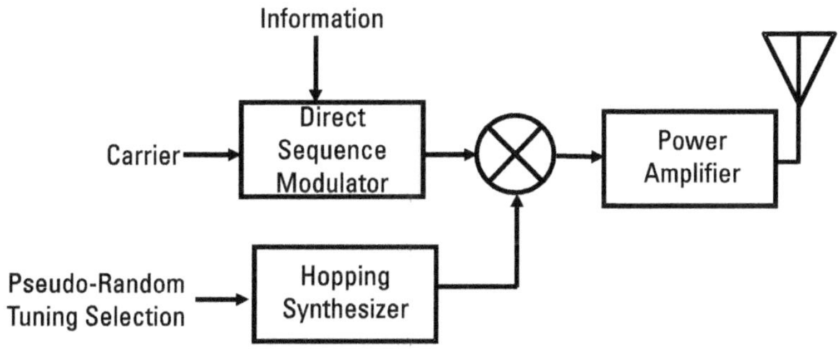

Figure 7.31 A hopped DSSS transmitter applies frequency hopping modulation to the digitally spread signal.

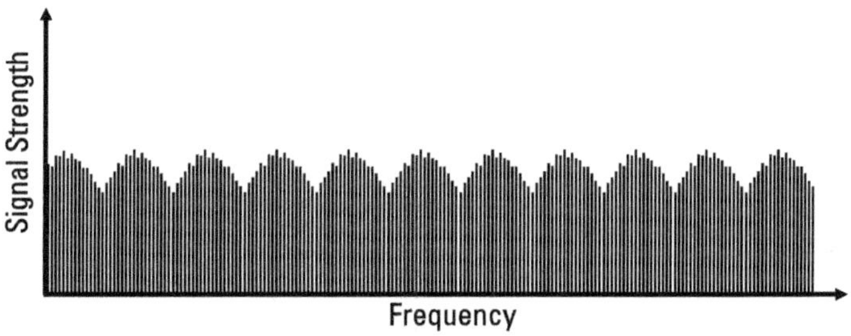

Figure 7.32 A signal with both DSSS and Frequency Hop has overlapping digital spectra centered on the hop frequencies.

7.7 Fratricide

Any situation in which communication jamming is employed has a potential for fratricide, the unintentional jamming of friendly communications. Particularly when broadband (barrage) jamming is used, friendly command and control communication, data links, and command links can suffer significant degradation.

There have been accounts of individuals who believe that because the effective range of a jammer is some specific distance, communication will be unaffected beyond that range. Figure 7.33 is intended to dramatically illustrate the danger of this misunderstanding. The analogy between the effective range of the jammer and the firearm is apt. The effective range of a firearm is the range at which it can be expected to hit and cause sufficient damage to a target when employed by an appropriately trained individual; the bullet travels much farther than the effective range. The effective range of a jammer is the distance at which it can cause sufficient J/S in an enemy receiver to prevent effective communication (with some safety margin); generally, full performance by a friendly link requires that the J/S in the receiver be far lower.

7.7.1 Fratricide Links

As shown in Figure 7.34, we consider four links in this analysis. The desired jamming operation causes a *J*/*S* in the target receiver defined by the following equation:

$$J/S = ERP_J - ERP_{ES} - LOSS_{JE} + LOSS_{ES}$$

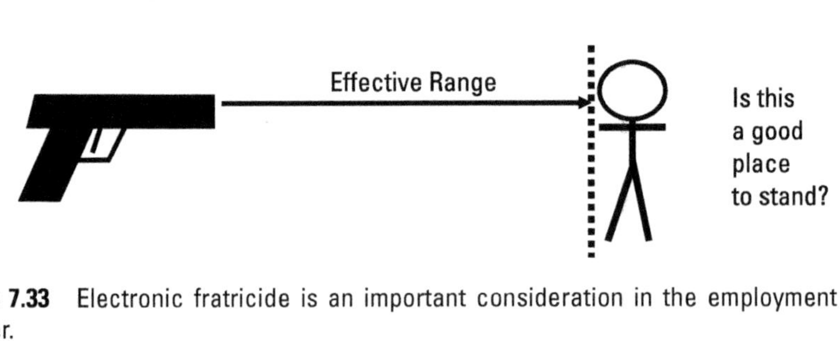

Figure 7.33 Electronic fratricide is an important consideration in the employment of any jammer.

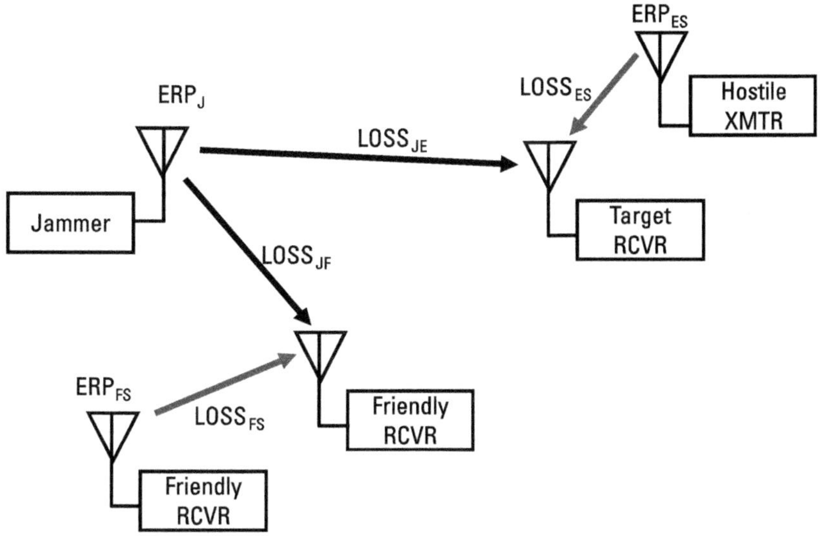

Figure 7.34 Fratricide vulnerability analysis requires calculation of J/S for both hostile and friendly communication links.

where ERP_J is the jammer ERP, ERP_{ES} is the hostile transmitter ERP, $LOSS_{JE}$ is the link loss between the jammer and the target receiver, and $LOSS_{ES}$ is the link loss between the hostile transmitter and the target receiver.

Now, consider the fratricide link. It is convenient to write a parallel equation for the unintended J/S of the friendly receiver.

$$J/S(\text{Fratricide}) = ERP_J - ERP_{FS} - LOSS_{JF} + LOSS_{FS}$$

where ERP_J is the jammer ERP, ERP_{FS} is the friendly transmitter ERP, $LOSS_{JF}$ is the link loss between the jammer and the friendly receiver, and $LOSS_{FS}$ is the link loss between the friendly transmitter and the friendly receiver.

Unfortunately, there is no magic rule of thumb for evaluating fratricide. If jamming is conducted at a frequency used for friendly communication, it is necessary to work both of these equations with the appropriate link loss models (i.e., line of sight, two-ray, or knife-edge diffraction), ERPs, link distances, and antenna heights or frequency (when appropriate). The effective J/S (fratricide) should generally be significantly below 0 dB (−15 dB is a reasonable target).

7.7.2 Minimizing Fratricide

Figure 7.35 summarizes the approaches to the minimization of fratricide. Each of them either reduces the jamming power received in the friendly receiver or enhances desired signals to reduce the effective J/S.

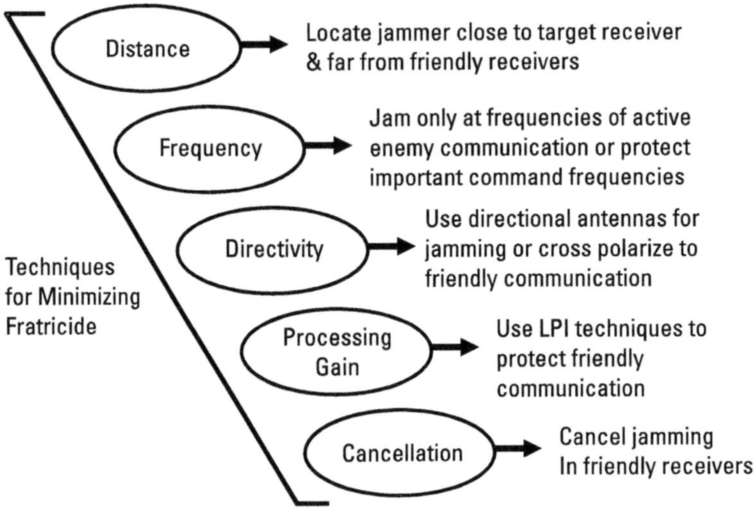

Figure 7.35 Several techniques can be used to minimize fratricide.

Minimize the jammer to target receiver distance and maximize the jammer to friendly receiver distance. Stand-in jamming involves the remote operation of a jammer as close to the enemy as practical. This includes jammers on UAVs, artillery-emplaced jammers, and hand-emplaced jammers. Remote jammers can be activated by command or timed to turn on in some optimum pattern. In general, they will be either barrage or swept spot jammers so they will be sure to cover the enemy's operating frequencies without direct operator intervention. The anti-fratricide advantage comes from the ratio of the link distances as shown in Figure 7.36. The advantage will be the square of the distance ratio for line-of-sight propagation or the fourth power of the distance ratio for two-ray propagation.

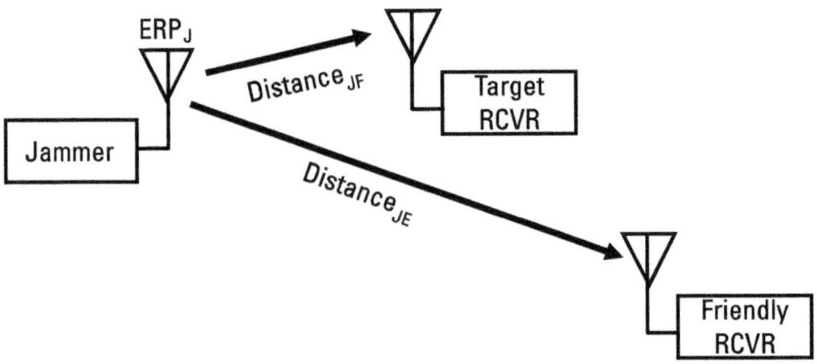

Figure 7.36 Relative distance to target and friendly receivers strongly impacts fratricide.

Use frequency diversity. It is best, whenever practical, to jam only on active enemy frequencies. Not only does this maximize the jamming effectiveness, but it also reduces the probability of fratricide. This assumes that command and control frequencies are chosen to be those not requiring jamming. It may also be practical to filter broadband jamming to protect friendly frequencies.

Note that where an enemy frequency hopper is jammed with a follower jammer, friendly communications will be minimally degraded because the jammer is seldom on a friendly frequency.

Use directional antennas for jamming as shown in Figure 7.37 when practical. If the jamming antenna is directed at the enemy's location, friendly receivers will most likely be in the lower gain side lobes of the jamming antenna. This will reduce the effective jammer ERP toward the friendly receiver by the side-lobe isolation ratio.

Another antenna consideration is polarization. Where practical, match the polarization of the jamming antenna to that of the enemy antennas and use cross-polarized antennas for friendly communication. Note that when everyone is communicating with whip antennas, friendly and enemy antennas will all be vertically polarized, so this technique does not apply.

Use LPI modulations for friendly communication. This will provide processing gain for desired signals in the friendly receiver, thus reducing the effective J/S from enemy or friendly jammers.

Signal cancellation techniques can sometimes be applied to reduce the effectiveness of jamming signals. As shown in Figure 7.38, an auxiliary antenna receives the jamming signal and passes it through a 180° phase shifter. When this phase-shifted signal is added to the signals from the normal communication antenna, the jamming signal will be cancelled (by some number of decibels). Note that the auxiliary antenna must typically have some advantage toward the jammer (10 dB in one case). The cancelling signal could also be hard connected to the jammer output, but this would only cancel the primary signal.

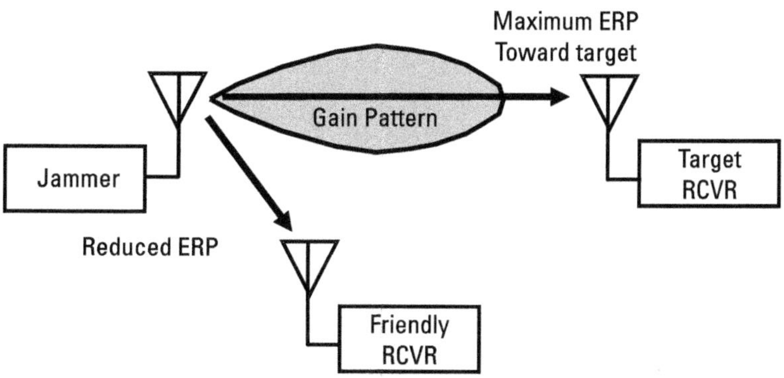

Figure 7.37 A directional jamming antenna will reduce the ERP toward friendly receivers.

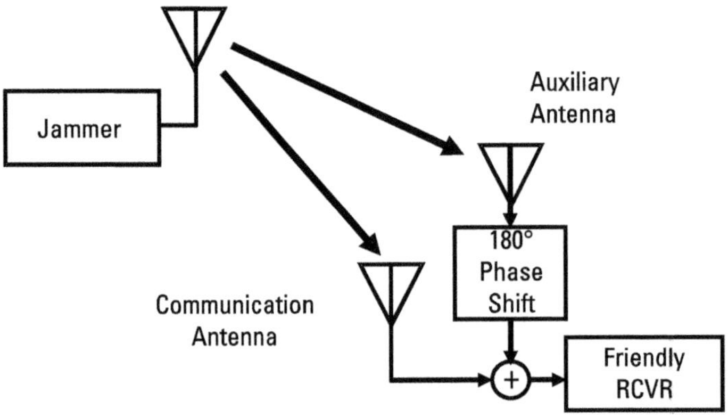

Figure 7.38 Injecting a 180° phase shifted version of the jamming signal significantly reduces it.

In virtually all situations, there will be multipath signals that add to form the signal actually received by the communication antenna. An auxiliary antenna should capture at least some of these multipath signals, improving the quality of the cancellation process.

7.8 Precision Emitter Location of LPI Transmitters

In general, all of the emitter location techniques described in Chapter 6 can be applied to LPI transmitters if the timing issues are properly handled; however, there are significant issues associated with the precision location of LPI emitters.

First, consider frequency hoppers. Time difference of arrival (TDOA) emitter location requires that many samples be taken with varying values of relative delay to determine the correlation peak. The delay values causing the correlation peak indicate the time difference of arrival. This process typically takes a large part of a second, so the short time that the hopper remains at one frequency (i.e., the hop duration) is very unlikely to allow enough time to determine the TDOA. Frequency difference of arrival (FDOA) requires only a measurement of frequency by each receiver, so if the emitter is in a fixed location, the receivers are airborne, and there is adequate SNR, FDOA may be practical.

Next, consider chirp spread spectrum signals. For TDOA, the rapidly changing frequency will present significant challenges in establishing a correlation peak, and accurate frequency measurement for FDOA is equally impractical.

Finally, consider DSSS signals. If the pseudo-random spreading code is known (for example, in a commercial communication system), it may be practical to perform either TDOA or FDOA emitter location. However, if the code is not known, the location of these signals requires energy detection approaches which will not provide adequate signal to noise ratio to support either TDOA or FDOA. An exception to this conclusion might occur for very strong DSSS signals using very short codes. Under these conditions, it may be practical to isolate a single spectral line and perform TDOA or FDOA analysis.

7.9 Jamming Cell Phones

In this section, we will discuss the jamming of cell phone links. First, we will discuss how various types of cell phone systems work and then we will consider a few jamming situations.

7.9.1 Cell Phone Systems

Figure 7.39 shows a typical cell phone system. A number of towers are connected to a mobile switching center (MSC) that controls the whole process. The MSC is also connected to a public switched telephone network so that cell phones can be connected to regular wired telephones.

Cell phone systems can be either analog or digital. This refers to the way that communication signals pass between the cell towers and cell phones. In analog systems, the communication channels are analog (frequency modulated), but there are also control channels, which are digital. Digital systems use digital channels for both control and communication. Each frequency in a digital cell

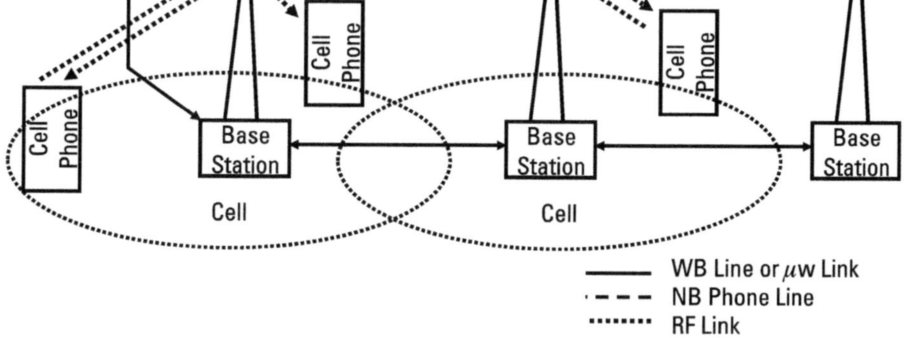

Figure 7.39 A cell phone system comprises several towers which are connected to a mobile switching center—which also connects to a public switched telephone network.

system has multiple communication channels. We will consider two important digital systems (GSM and CDMA) as typical.

7.9.2 Analog Systems

In analog cell phone systems, duplex operation is provided by assignment of two RF channels to each cell phone, one from the tower to the phone (the downlink) and one from the phone to the tower (the uplink). One user continuously occupies two RF channels during a call. Each channel carries the transmitted signal most of the time, but interrupts that signal for short periods to send digital control data. In some systems, the control data is modulated onto the voice signal so no interruption is required. Figure 7.40 shows the way that signals are carried in analog cell phone channels for a typical system. A few of the RF channels carry digital signals for access and control functions, these are control channels.

When a cell phone is activated, it searches control channels to find the strongest tower signal (i.e., the closest cell tower). After the cell system validates the cell phone as an authorized user, the cell phone enters the idle mode, monitoring the control channel for incoming calls. When the cell phone is called, the tower sends a control message assigning a pair of RF channels. When the cell phone initiates a call, the tower sends a control message to assign the RF channels. When no channels are available, the system delays by a randomized period before retrying. To prolong cell phone battery life, the cell phone transmitter is

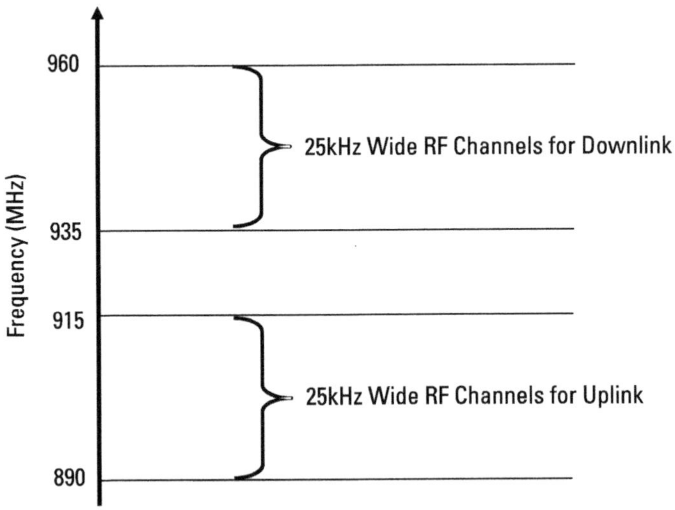

Figure 7.40 An analog cell phone system carries one conversation per RF channel. Up and down link channels for one phone are 45 MHz apart.

turned off when the user is not talking. The digital control signals in the voice channels allow the system to change the RF channel assignment and to turn down the transmit power from the cell phone to the minimum acceptable level (to further prolong battery life and to avoid interference).

Analog cell systems typically operate at about 900 MHz and can have up to 50W of transmit power on each RF channel from the cell towers. Cell phones have maximum transmit power of 0.6W to 15W, but are turned down to minimum required power by command from the tower. Minimum cell phone transmit power is usually 6 mW.

7.9.3 GSM Systems

The Global System for Mobile Communication (GSM) has eight time slots per 200-MHz-wide RF band, allowing eight users to share the same RF band. A system will have many RF bands. Digitized voice data from each user is carried in one digital data block per frame as shown in Figure 7.41. The frame repeats at 33,750 frames per second for a total bit rate per RF channel of 270 kbps. Some systems operate in the half-rate mode in which each user occupies the assigned slot in every second frame, so that 16 users share each frequency band. At the receiver, the bits in one time slot are passed through a digital-to-analog converter (DAC) to reproduce the signal that was digitized at the transmitter.

Some of the user time slots in the cell system are occupied by control channels for paging and assignment of RF channels and time slots.

Operation is very similar to that of analog cell systems. When a cell phone is activated, it searches control channels to find the strongest tower signal, and after authorization enters the idle mode, monitoring the control channel for incoming calls. When the cell phone is called or initiates a call, the tower sends a control message assigning a pair of RF channels (one each for uplink and downlink). However, in a GSM system, it also assigns a time slot in each assigned RF channel.

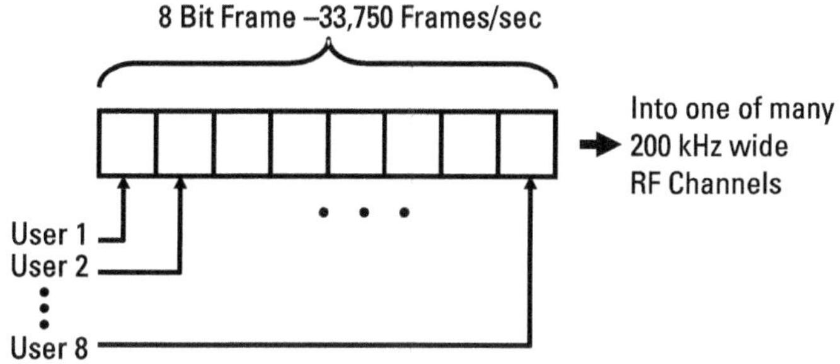

Figure 7.41 GSM cell phones carry digital user data in one RF channel for the uplink and another RF channel for the downlink.

The randomized delay before retry when no channel/time slot is available and the control of cell phone transmitter power to maximize battery life is the same as described above for analog systems.

GSM systems operate at 900, 1,800, and 1,900 MHz. Separate RF channels are used for the uplink and downlink to each cell phone for full duplex operation. Note that different time slots are used for the uplink and downlink so that a cell phone is not transmitting and receiving at the same time. The transmitted power from cell phones and towers are similar to those in analog systems.

7.9.4 CDMA Systems

Code division multiple access (CDMA) cell phone systems use DSSS modulation as described earlier in this chapter. Each user voice input signal is digitized. A high rate digital modulation carrying a pseudo-random code is applied to each digitized user voice signal in the transmitter. This spreads the signal power over a wide frequency spectrum, thereby reducing its power density. When the same pseudo-random code is applied to the received signal at the receiver, the signal is returned to its original form. When passed through a DAC, the signal can be heard by the user for whom it is intended. If the correct code is not applied to the received signal, it remains so faint that it cannot even be detected by a listener. By using 64 different codes, which have been selected for optimum signal isolation, voice signals from 64 different users can be carried on the same 1.23-MHz-wide RF channel as shown in Figure 7.42. A CDMA cell system has multiple RF channels. Some of the access channels (code and RF channel) in the system are used for control functions.

Operation is very similar to that of GSM cell systems as described above. However, the control signals to cell phones assign spreading codes rather than time slots. The IS-95 CDMA system operates throughout the United States at 1,900 MHz using tower and cell phone transmit powers like those described above for analog cell phone systems.

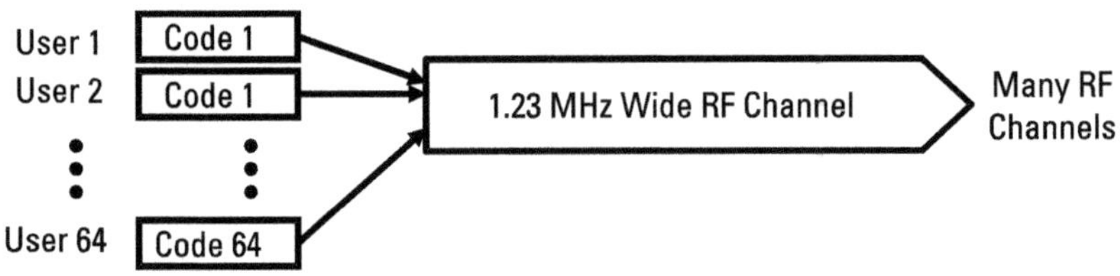

Figure 7.42 CDMA cell phones carry up to 64 digital user signals, each using a different spreading code, on each RF channel.

7.9.5 Cell Phone Jamming

We will now consider some cell phone jamming situations. We will use the propagation and jamming formulas discussed in Chapter 6.

Because any propagation loss model might be appropriate to any link, it is necessary when approaching a communication jamming problem to first determine the appropriate loss model for each of the links involved. Because cell phones and cell towers are near the ground, the uplink (i.e., cell phone to tower) and the downlink (i.e., tower to cell phone) will be either line of sight or two-ray depending on the range, frequency, and antenna heights. This also applies to the link from a jammer (regardless of its location) to the cell phone or to the cell tower. Thus, the first step in analyzing cell phone jamming is to determine the Fresnel zone distances for the cell phone and jamming links. Then, the J/S can be calculated.

We will consider four cases: jamming from the ground and from the air against the uplink and the downlink. In each of these cases, the cell system is operating at 800 MHz and we are jamming the whole RF channel. If the cell system is analog, this will jam one signal. If the system is digital, this will jam all user channels using that RF channel. To jam only one user channel in a digital system, it is necessary to limit the jamming to the appropriate time slot (for GSM systems) or apply the code for one user (for CDMA systems).

7.9.6 Uplink Jamming from the Ground

As shown in Figure 7.43, the cell phone is 1m from the ground, 2 km from a 30-m-high cell tower. The cell phone has a maximum ERP of 1W. The jammer is 4 km from the cell tower, 3m above the ground, and generates 100-W ERP.

As the uplink goes from the cell phone to the tower, we must jam the link receiver, which is in the tower. The cell phone transmit power can be reduced to as little as 6 mW, providing only the amount of power needed for adequate

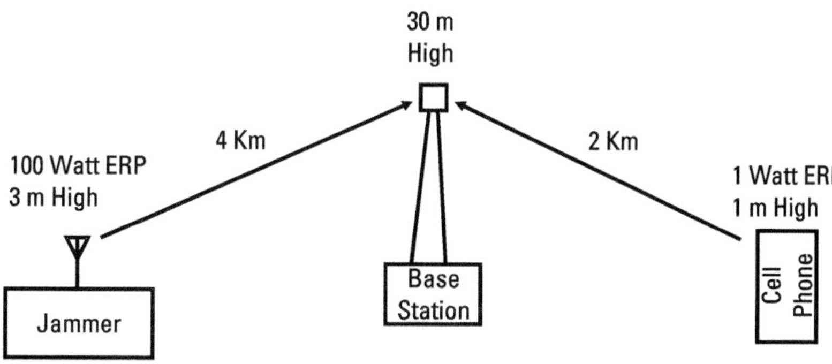

Figure 7.43 Jamming a cell phone uplink requires broadcasting to the cell tower.

SNR in the tower receiver. However, we can assume that our jamming will cause very low SNR in the jammed link, so the cell phone would remain at its maximum power during jamming.

First, let us calculate the Fresnel zone distance for the cell phone and jamming links, using the formula:

$$FZ = \left(h_T \times h_R \times F\right)/24{,}000$$

where FZ is the Fresnel zone distance (in kilometers), h_T is the transmitter height (in meters), h_R is the receiver height (in meters), and F is the link frequency (in megahertz).

The Fresnel zone distance for the cell phone to tower link is:

$$FZ = (1 \times 30 \times 800)/24{,}000 = 1\,\text{km}$$

The cell phone is 2 km from the tower, which is greater than the Fresnel zone distance, so two-ray propagation applies to the cell phone link.

For the jamming link:

$$FZ = (3 \times 30 \times 800)/24{,}000 = 3\,\text{km}$$

Because the link distance is greater than the Fresnel zone distance, the propagation is two-ray.

As in all communication jamming when the receiving antenna has approximately the same gain in all directions, the J/S is calculated from:

$$J/S = ERP_J - ERP_S - LOSS_J + LOSS_S$$

where ERP_J is the ERP of the jammer (dBm), ERP_S is the ERP of the desired signal transmitter (dBm), $LOSS_J$ is the loss from the jammer to the receiver (in decibels), and $LOSS_S$ is the loss from the desired signal transmitter to the receiver (in decibels).

Converting the two ERP values to dBm, 100W = 50 dBm and 1W = 30 dBm. The loss from the jammer (two-ray propagation model) is:

$$\begin{aligned} LOSS_J &= 120 + 40\log(4) - 20\log(3) - 20\log(30) \\ &= 120 + 24 - 9.5 - 29.5 = 105\,\text{dB} \end{aligned}$$

The loss from the cell phone to the tower (two-ray propagation model) is:

$$LOSS_S = 120 + 40\log(2) - 20\log(1) - 20\log(30)$$
$$= 120 + 12 - 0 - 29.5 = 102.5\text{dB}$$

So the *J*/*S* is:

$$J/S = 50 \text{ dBm} - 30 \text{ dBm} - 105 \text{ dB} + 102.5 \text{ dB} = 17.5 \text{ dB}$$

7.9.7 Uplink Jamming from the Air

As shown in Figure 7.44, the cell phone link is the same as in the previous case, but now the 100-W jammer is in an aircraft flying at 2,000-m altitude 15 km from the cell tower.

The cell phone tower link is the same, but we must calculate the Fresnel zone distance for the jammer to tower link:

$$FZ = (2{,}000 \times 30 \times 800)/24{,}000 = 2{,}000 \text{ km}$$

The jammer to tower link is much shorter than *FZ*, so it definitely uses line of sight propagation. The jamming link loss is then:

$$LOSS_J = 32.4 + 20\log(d) + 20\log(F)$$

where *d* is the link distance in kilometers and *F* is the operating frequency in megahertz.

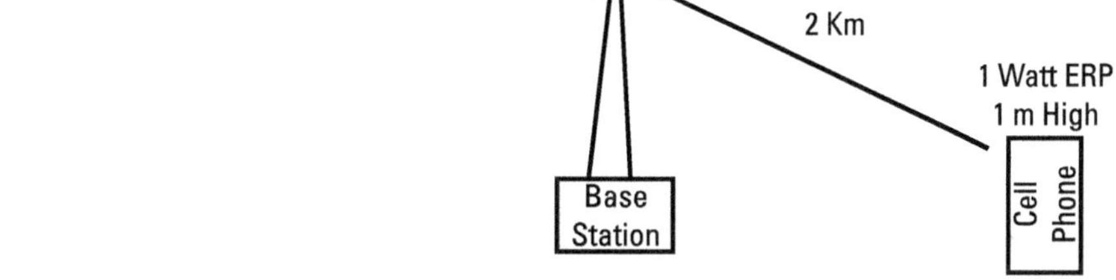

Figure 7.44 An airborne uplink jammer can achieve good J/S even at long range because of its elevation.

$$LOSS_J = 32.4 + 23.5 + 58.1 = 114 \text{ dB}$$

The other link values (ERP_S, ERP_J, and $LOSS_S$) are the same, so J/S is calculated as:

$$J/S = 50 \text{ dBm} - 30 \text{ dBm} - 114 \text{ dB} + 102.5 \text{ dB} = 8.5 \text{ dB}$$

It is interesting to note that if the jammer were 3m from the ground rather than 2,000m, the J/S would be 14 dB less.

7.9.8 Downlink Jamming from the Ground

It is interesting to note that downlink jamming has an operational advantage, even though the large effective radiating power of the transmitter in the cell tower reduces the J/S that can be produced. That advantage comes from the way cell towers are selected for the uplink. If we are jamming the uplink (i.e., jamming the receiver in a cell tower), the received signal quality will be low, causing the system to choose a different tower.

The downlink jamming problem is as shown in Figure 7.45. The 30-m-high cell tower ERP is 10W, the 1-m-high cell phone is 2 km from the tower, and the 100-W, 3-m-high jammer is 1 km from the cell phone.

Because we are jamming the downlink, the jamming link is from the jammer to the phone. The FZ calculation for the downlink is the same as for the uplink above (i.e., 1 km), so the downlink uses two-ray propagation. The jammer FZ is:

$$FZ = (3 \times 1 \times 800)/24{,}000 = 100\text{m}$$

The phone link is longer than FZ, so it uses two-ray propagation. The jamming link loss is:

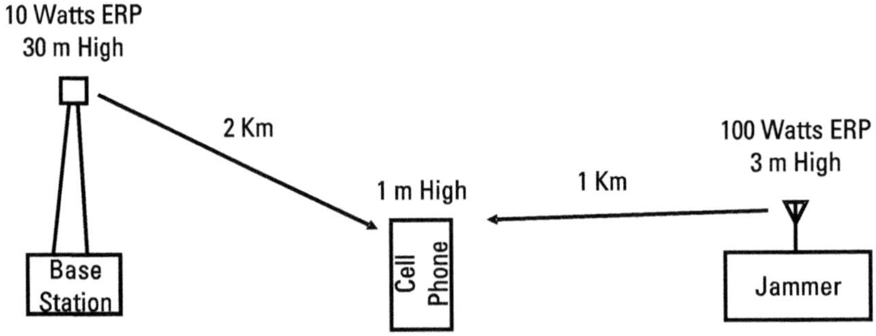

Figure 7.45 A down link jammer broadcasts to the cell phone and must overcome the high power of the cell tower transmitter.

$$LOSS_J = 120 + 40\log(1) - 20\log(3) - 20\log(1) = 120 + 0 - 9.5 - 0 = 110.5 \text{ dB}$$

The 10-W ERP from the tower is 40 dBm. The other parameters (ERP_J and $LOSS_S$) are the same as for the uplink jamming from the ground case, so the J/S is thus:

$$J/S = 50 - 30 - 110.5 + 102.5 = 12 \text{ dB}$$

7.9.9 Downlink Jamming from the Air

The jammer is now at 2,000m, 15 km from the receiver. The jamming link *FZ* is:

$$FZ = (2{,}000 \times 1 \times 800)/24{,}000 = 66 \text{ km}$$

which is greater than the jamming link distance, so the jammer link is line of sight, and has the same loss as for the uplink jamming from the air case.

The cell phone down link ERP is 10W (40 dBm), but the other parameters (ERP_J and $LOSS_S$) are the same as the uplink jamming from the air case. Thus, the J/S is:

$$J/S = 50 \text{ dB} - 40 \text{ dB} - 110.5 \text{ dB} + 102.5 \text{ dB} = 2 \text{ dB}$$

Again, the J/S would be 14 dB less if the jammer were 3m rather than 2,000m high.

Reference

[1] *Journal of Electronic Defense,* EW101 Column, December 2006.

8

Digital RF Memories

The digital radio frequency memory (DRFM) is an important development supporting electronic countermeasures. It allows the rapid analysis of complex received waveforms and generation of countermeasure waveforms. It can increase the effectiveness of a jamming system against complex waveforms by many decibels.

8.1 DRFM Block Diagram

As shown in Figure 8.1, the DRFM downconverts received signals to the appropriate intermediate frequency (IF) for digitization. Then it digitizes the bandwidth of the IF signal. The digitized signal is placed into a memory for transmission to a computer. The computer makes any necessary analysis and modifications to the signal to support the jamming technique being employed. Then the modified digital signal is converted back to analog RF. This signal is frequency converted back to the received frequency using the same local oscillator used in the original frequency conversion. The use of a single oscillator maintains the phase coherence of the signal through the downconversion and upconversion processes.

The key element of the DRFM is the analog-to-digital converter (ADC). It must support the digitization rate of about 2.5 samples per hertz of the frequency band it digitizes, and it must output an I&Q (in-phase and quadrature) digital signal. As shown in Figure 8.2, the I&Q digitization has two samples per hertz of the digitized RF signal that are 90° apart in phase. This captures the phase of the digitized signal. Note that the 2.5 samples per hertz is greater than the Nyquist rate of two samples per hertz that is required in a digital receiver.

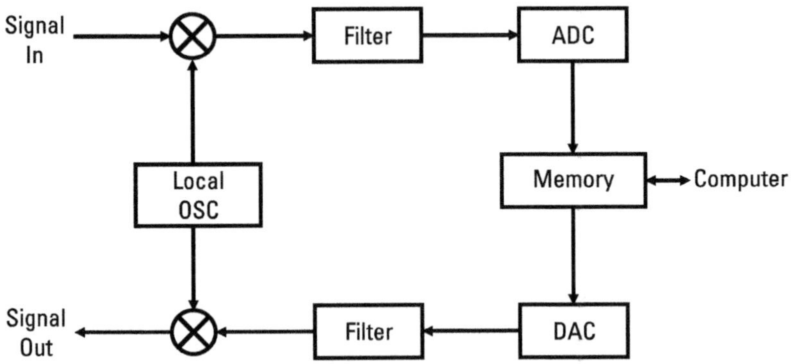

Figure 8.1 The DRFM digitizes a received signal, passes it to a computer for modification, and coherently regenerates the modified signal for rebroadcast.

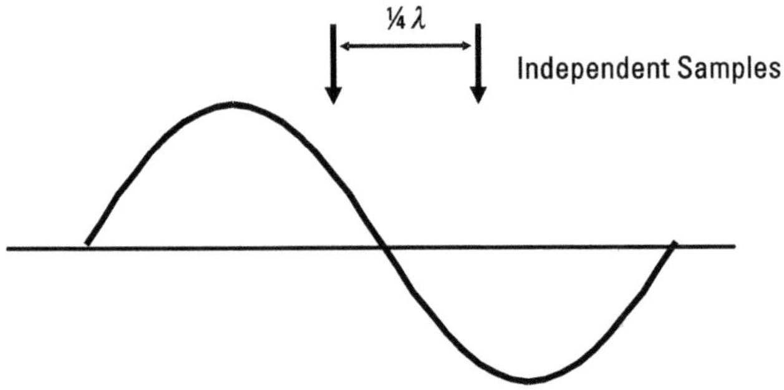

Figure 8.2 An I & Q digitizer digitizes a signal at two points one quarter wavelength apart to capture the frequency and phase of the signal.

This oversampling is required because the signal is being reconstructed. The digital signal must typically have several bits per sample, although there are cases in which 1-bit digitization or phase-only digitization is used.

The computer performs analysis of the captured signal, including determination of its modulation characteristics and parameters. The computer can typically analyze the first pulse received by the system and generate subsequent pulses with the same or systematically varied modulation parameters.

The digital-to-analog converter (DAC) that generates the RF output signal will have more bits than the ADC to assure that the signal quality is not degraded in the reconstruction of the RF signal.

8.2 Wideband DRFM

A wideband DRFM digitizes a wide IF bandwidth that may include several signals. The jammer system tunes across the frequency range of threat signals it must jam and outputs an IF signal with the bandwidth the DRFM can handle. As shown in Figure 8.3, the frequency conversion and the later reconversion to the received frequency are done using a single system local oscillator to preserve phase coherence. The DRFM bandwidth is limited by the digitization rate of its ADC. Because there can be expected to be multiple signals present in the bandwidth, a significant spurious free dynamic range is required, so the ADC requires the maximum practical number of digitization bits.

Dynamic range is discussed in detail in Chapter 6. The dynamic range of a digital circuit is: 20 $\log_{10}(2^n)$, where n is the number of digitizing bits. It is important to remember that the analog circuitry before the ADC must have as much dynamic range as the digital circuitry. Analog dynamic range is also discussed in Chapter 6.

Wideband DRFMs are highly desirable, because they can handle signals with wide frequency modulations and frequency agile threats. We will be discussing the implications of frequency agile threats in detail later in this chapter.

Simply put, as the state of the art in digitizers improves, wideband DRFMs can be expected to be wider and more plentiful. There is an inverse relationship between the digitizing speed and the number of bits that can be provided; the driving requirement for future DRFMs is more samples per second with more bits per sample.

There are a number of approaches to the generation of faster sampling with more bits than can be produced by a single ADC. Here are two typical approaches:

- One technique is the use of several single-bit digitizers at different voltage levels. These do not require computers and can thus be very fast.

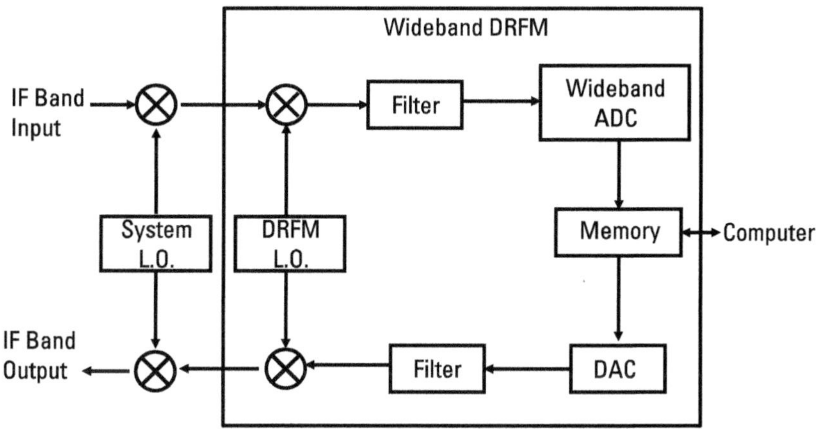

Figure 8.3 A wideband DRFM handles a frequency range containing multiple signals.

Their outputs are combined to create very high rate multiple-bit digital words.

- Another technique is to place several multiple bit digitizers on the outputs of a tapped delay line. The delays between the taps allow these (slower) digitizers to sample at time-spaced intervals during each cycle of the signal being digitized. The outputs are combined to form high-speed multiple-bit digital outputs.

8.3 Narrowband DRFM

A narrowband DRFM need only be wide enough to capture the widest signal the jammer must handle. This means that a narrowband DRFM can operate with an ADC that is reasonably within the state of the art.

As shown in Figure 8.4, the jammer system converts a frequency range of interest into the frequency range covered by multiple narrowband DRFMs. The DRFM input signal is power divided to the individual DRFMs. Each of the DRFMs is tuned to an individual signal and performs its function in support of jamming operation. Then the analog RF outputs from the DRFMs are combined and converted (coherently) back to the original frequency range.

It should be noted that spurious responses are less a problem in narrowband DRFMs because each contains only one signal.

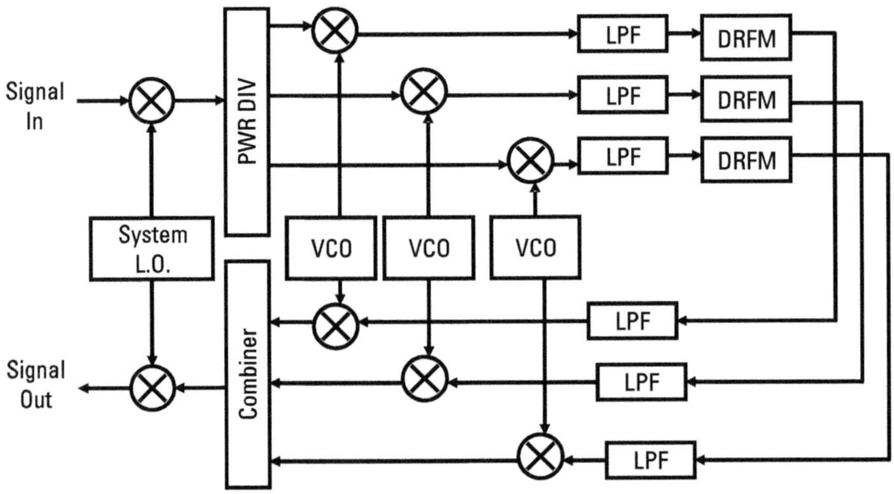

Figure 8.4 A narrowband DRFM handles only one signal. Multiple narrowband DRFMs are required to handle a multiple signal environment.

8.4 DRFM Functions

DRFMs are particularly valuable in dealing with pulse compressed radars. Chapter 4 describes radars that improve their range resolution through pulse compression. This chapter has some illustrations that will help your understanding of pulse compression if you are not familiar with it. The two techniques discussed are chirp and Barker code.

- Chirp involves adding a linear frequency modulation across each transmitted pulse. In the radar's receiver, a compressive filter reduces the effective pulse width by the ratio of the FM sweep range to the radar's coherent bandwidth. If a jammer produces signals that do not have this frequency modulation, the effective jamming-to-signal ratio (J/S) is reduced by the compression factor. By generating chirped jamming pulses, the DRFM maintains the full J/S.
- Barker code pulse compression involves the binary phase shift-keyed modulation of each pulse with a code. In the radar receiver, there is a tapped delay line with as many stages as the number of bits in the code. Some of the outputs have 180° phase-shifted outputs, so that when the pulse exactly fills the shift register, all of the bits add constructively. When the pulse is not exactly aligned with the shift register, the output is approximately zero. This, in effect, shortens the received pulse to the duration of 1 bit of the code. This compresses the range resolution by the ratio of the number of bits across each pulse. Because a jamming pulse without the Barker code is not compressed, the effective J/S is reduced by the number of bits in the code. The DRFM can create jamming pulses with the correct Barker code, maintaining the full J/S.

8.5 Coherent Jamming

One of the advantages of using a DRFM is that it can generate a coherent jamming signal. This is particularly important when jamming a pulse-Doppler (PD) radar. Figure 8.5 shows a range versus velocity matrix that is part of the PD radar processing for all signals entering the radar receiver. The velocity dimension of the matrix is generated by a bank of narrow filters, usually implemented in software. Because the transmitted signal is coherent, the legitimate return signal from a target will fall into one of many filters, and those filters are quite narrow. However, a noncoherent jamming signal like barrage or spot

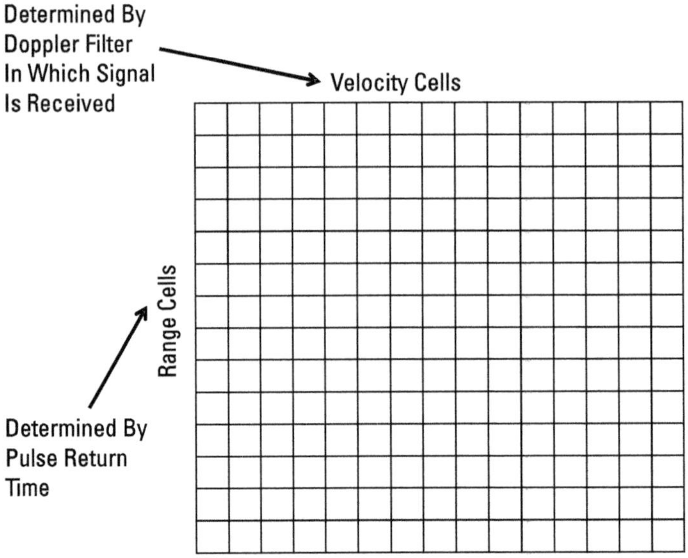

Figure 8.5 The processing hardware and software of a PD radar includes a matrix of time vs. radial velocity for each received pulse.

noise will enter several filters. This allows the radar to reject jamming in favor of the return of its own coherent signal.

8.5.1 Increased Effective J/S

Noise jamming can have its effective J/S against a PD radar reduced by many decibels by the radar's processing gain. Consider the case of an acquisition radar that has its coherent processing interval (CPI) equal to the time that its scanning beam will illuminate a target. The radar has a circular scan with a period of 5 seconds, a beamwidth of 5°, and a pulse repetition frequency (PRF) of 10,000 pulses per second. The beam illuminates the target (a time equal to the CPI) for 69.4 ms, calculated from the following formula (see Figure 8.6):

$$\text{Illumination Time} = \text{Scan Period}\left(\text{Bandwidth}/360^\circ\right)$$
$$= 5 \text{ Seconds } \left(5^\circ/360^\circ\right) = 69.4 \text{ ms}$$

The processing gain of a PD radar is its CPI multiplied by its PRF, so the processing gain is:

$$\text{Processing Gain} = 0.0694 \times 10{,}000/\text{sec} = 964$$

which equals 28.4 dB.

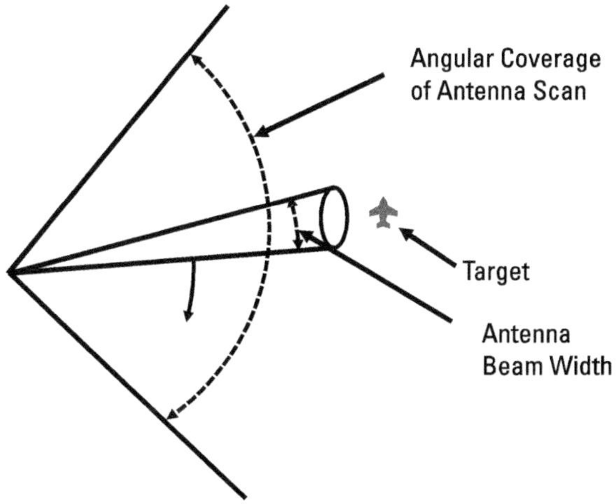

Figure 8.6 The amount of time a scanning radar illuminates a target depends on its beam width, scan rate, and angular scan coverage.

The bandwidth of a single Doppler filter can be as narrow as the inverse of the CPI or 14.4 Hz. This means that the skin return from the radar's own signal will be enhanced by 28.4 dB, but that a noncoherent jamming signal will not be enhanced. Hence, a coherent jamming signal (generated by a DRFM) that falls into a 14.4-Hz filter will provide 28.4-dB better jamming than a noncoherent noise jamming signal of the same jammer effective radiated power.

8.5.2 Chaff

Radar signals reflected from chaff are spread in frequency by the movement of the many chaff elements as shown in Figure 8.7. With proper analysis, the PD radar can discriminate against chaff returns, precluding the ability of chaff to break the lock of the radar on the true target, and allowing the radar to select and process the valid target return in the presence of chaff. This reduces or eliminates the effectiveness of chaff as a radar countermeasure against PD radars. However, if coherent jamming signals (from a DRFM) are used to illuminate chaff, it can be effective in breaking the lock of the radar.

8.5.3 RGPO and RGPI Jamming

The bank of Doppler filters allows the determination of the rate of change of range to a target. As shown in Figure 8.8, a PD radar can discriminate separating targets, each associated with its Doppler shift. The radar processing can look at the range versus time history of signals and calculate the radial velocity of each of the separating targets. For a legitimate target return, the rate of change of range will be the same as the Doppler-derived velocity. If a range gate pull-

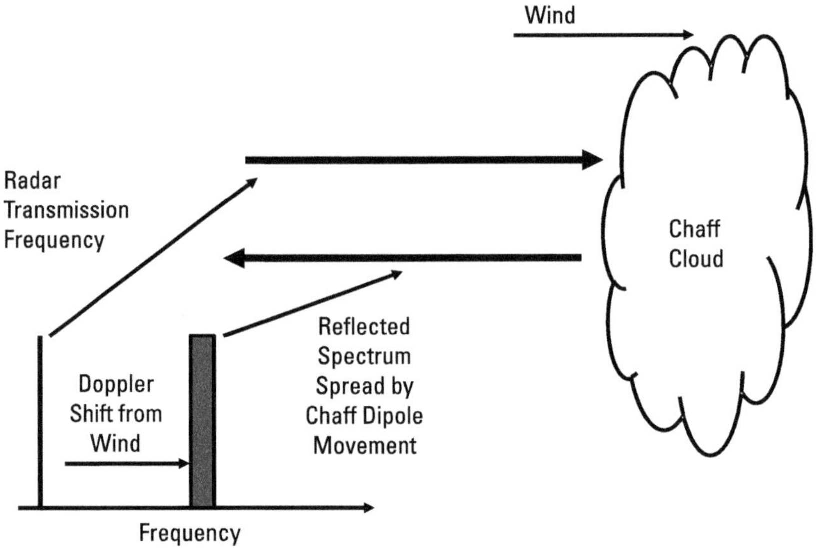

Figure 8.7 The random movement of the dipoles in a chaff cloud causes spreading of the frequency in radar signals reflecting from the cloud. Wind movement of cloud causes frequency shift.

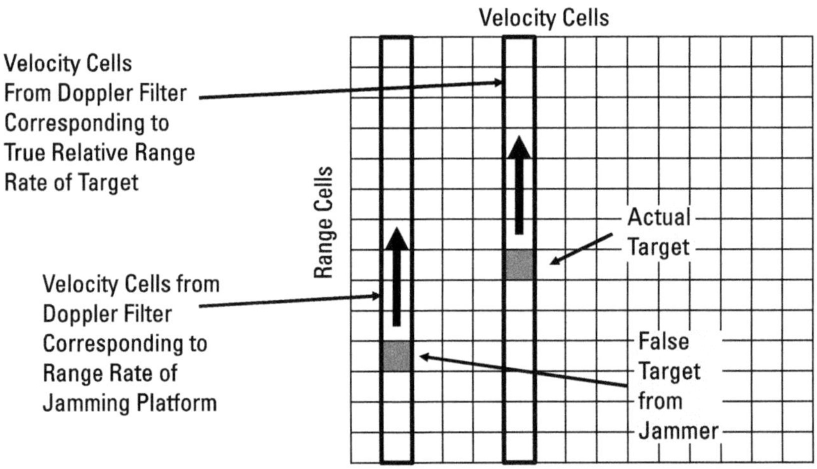

Figure 8.8 A PD radar places separating target pulses in its time/velocity matrix with the appropriate Doppler shift for each.

off (RGPO) or range gate pull in (RGPI) jamming technique is used against the radar, the Doppler shift will not be consistent with the rate of change of range. This is because the jammer only delays or advances the pulses at their transmitted frequency. This will allow the radar to reject the jamming pulses and continue to track the true target.

A DRFM can change both the time and frequency of a radar pulse before coherently rebroadcasting it. This makes the jamming signal appear, to the radar, to be a legitimate target return, so the jamming can break the radar's lock on the target.

8.5.4 Radar Integration Time

A radar receiver is optimized to its own signal. Thus, a pulse of exactly the right length will have the same integration characteristics as the radar's own signal. This enhances the processing gain of the jamming signal as compared to a jamming pulse of a different pulse width. The DRFM can generate jamming pulses of exactly the right pulse duration, maximizing the achieved J/S.

8.5.5 Continuous-Wave Signals

A DRFM continuously records the continuous-wave (CW) signal, converting it to sequential digital data that is then stored in digital memory. This stored data is then replayed out and converted back to an analog signal after a delay, as long as the CW signal is present. To determine the range to its target, a CW radar must place a frequency modulation (FM) on its signal as shown in Figure 8.9. Many FM waveforms can be used. With the FM modulation waveform shown, the first part of the waveform holds constant frequency to allow radial velocity determination. The second part allows the range to be determined by comparing the transmitted and received signal frequencies (with the Doppler shift removed). As the DRFM records the CW signal, any frequency modula-

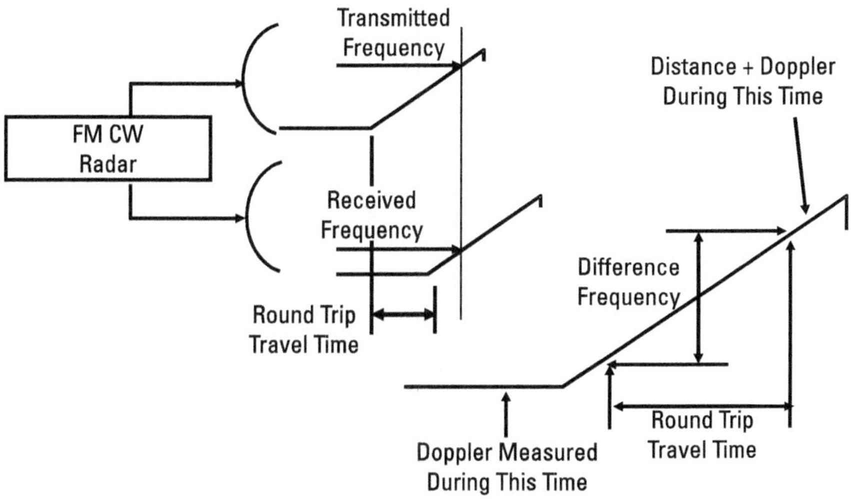

Figure 8.9 A frequency modulation on a CW radar allows determination of the range to the target by comparing the frequency of the transmitted and received signals.

tion is also recorded and subsequently replayed. By mixing in additional frequency modulation, the DRFM can simulate any desired target velocity (i.e., Doppler shift).

8.6 Analysis of Threat Signals

One of the important advantages that DRFMs (and their associated processors) provide to electronic warfare (EW) operations is the ability to very quickly analyze intercepted threat signals. One issue is the threat radar frequency. Measurement and replication of transmission frequency are important because of the issue of frequency diversity in modern threat radars.

8.6.1 Frequency Diversity

One of the electronic protection (EP) measures that a radar can use is frequency diversity. A radar can have an operator selectable frequency or with more complexity can change frequency periodically. In both of these cases, the DRFM can analyze the first pulse it sees and coherently transmit subsequent pulses at the same frequency. This requires a DRFM system throughput latency short enough to receive, analyze, set jamming parameters, and rebroadcast during the interpulse period (from many microseconds to a millisecond or so. This is well within the state of the art of wideband and narrowband DRFMs.

8.6.2 Pulse-to-Pulse Frequency Hopping

A more challenging situation is a radar that has pulse-to-pulse frequency hopping as shown in Figure 8.10. This radar will have an array of transmit frequencies that are pseudo-randomly selected. The total frequency range can be up to about 10% of the nominal transmit frequency. This is to avoid the loss of

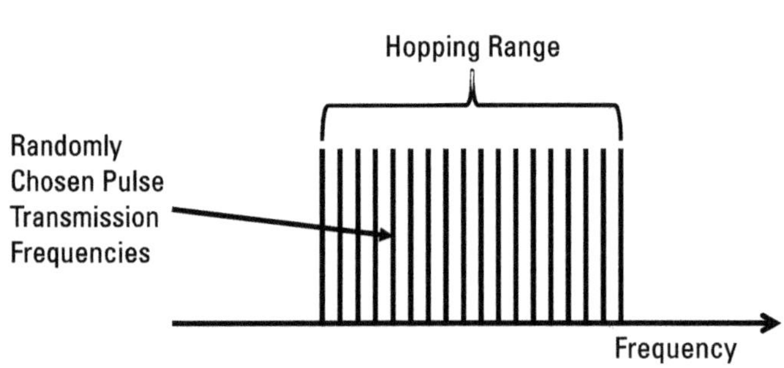

Figure 8.10 A radar with pulse to pulse frequency hopping pseudo-randomly selects one of several frequencies on which to transmit each pulse.

antenna and transmitter efficiency that occurs when operating over wider frequency ranges.

The frequency hopping radar not only choses a random frequency in its hopping range for each pulse, but can have a least jammed feature in which the radar skips frequencies at which jamming causes reduced quality of skin return signals. Every pulse is broadcast, but as shown in Figure 8.11, the jammed frequencies are not chosen.

8.7 Noncoherent Jamming Approaches

Pulse-to-pulse frequency hopping gives a noncoherent jammer two options to jam all of the pulses: either divide the jamming power among the observed frequencies as in Figure 8.12 or spread jamming across the whole hopping range as in Figure 8.13. If there are, let us say, 25 frequencies used by a radar hopping

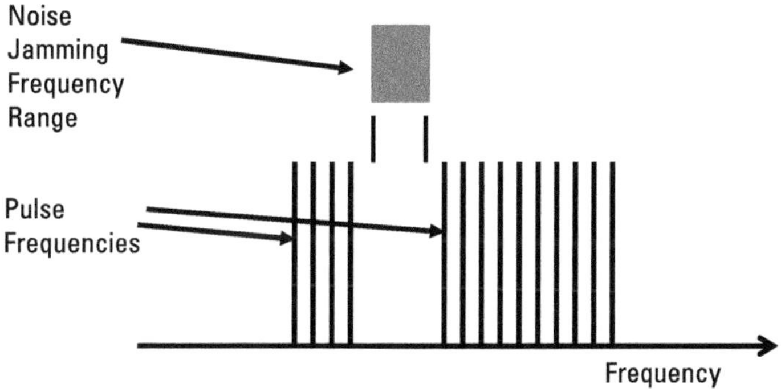

Figure 8.11 A least jammed frequency capability allows a frequency hopping radar to skip frequencies at which jamming is present.

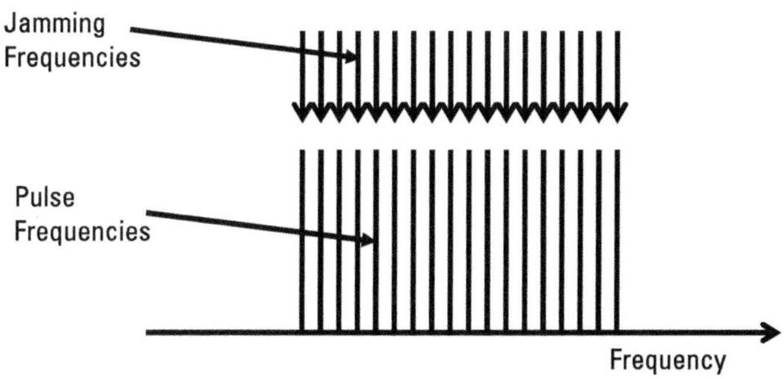

Figure 8.12 If a jammer can be configured to transmit a bandwidth matched jamming signal on each frequency at which a pulse can be transmitted, the jamming power at each frequency is reduced by the number of frequencies.

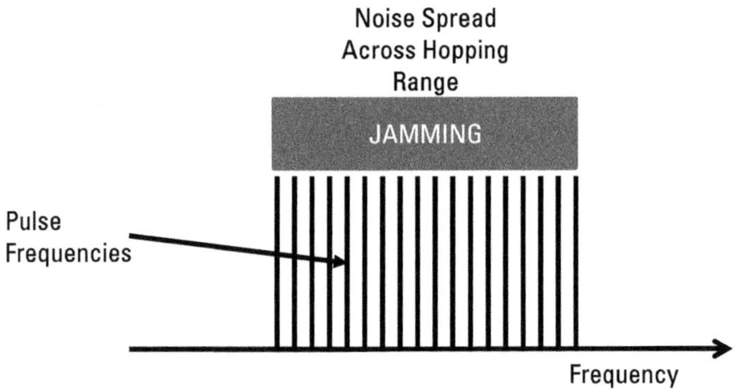

Figure 8.13 If a jammer spreads its power over the full hopping range, the jamming power at each hop frequency is reduced by the ratio of the jamming range to the radar receiver coherent bandwidth.

over a band at 4 GHz, they would most likely be spread over 400 MHz. This is 10% of the radar's RF frequency. (Note that an operating frequency range of less than 10% allows optimum radar antenna and amplifier performance.) If we can split our jamming into 25 transmissions, we can jam every pulse. However, this would reduce the effective jamming at each frequency by a factor of 25, which will reduce the J/S by 14 dB: 10 $\log_{10}(25)$ is 14 dB.

To consider the effect of spreading the jamming signal across the whole hopping range, we first need to determine the coherent bandwidth of the radar receiver. The coherent bandwidth will be the inverse of the pulse width. If the pulse width is 1 μs, the coherent bandwidth would be 1 MHz. It would be optimum to place spot jamming within the radar receiver bandwidth; however, with noncoherent jamming the jamming bandwidth would typically be a little wider, let us say, 5 MHz. This means that spreading our jamming over the full hopping range (i.e., 400 MHz) reduces the jamming power at each hop frequency by a factor of 80. This reduces the J/S by 19 dB:10 $\log_{10}(80)$ is 19 dB.

We could cover only some of the frequencies at a higher jamming level, but a least jammed capability defeats this strategy by not transmitting where we are jamming. Becausr every radar pulse is broadcast at some frequency, the skin return energy to the radar would remain the same, so the jamming would be completely ineffective.

8.8 Follower Jamming

However, if we could measure the frequency on each pulse and jam on that frequency, we would achieve the full J/S (i.e., 14 to 19 dB more J/S than would be provided by the two approaches we have discussed).

To enable pulse-to-pulse follower jamming, a DRFM (along with its associated processing components) must determine the transmission frequency and set the jamming to that frequency during a small part of the pulse. Let us consider a threat radar with 1-μs pulse width. If the latency in the DRFM, including both signal propagation and processing time is less than 100 ns, the jammer could jam the last 90% of the pulse as shown in Figure 8.14. This is a reduction of 11% in the energy of the jamming pulse compared to what it would be if there was no latency in the DRFM. An 11% change translates to 0.5 dB, so this follower jammer with 100 ns of process latency will lose only 0.5 dB of effective J/S. Note that this is based on noncoherent noise jamming, and the set on accuracy will be limited. Also, if the radar has leading edge tracking, it can still track its target in the time before jamming at the new frequency starts.

If the pulse is longer, so that the DRFM can process the signal longer, the frequency will be known to be much better accuracy. Assuming that the radar's hop frequencies are known, the jamming can be accurately set to the hop frequency.

DRFMs, along with associated digital signal processors (DSP), make it possible to create jamming signals with some subtle radar waveform features. The J/S created by jamming signals without these subtle features can be significantly reduced. The first radar feature we will discuss is pulse compression (PC).

8.9 Radar Resolution Cell

The resolution cell is the physical volume in which a radar cannot determine that multiple targets are present. This is shown in Figure 8.15. The cross-range dimension of this cell is the distance over which the radar cannot distin-

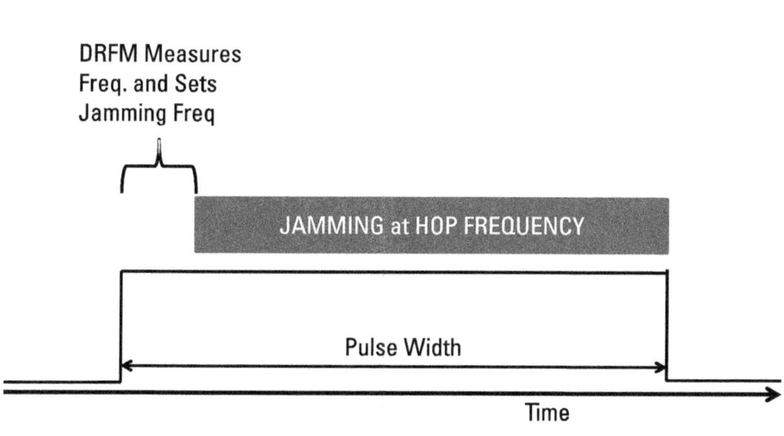

Figure 8.14 A DRFM can measure the frequency of each pulse and set the jamming to that frequency with process latency much less than the threat pulse width. The rest of the pulse is jammed, reducing the skin return energy available to the radar.

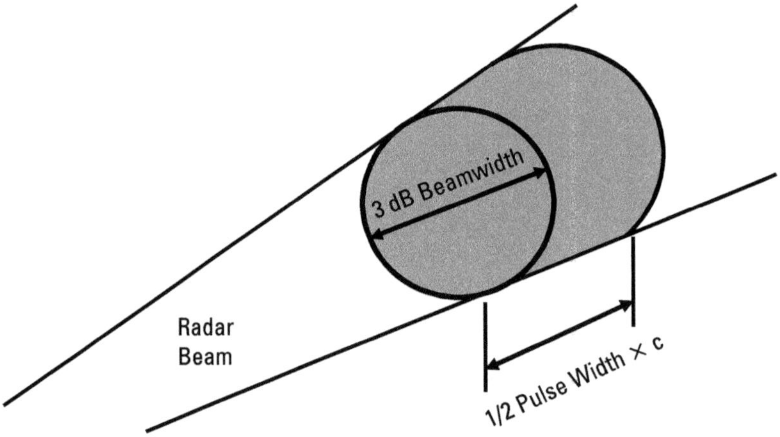

Figure 8.15 A radar resolution cell is the volume in which a radar cannot determine the presence of multiple targets. It is a segment of the 3 dB beam as long as half the pulse duration x the speed of light.

guish multiple targets which are separated in angle. It is determined from the expression:

$$\text{Range} \times 2\sin\left(BW/2\right)$$

where range is the distance from the radar to the target and BW is the 3-dB beamwidth of the radar's antenna.

For example, if the range is 10 km and the beamwidth is 5°, the cross-range dimension of the resolution cell is:

$$(10{,}000\text{m})(2)(0.0436) = 873\text{m}$$

The depth of this cell is the distance increment over which the radar cannot distinguish multiple targets that are separated in range. The depth is determined from the expression:

$$(\text{PD}/2) \times c$$

where PD is the pulse duration and c is the speed of light.

For example, if the pulse duration is 1 μs, the depth of the resolution cell is:

$$\left(10^{6}\,\text{sec}\right)(0.5)\left(3\times10^{8}\text{ m/s}\right) = 150\text{m}$$

Multiple targets within the resolution cell could include any of the following:

- Multiple valid targets;
- A valid target and a decoy;
- A valid target and a false target generated by a jammer.

Any of these situations make it difficult or impossible for a radar to track (and thus attack or hand off) valid targets. This is particularly problematic in long-range acquisition radars which typically have long pulse duration to increase the energy in each pulse. (Note that the effective range of a radar is a function of its effective radiated power and the time its signal illuminates its target.)

8.9.1 Pulse Compression Radar

As discussed above, pulse compression involves the addition of modulation to radar pulses. This modulation is processed in the radar receiver to reduce the depth of the radar's resolution cell. This modulation can be either linear frequency modulation on pulse (LFMOP) called chirp or binary phase modulation on pulse (BPMOP) called Barker code. In either case, the depth of the resolution cell can be reduced by a small or large amount depending on the specific modulation placed on the pulse. The compression ratio achieved by either technique can be up to the order of 1,000.

8.9.2 Chirp Modulation

As shown in Figure 8.16, chirp modulation is a frequency modulation through the duration of the pulse. Note that the chirp waveform can also be nonlinear if it is monotonic. The amount of compression achieved is determined from the expression:

FM Width/Coherent Radar Bandwidth

where the FM width is the range over which the frequency is swept during the pulse and the coherent radar bandwidth is 1/Pulse duration.

For example, if the width of the frequency modulation is 5 MHz and the pulse duration is 10 μs, the compression ratio would be:

5 MHz/100 kHz or 50

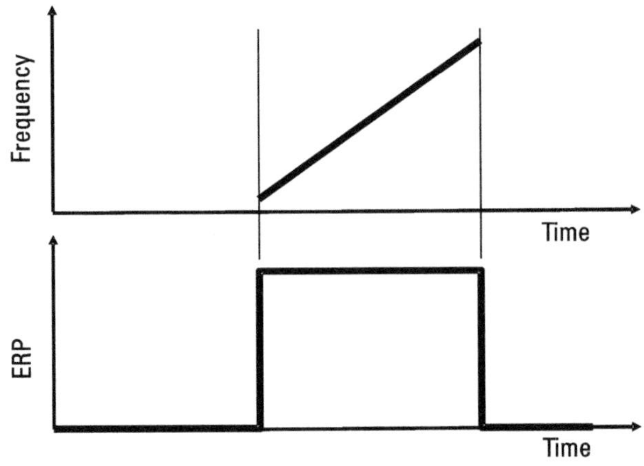

Figure 8.16 A chirped pulse has a linear (or monotonic) frequency modulation across its pulse duration.

The resolution cell with pulse compression is now modified as shown in Figure 8.17. Note that this figure shows the range compression in two dimensions for clarity, but the reduced resolution cell is actually a volume as shown in Figure 8.15.

The impact on jamming is as shown in Figure 8.18. The FM modulated skin return pulse is compressed while the jamming pulse (which does not have the FM modulation) is not. The radar processes both signals over the duration of the compressed pulse. The energy of the jammer signal is reduced (during this processing time) by the compression factor. Thus, the effective J/S is reduced by the amount of the compression. In the above example, the J/S reduction is 50 or 17 dB.

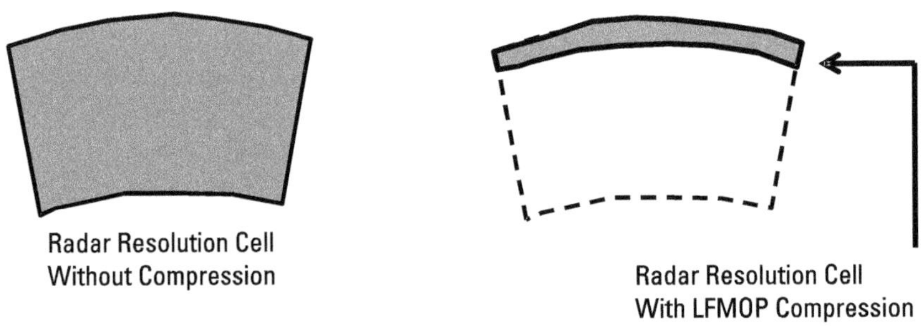

Figure 8.17 With chirp pulse compression, the range dimension of the resolution cell is reduced by the compression ratio.

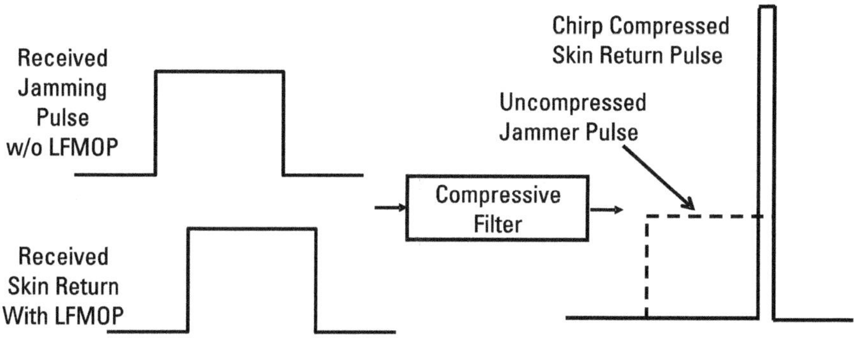

Figure 8.18 During processing in the radar receiver, the skin return pulse is compressed by the compressive receiver as shown. The jamming pulse is not compressed because it does not have the LFMOP modulation.

8.9.3 Role of DRFM

Figure 8.19 is a flowchart of the process for matching the pulse compression characteristics of the jamming pulse to the skin return:

- The received radar signal is converted to the operating frequency of the DRFM.
- The DRFM digitizes the first threat pulse received.
- This digitized pulse is passed to the DSP where the frequency history of the pulse is determined.

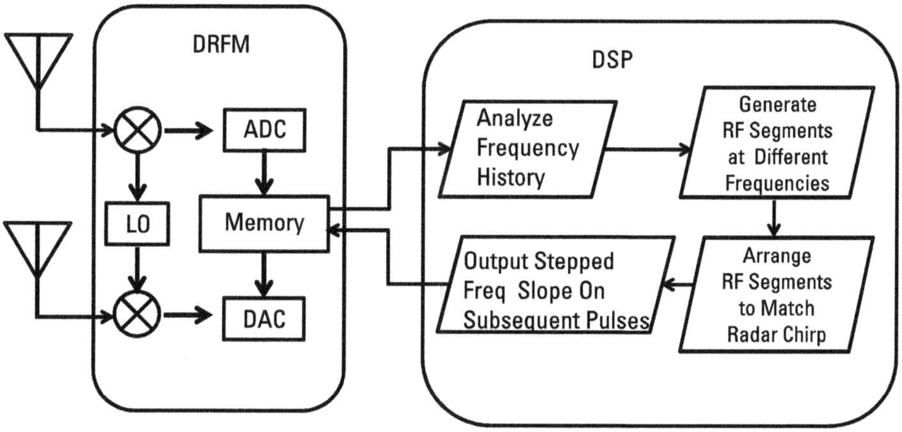

Figure 8.19 The DRFM converts received signals to the DRFM operating frequency, digitizes it and passes it to the DSP. The DSP determines the frequency history of the first received pulse and generates a stair step frequency slope for subsequent pulses. The DRFM generates subsequent jamming pulses with this stair stepped frequency slope.

- A set of signal segments with a progression of different RF frequencies is passed back to the DRFM for subsequent pulses.
- The DRFM produces jamming pulses with a stair step approximation of the radar chirp.
- The DRFM output is coherently returned to the frequency at which the radar pulse was received and is broadcast as properly chirped jamming pulses.

Note that if the radar pulse has a linear frequency modulation, this process can be performed without a DRFM. An instantaneous frequency measurement (IFM) receiver can determine the frequency modulation and a serodyne circuit can generate a jamming signal with a matching frequency modulation. However, a DRFM will provide more accuracy and will also produce jamming signals with nonlinear frequency modulation if required.

8.9.4 Barker Code Modulation

The other technique for pulse compression is the addition of a binary phase shift keyed (BPSK) digital modulation to each pulse as previously discussed. There are a fixed number of bits in this code during each pulse, and when the pulse is received by the radar as a skin return, it is passed to a tapped delay line assembly as shown in Figure 8.21.

The code, which can be a Barker code or one of several other codes, is a maximal length code. This means that it is pseudo-random, and if the number of zeros is subtracted from the number of ones, the sum will be zero or minus one. The code on the pulse at the top of Figure 8.21 is a 7-bit Barker code, in which "+" indicates a one and "–" indicates a zero. Note that this is a short

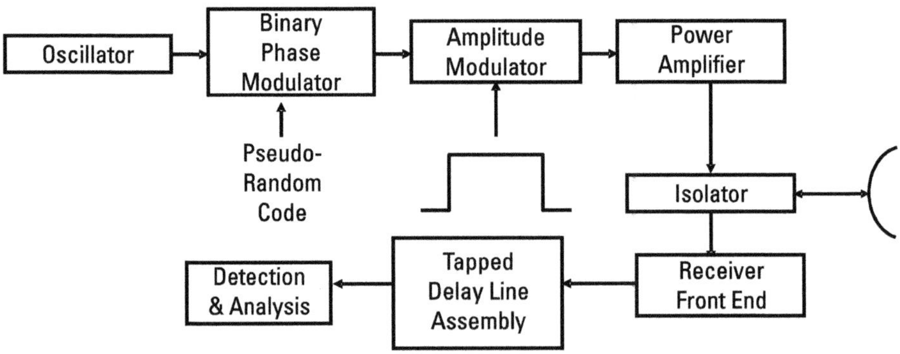

Figure 8.20 A radar with Barker code places a BPSK modulation on each transmitted pulse and compresses the received skin return pulses by passing them through a tapped delay line.

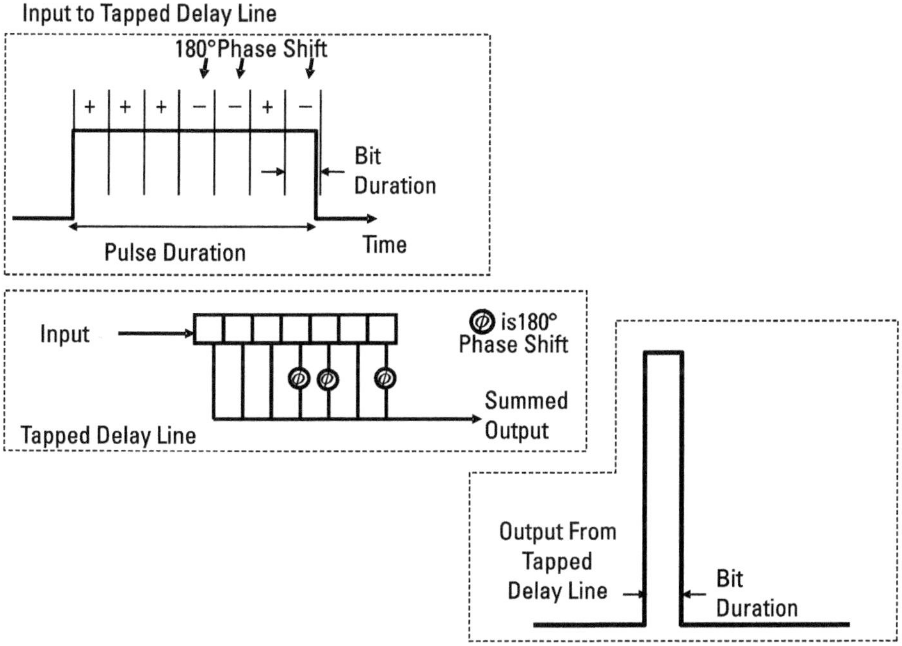

Figure 8.21 The coded pulse produces a large output from the delay line only when all of its bits align to the taps. This reduces the post processing pulse width to the duration of one bit of the code.

code, but the typical codes used in pulse compression radars are much longer (up to the order of 1,000 bits). There are 180° phase shifts on some of the taps. They are designed so that when the pulse exactly fills the delay line, each of the zero bits is at a tap with a phase shifter. Thus, when the pulse fills the delay line and the taps are summed, the pulse has is full amplitude. At any other time, the summed output is significantly less. With a 7-bit code as shown, the summed output will be either 0 or −1 when the pulse is not aligned with the delay line. For longer codes, there will be some time periods during which the sum is larger, but still significantly below the full pulse amplitude. When the pulse leaves the delay line and summing process, the pulse duration is effectively 1 bit wide (i.e., the effective postprocessing pulse duration is the time that the pulse exactly fills the tapped delay line).

Figure 8.22 shows the effect of this pulse width reduction on the radar's resolution cell. The cross-range dimension of the resolution cell is still the 3-dB beamwidth of the radar's antenna, but the depth of the cell is now on half of a code bit duration multiplied by the speed of light. Thus, the range resolution is improved by the factor equal to the number of bits transmitted during each pulse.

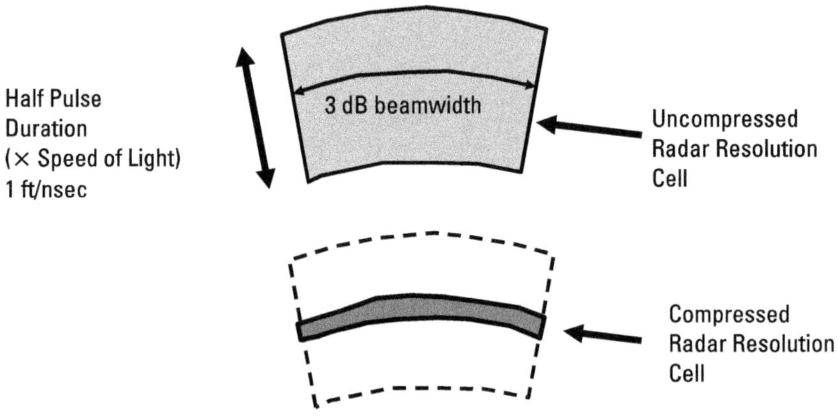

Figure 8.22 With Barker code compression, the resolution cell depth is reduced to one half of the period of a code bit multiplied by the speed of light.

8.9.5 Jamming Barker Coded Radars

Now consider a noncoherent jammer operating against a radar that has Barker coded pulses (as shown in Figure 8.23). The skin return pulse has coding which is matched to the tapped delay line configuration. This means that postprocessing pulse width will be effectively reduced to the bit duration. For example, if there are 13 bits in the Barker code, the pulse will be reduced by a factor of 13. However, a jamming signal that does not have the Barker code modulation will not be shortened. Because the radar processing is optimized for the much shorter compressed skin return pulses, it will process jamming pulses only during one-thirteenth of the time. This reduces the effective jamming power by 11

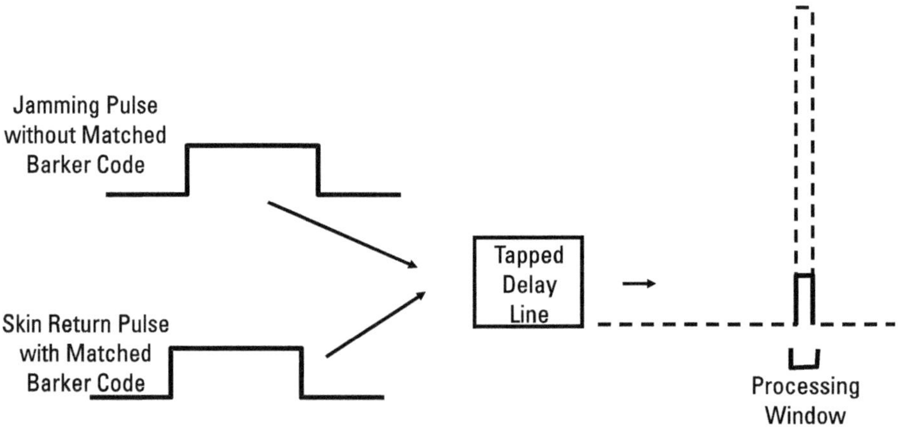

Figure 8.23 Unless jamming has the correct BPSK modulation, the effective J/S is reduced by the compression factor.

dB relative to the skin return power at this point in the processing, so the J/S is reduced by 11 dB. (Note that the ratio 13 converts to 11 dB.) If there were 1,000 bits in the code, the J/S reduction would be 30 dB.

The answer to this problem is to add Barker code to the jamming pulses. The only practical way to accomplish this is by use of a DRFM in the jammer.

As shown in Figure 8.24, pulses from the radar are input to a DRFM, which digitizes the first pulse received and passes it to a processor. The processor determines the code bit duration and the sequence of ones and zeros in the code. The processor generates a digital representation of a one bit and a zero bit. It outputs these code blocks in the correct sequence to form a digital representation of the Barker coded radar pulse back to the DRFM. This output can be delayed or frequency shifted as required to perform the desired jamming functions.

The DRFM produces RF jamming pulses and coherently converts them to the operating frequency of the jammed radar. The jamming pulses are modified from the received radar pulses in amplitude, Doppler shift frequency and timing to produce the desired jamming techniques.

8.9.6 Impact on Jamming Effectiveness

When the BPSK coded jamming pulses are received by the jammed radar, the radar's processing circuitry compresses them just as it does the skin return pulses. This means that the J/S is not reduced by the compression factor, which can improve the jamming effectiveness (relative to noncoherent jamming) by many decibels.

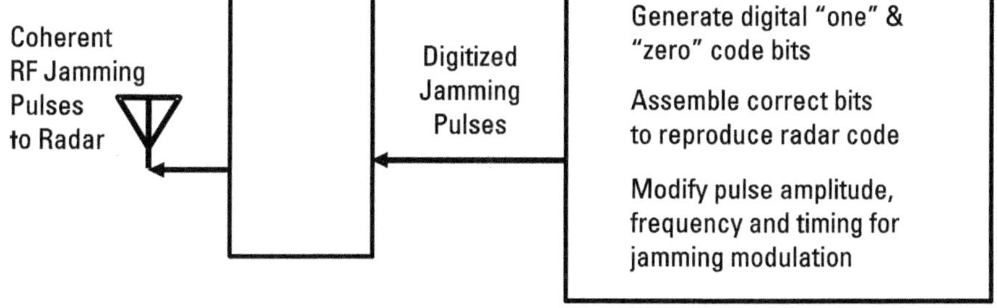

Figure 8.24 A jammer with a DRFM can produce jamming pulses which have the same Barker code as the skin return pulses and thus produce the full J/S ratio in the jammed radar.

Another benefit of jamming with a DRFM is that the constructed jamming pulses have exactly the correct pulse duration. The radar receiver's processing circuits are optimized for pulses with a specific pulse duration, so the jamming pulses benefit from the same processing features as skin return pulses.

8.10 Complex False Targets

Modern radars, particularly synthetic aperture radars (SAR) and radars with active electronically steered arrays (AESA), can characterize a complex target with a radar cross section (RCS) that includes many scattering points caused by the shapes of various parts of the target. Each of these scattering points generates a return with its own phase, amplitude, Doppler shift and polarization characteristics. These multiple returns combine to form a complex skin return that a modern radar can analyze to support accurate target identification. A simple false target from a noncoherent jammer will be received by the jammed radar with a waveform that is significantly different from that of the true skin return.

This allows a radar with the latest processing capabilities to reject false targets with incorrect RCS characteristics. Thus, effective jamming of modern radars requires that false targets produced in the application of such techniques as range gate pull-off, range gate pull-in, and others have correct, complex waveforms.

8.10.1 The Radar Cross Section

Figure 8.25 shows a few examples of the points on an aircraft that contribute to its composite RCS. In addition, there are contributions from the engine inlet

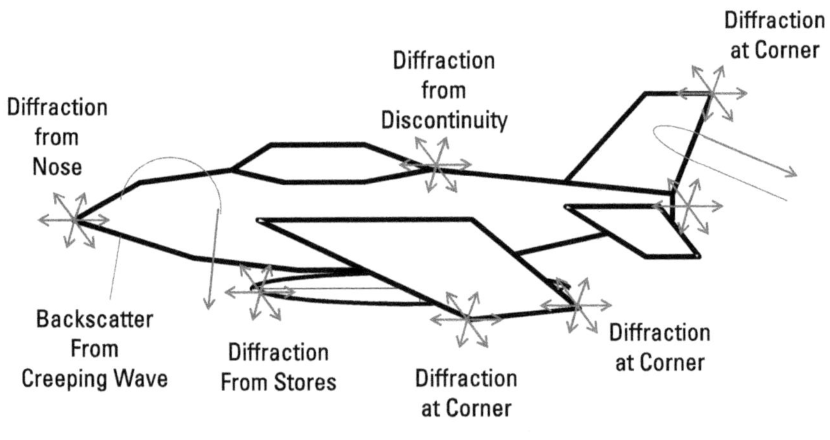

Figure 8.25 There are many contributing factors to the RCS of an aircraft. Together, they cause an RCS with complex amplitude and phase components.

and output openings and (in some aircraft) from the moving internal parts of the engines. The combination of all of these factors creates a very complex RCS which changes with the aspect angle as the target maneuvers.

There are also target characteristics such as jet engine modulation (JEM) and rotor blade modulation (RBM). JEM causes a complex compression pattern ahead of the aircraft that causes a strong spectral component in the radar return.

Radar reflections from helicopter targets have spectral characteristics related to the number of blades and their rotation rate.

The RCS has time-varying characteristics as the target maneuvers. Modern radars can analyze this time-varying characteristic to detect and reject false targets.

8.10.2 Generating RCS Data

The detailed RCS of a target can be determined either by measurement in an RCS chamber or by computer analysis. As shown in Figure 8.26, the RCS chamber is an anechoic chamber in which a low-power radar illuminates either an actual object or a scale model of that object. The surfaces of the chamber are covered with radio absorptive material so that there are no reflections. Over most of the chamber surfaces, the radio absorptive material is formed into pyramids with a steep internal angle between adjacent spikes so that signals reflected

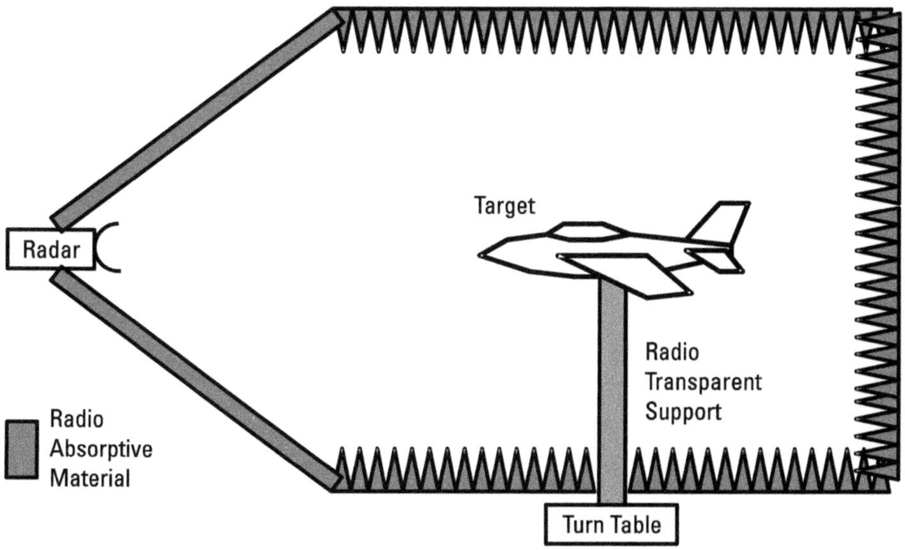

Figure 8.26 A radar cross section chamber is an anechoic chamber with a low power radar aimed at a model mounted in the middle of the chamber. As the model is rotated, the radar measures the skin return signals from which the radar cross section can be determined.

from the model are directed into the material. This allows the radar to get a clean radar skin return just as though the model were in a free space environment. If the target is small, the actual object can be used in the chamber. If the target is too large to fit into the available chamber (for example a large aircraft), a scale model is used. The operating frequency of the radar is increased by the same scale factor as the size reduction of the model. For example, a one-fifth scale model requires testing at five times the frequency. This makes the ratio between the target dimensions and wavelength of the radar signal reflections correct. Because the RCS data being measured is fine scale, the important surface features of the model must be very accurate to produce the correct RCS results.

The target is placed in the center of the chamber and rotated to generate RCS data from all important aspect angles. This data is then analyzed and characterized to develop target ID tables for radars.

8.10.3 Computed RCS Data

The other way to generate an RCS table is by computer analysis. The target (for example, an aircraft) is characterized by a set of flat or curved surfaces. Figure 8.27 shows a few of the various characteristic shapes that make up an aircraft. The actual model used for computer analysis will be much more complex.

There is a formula for the RCS of each type of surface, so a composite model of the aircraft can be computer generated. The formula for each surface component will characterize the RCS of that component in amplitude and phase as a function of its size and orientation relative to the radar. The material

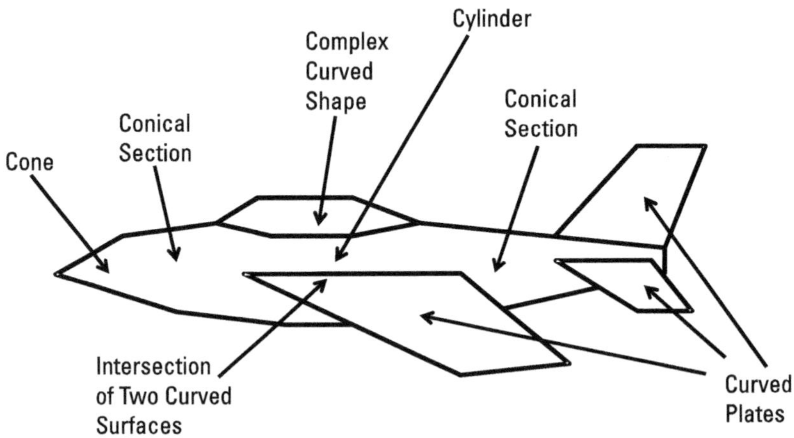

Figure 8.27 An aircraft can be characterized as a very large number of individual shapes. A formula can be developed for the RCS of each shape component as a function of aspect angle, material, surface, and frequency. The radar cross section of the aircraft is calculated from the combination of these formulas.

in each component (type of metal, glass, plastic, and so forth) and the nature of its surface will also be considered. The computer model of the target will combine the formulas for all of these surfaces in their positions relative to the other components.

Note that this section includes information from [1], which is recommended reading for more detail on the subject.

8.11 DRFM-Enabling Technology

The primary limitation on DRFM performance has always been the analog-to-digital converter (ADC). The bandwidth over which the DRFM can operate is limited by the digitizing speed and the accuracy with which signals can be reproduced is a function of the number of bits. The number of bits per sample also determines the level of spurious responses that will be present in output signals.

At the time of this writing, the state of the art was well over 2 GHz of sample rate with 12 bits of quantization. Note that so much development effort is underway that these are moving targets. ADC performance is on a significantly positive slope.

Another significant supporting technology is the field programmable gate array (FPGA). These have allowed significantly more processing to be performed on a single DRFM board. As a result, the programmability and speed of necessary DRFM functions have increased significantly.

8.11.1 Capturing Complex Targets

The range and aspect angles of radar targets are constantly changing during an engagement. This, along with the fact that there are multiple scattering points on the target, means that a modern radar receives a constantly changing and very complex skin return. Modern radars can determine that a false skin return generated by a jammer is different from the skin return it has been receiving. Thus, to perform successful deceptive jamming against such a radar, the jammer must be able to make its false returns reasonably close to valid skin returns.

As discussed above, accurate (complex) RCS data can be acquired either by measurements in an RCS chamber or by computer simulation. Such data can also be measured in an operational environment, but like all such open-air data gathering, it is challenging to separate the desired data from the environmental conditions.

As shown in Figure 8.28, the collected data is processed in special software to determine the dominant scattering points. The return from each of these

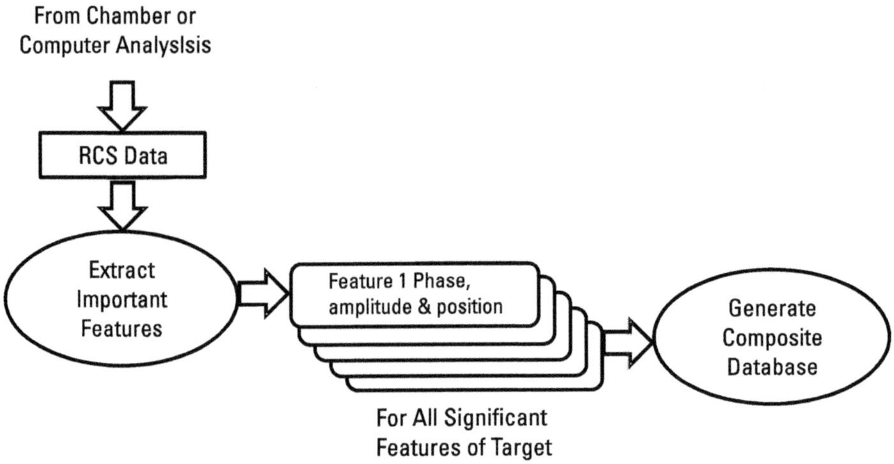

Figure 8.28 A computer model of a target is analyzed to extract the important features. The phase, amplitude, position, and Doppler shift of each is incorporated into a composite data base.

scattering points is characterized in terms of its phase, amplitude, Doppler shift, and position as a function of the aspect angle. This data is stored in a database from which DRFM channels can be driven to generate an accurate, dynamic target return.

8.11.2 DRFM Configuration

Figure 8.29 is a block diagram of an older DRFM system to generate complex false targets. There are multiple DRFM cards, each of which can generate one or two returns. Each DRFM digitizes the signal input from a receiver and modifies it to represent the return from on the target's scatterers. The output has the proper amplitude, phase, and Doppler shift for its assigned scattering point (see Figure 8.30). It also has a time delay appropriate to the distance from the radar with the current target aspect angle. The RF outputs of the DRFMs are combined and coherently retransmitted to the target radar.

With the introduction of FPGAs, a single DRFM board can generate returns for 12 scattering points. Each scattering point signal has a unique modulation that has the Doppler shift and range delay appropriate to its relative location in the current target orientation with the current velocity and three-dimensional angular velocities.

Each of these scattering point channels also applies the modulation required to perform the applied deceptive jamming technique, which must be different for each point to fool the jammed radar.

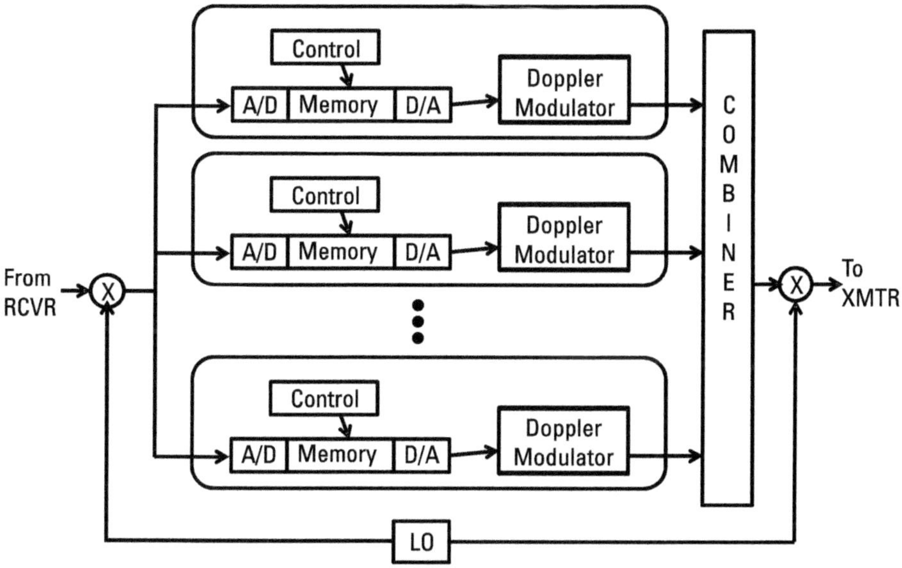

Figure 8.29 Older systems can generate complex targets with multiple DRFMs, each of which can replicate one or two scatterers.

8.12 Jamming and Radar Testing

This discussion has been presented in terms of deceptive jamming, but it can be equally important for the testing of modern radars.

To test a radar that has processing capable of detecting complex radar returns, it is necessary to have accurate dynamic scenarios depicting various targets through many typical engagements. These testing scenarios must include realistic multipoint scattering returns with the proper amplitude, phase, and position characteristics to test all of the radar's hardware and software features.

8.13 DRFM Latency Issues

In Sections 8.9.2 and 8.9.5, we discussed the reproduction of chirped and Barker coded pulses. In both cases, the DRFM and its associated DSP captured and analyzed the first pulse received and copied the characteristics of that pulse when rebroadcasting subsequent pulses. This assumes that all of the pulses in a received radar transmission will be identical. The rebroadcast pulses are coherent with the received pulses and there are other modulation elements applied to support the jamming technique being employed. For example, each of the subsequent pulses may be delayed and/or frequency shifted.

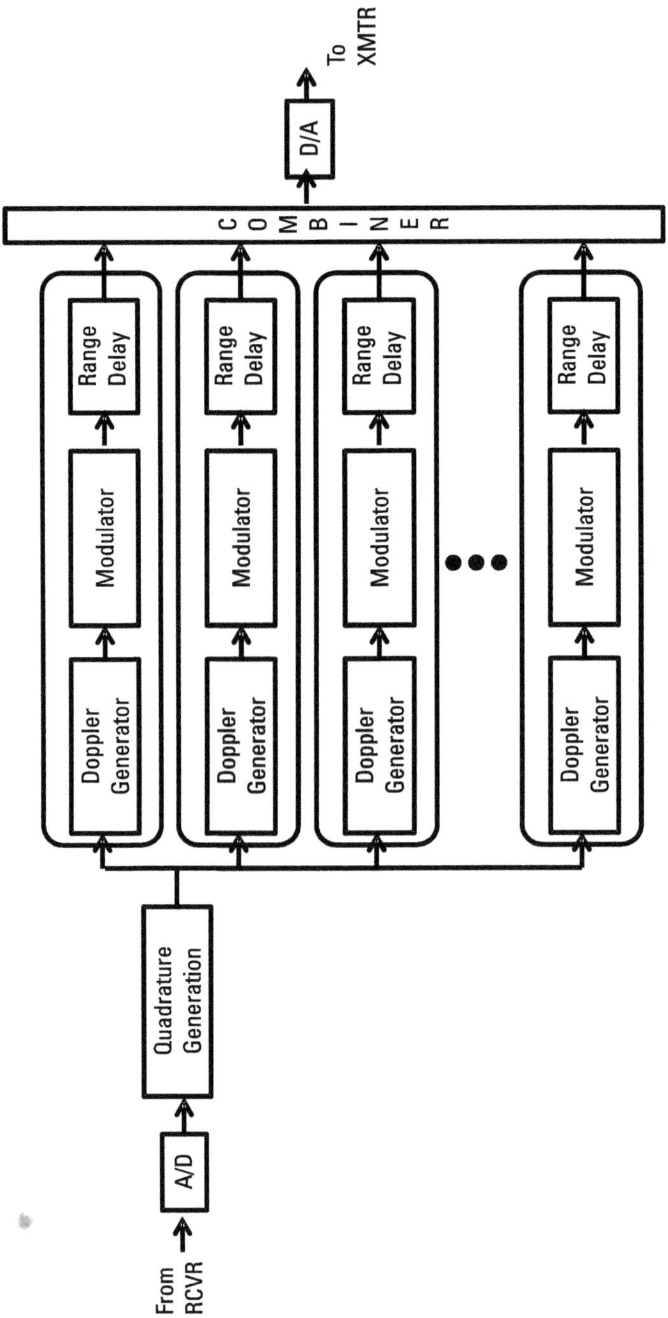

Figure 8.30 A single DRFM unit with FPGA technology can emulate 12 scatterers, with control and Doppler shift functions on the board.

8.13.1 Identical Pulses

When all of the pulses in a received radar transmission are identical, the DRFM and its associated processor analyze the first pulse received and generate jamming pulses with the proper modulation to jam each subsequent pulse in the transmission. The process latency required is short enough to complete the necessary processing during the interpulse period. This is a few tens of microseconds to a few milliseconds.

8.13.2 For Identical Chirped Pulses

As shown in Figure 8.31, the analysis of the first pulse must be accomplished during the interpulse period. Consider a tracking radar with a pulse width of 10 μs and a 10% duty cycle. Remember that the pulse interval is the time between the leading edges of subsequent pulses. This means that the interpulse time available for the DRFM to make calculations is 90 μs. The process throughput latency time, during which the DRFM processor accepts the digitized pulse data from the DRFM, determines the pulse modulation parameters, generates the desired jamming pulse, and returns the modified signal (in digital form) to the DRFM, must thus be less than 90 μs.

For a chirped pulse, the first received pulse must be digitized in the DRFM and passed to the processor. As shown in Figure 8.32, the slope of the frequency modulation must be measured. Note that the frequency modulation on the pulse can be either linear or nonlinear. Blocks of code are generated for each of many time increments over the pulse width. A digital representation of the whole pulse is then generated with the frequency during each increment determined from the measured frequency in the received pulse and the desired

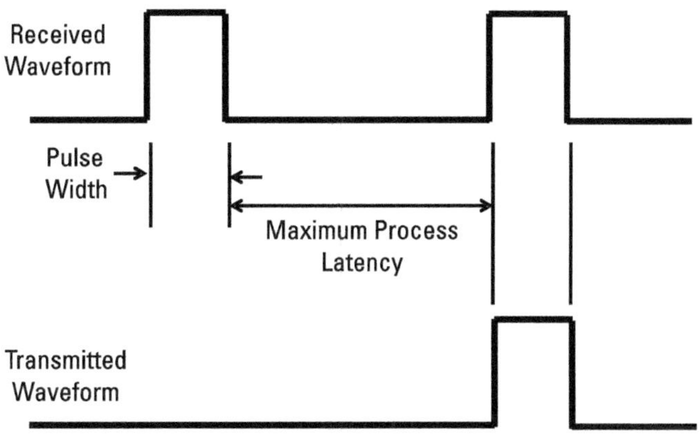

Figure 8.31 In order to copy the first received pulse in subsequent pulses, the DRFM and processor must complete the full process within the inter-pulse time.

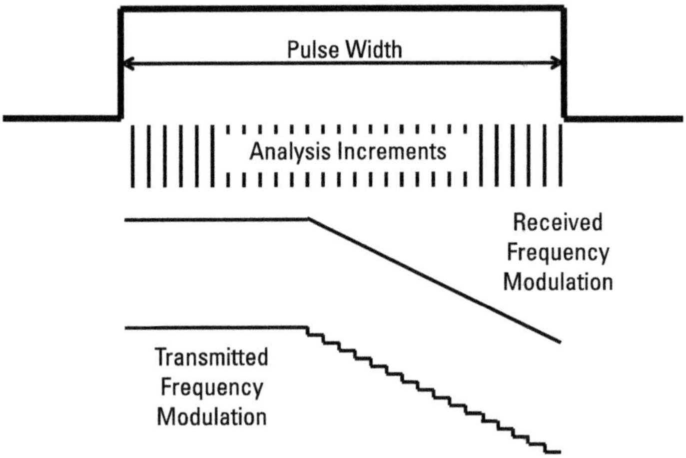

Figure 8.32 During the inter-pulse time, the received chirped signal is analyzed to determine its frequency at each analysis increment. Then a digital return signal is formulated with a stair-step approximation of the modulating frequency. The return signal is off-set in frequency and time to support the chosen jamming technique.

Doppler shift offset. The digital signal returned to the DRFM for rebroadcast will have a stair stepped representation of the frequency modulation of the received pulse and will be offset in frequency and time by the amount dictated by the jamming technique employed.

8.13.3 For Identical Barker Coded Pulses

When there is a binary phase shift keyed (BPSK) signal on a received radar signal, the DRFM digitizes the first pulse received. Then the processor determines:

- The clock rate of the code (a Barker code or some longer maximal length code);
- The sequence of ones and zeros in the code;
- The received frequency;
- The time of arrival of the pulse.

Then the processor develops digital signals for a one bit and for a zero bit. Finally, as shown in Figure 8.33, the processor outputs a digital representation of the BPSK modulated pulse for each subsequent pulse in the received signal. The generated signal has the correct frequency from the received frequency and the Doppler shift appropriate for the jamming technique chosen. The signal is delayed by an amount that will place the pulse at the proper time for

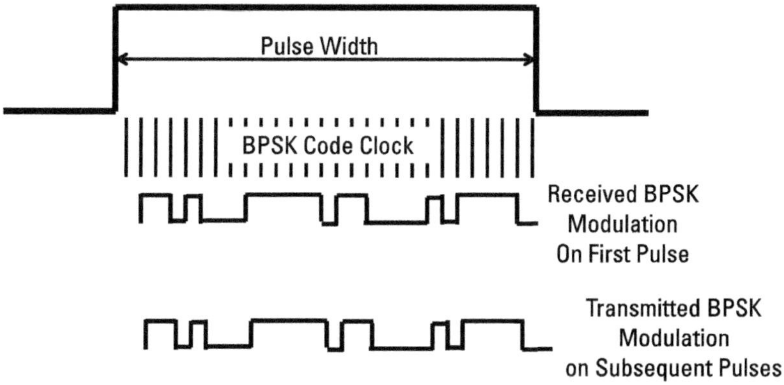

Figure 8.33 After receiving the first BPSK modulated pulse, the processor determines the code clock and the sequence of ones and zeros in the code. Then, it creates a digital model for a one and for a zero. Finally, it outputs a digital signal with the correct code for each subsequent pulse and outputs this signal to the DRFM for retransmission with the time and frequency shift appropriate to the jamming technique employed.

each subsequent pulse considering both the pulse repetition interval (PRI) of the received signal and the time offsets required for the chosen jamming technique. The DRFM then coherently rebroadcasts each pulse after the first pulse received.

8.13.4 For Unique Pulses

Now consider the more challenging requirement to reproduce radar signals that change on a pulse-to-pulse basis. The primary example is a pulse-to-pulse frequency-hopping radar. There will be multiple frequencies that are pseudo-randomly selected by the radar. Also, it is reasonable to assume that the radar can sense when it is being jammed and will have a least jammed frequency mode. Frequencies at which jamming or other interference is detected will be skipped in the hopping sequence. This means that a jammer without the ability to measure the frequency of each pulse must cover the whole frequency-hopping range and cannot maximize its J/S by concentrating its power in part of the covered range (a technique called partial-band jamming).

If the jamming frequency bandwidth is spread, the achieved J/S is reduced by the ratio of the radar's receiver bandwidth to the pulse hopping range. Take, for example, a radar with a 3-MHz receiver bandwidth operating at 6 GHz. The radar's hopping frequency range would typically be 10% of its operating frequency (i.e., 600 MHz). The ratio of the hopping range to the receiving bandwidth is thus:

$$600 \text{ MHz}/3 \text{ MHz} = 200$$

This reduces the effective RCS by 23 dB.

Now consider a jammer with a DRFM that can measure the frequency of each received pulse. By knowing the frequency of each received pulse, it can jam that pulse at the correct frequency, avoiding this loss of effective J/S.

Because the frequency of each pulse is not known until it is received at the jammer, a DRFM and its associated processor must:

- Determine the radar transmitting frequency.
- Generate a digital representation of the pulse with the correct frequency and timing (including any frequency and time offset for the chosen jamming technique).
- Begin coherently rebroadcasting at that frequency.

All during a small part of the radar's pulse width, as shown in Figure 8.34.

The energy of the jamming pulse is reduced by the ratio of the duration of the jamming pulse (i.e., the radar's PW less the processing latency time versus the original PW). For example, if the pulse width is 10 μs and the processing latency time is 100 ns, the reduction in jamming energy is:

$$9.9\mu s/10\mu s = 0.99$$

which is only 0.04 dB.

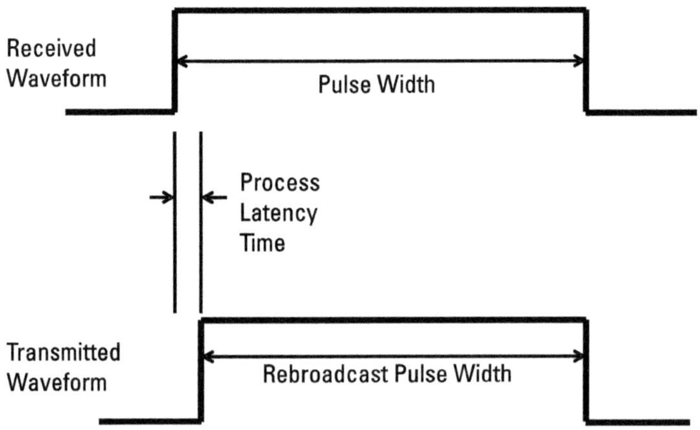

Figure 8.34 If every pulse is unique, the width of rebroadcast pulses will be reduced by the processing latency time.

8.14 A Summary of Radar Techniques That Call for DRFM-Based Countermeasures

Several radar techniques are difficult for traditional jammers to counter, including:

- Coherent radar;
- Leading edge tracking;
- Pulse-to-pulse frequency hopping;
- Pulse compression;
- Range rate/Doppler shift correlation;
- Detailed analysis of target RCS.

8.14.1 Coherent Radars

Coherent radars expect their skin returns to fall within a single frequency cell as shown in Figure 8.35. This refers to a pulse Doppler radar that has a bank of filters in its processing circuitry. Because noncoherent jammers, even in spot jamming modes, spread their power across multiple filters, the radar can detect jamming and can go into a home-on-jam mode. It will also reduce the achieved J/S by coherent pulse processing gain.

Because DRFM-equipped jammers can generate coherent jamming signals, the pulse-Doppler radar provides the same processing gain to jamming signals and cannot detect the presence of jamming. This both improves the J/S and prevents the activation of home-on-jam modes.

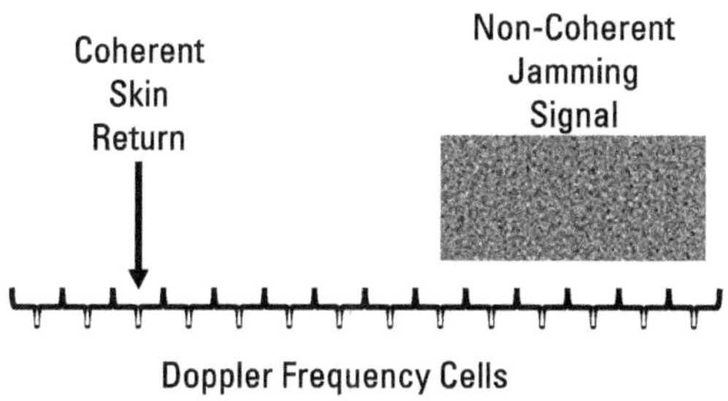

Figure 8.35 A coherent radar produces a skin return in a signal frequency cell, whereas a non-coherent jamming signal occupies several.

8.14.2 Leading-Edge Tracking

Leading-edge tracking makes range gate pull off jamming ineffective because the jamming pulses are progressively delayed from the radar's skin return. The radar tracks targets using only the leading edges of pulses. Because the leading edges of the jamming pulses are later than those of the skin return, the radar continues to track its skin return pulses, ignoring the jamming pulses as shown in Figure 8.36. Depending on the geometry, leading-edge tracking may also allow the radar to ignore the jamming pulses of terrain bounce jamming that are delayed because of their longer transmission path.

Modern DRFMs, because they have very short latency times (of the order of 50 ns), can generate jamming pulses quickly enough to capture the leading-edge tracker. This will make both range gate pull off and terrain bounce jamming effective.

8.14.3 Frequency Hopping

Frequency hopping, either from coherent processing interval (CPI) to CPI or from pulse to pulse, requires a conventional jammer to cover the whole hopping range of the radar. (The radar uses only one frequency at a time, but the jammer does not know which.) This reduces the J/S that the jammer can produce.

By measuring the radar's frequency during the first 50 ns of each pulse (as in Figure 8.37), a DRFM-equipped jammer can produce jamming signals that follow the frequency hopping and cover an extremely large percentage of the skin return pulses.

8.14.4 Pulse Compression

In addition to improving the range resolution, a radar also reduces the J/S that a jammer can produce by the same amount as the compression ratio. This assumes that the jammer pulses do not have the proper pulse compression modu-

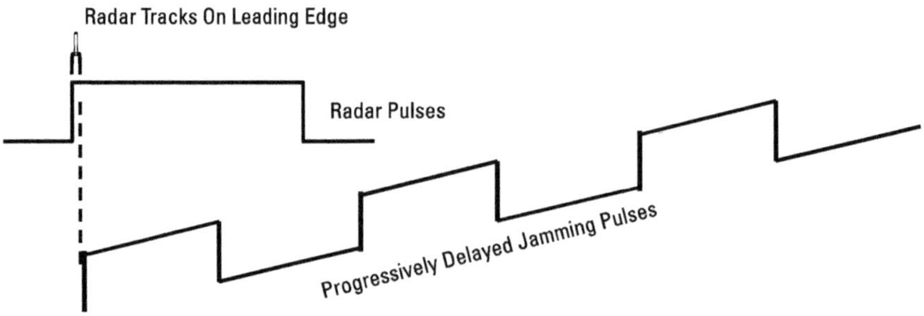

Figure 8.36 If a radar uses leading edge tracking, a DRFM equipped jammer can generate jamming pulses with leading edges matching the skin return within 50 nsec.

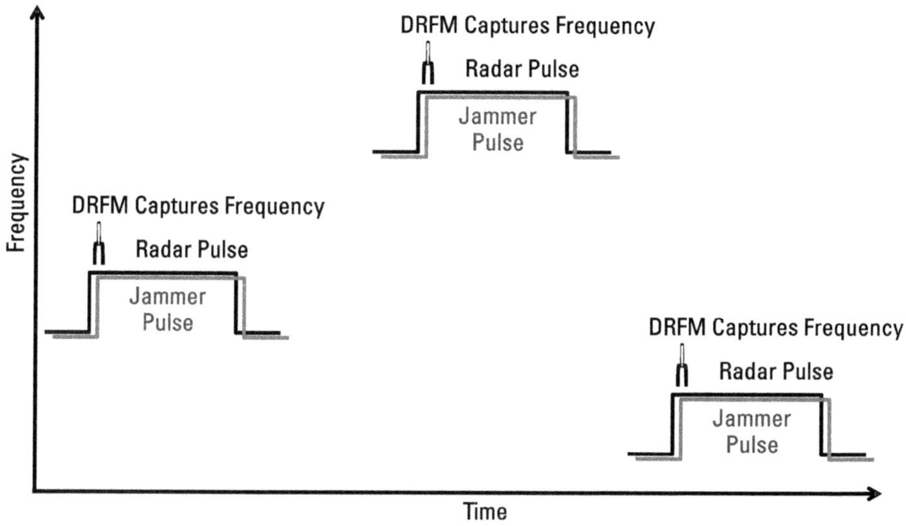

Figure 8.37 A DRFM equipped jammer captures the frequency of a frequency hopped pulse in the first 50 nsec and matches the frequency in the corresponding jamming pulse.

lation. Pulse compression can be achieved by either chirping (i.e., frequency modulating) the pulses or applying a Barker code. In either case, the J/S that a jammer can produce is reduced by the same factor as the compression ratio. This can reduce jamming effectiveness by multiple orders of magnitude.

There are other ways to produce linearly chirped jamming pulses, but a DRFM-equipped jammer can measure the frequency modulation on the radar's pulses (either linear or nonlinear modulation). It can then generate jamming pulses with frequency modulation which very closely approximates that on the skin return pulses.

A DRFM-equipped jammer can determine the bit rate and the exact digital code on the first Barker coded pulse received. It can then produce jamming pulses for all subsequent skin return pulses which have the proper Barker code as shown in Figure 8.38.

In either case, the DRFM improves the achieved J/S by multiple orders of magnitude against pulse compression radars.

8.14.5 Range Rate/Doppler Shift Correlation

Pulse Doppler radars can detect separating targets and capture the range history along with the Doppler frequency history of each of those targets. By correlating the rate of change of range with the Doppler shift, the radar can discriminate against false targets, allowing it to continue tracking targets with its true skin returns (see Figure 8.39).

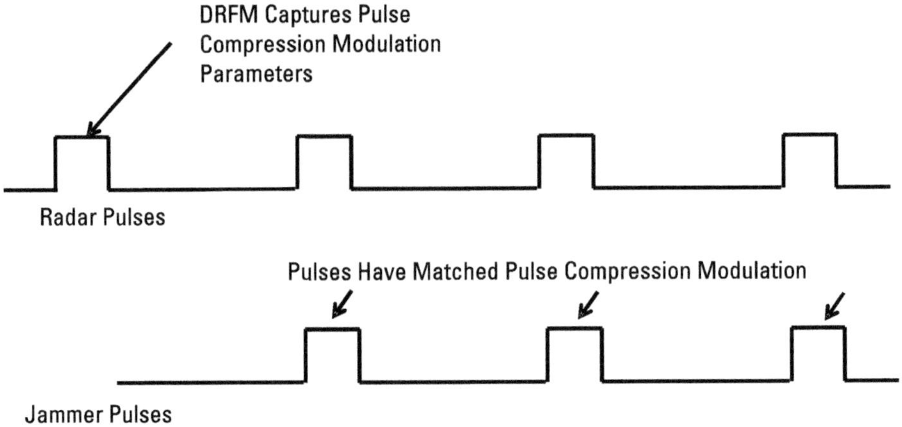

Figure 8.38 A DRFM equipped jammer captures the compression modulation on the first pulse and generates subsequent pulses with matched modulation.

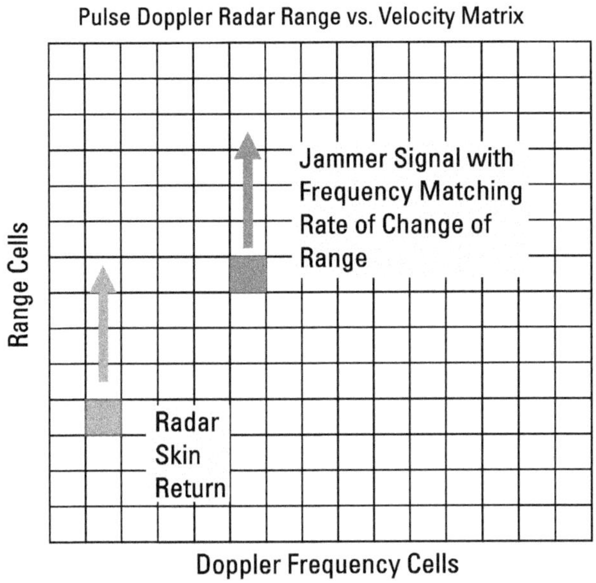

Figure 8.39 A DRFM equipped jammer can generate false target pulses at frequencies which simulate the Doppler shift matching their rate of change of range.

DRFM-equipped jammers can set both the pulse timing and the frequency of jamming pulses so that they are consistent with true skin returns, and thus will allow range gate pull off, range gate pull in, and other false target jamming techniques to be effective.

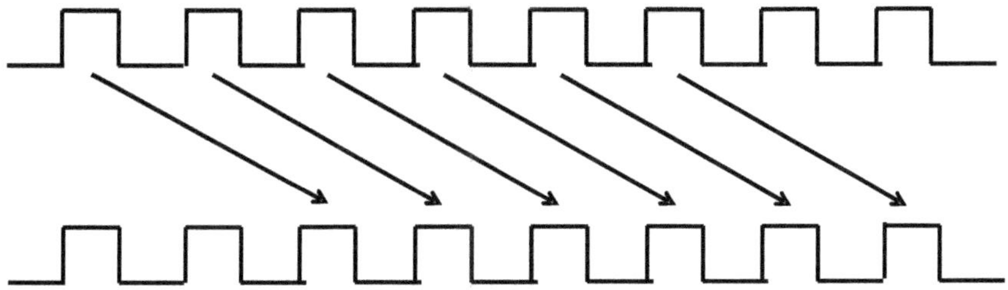

Figure 8.40 In pipe-lining mode, a DRFM takes more than one pulse interval to complete the processing required to produce a matched jamming pulse.

8.14.6 Detailed Analysis of Radar Cross Section

The detailed analysis of an RCS allows a radar to detect changes in the received return from a target when false target pulses are generated by a jammer. By noting the change, the radar can reject the newly introduced jamming signals and reacquire the true skin return.

Because the most modern DRFM equipped jammers can create very complex pulses which incorporate multisurface RCS patterns, they can generate false targets that a radar has great difficultly identifying as false.

8.14.7 High Duty-Cycle Pulse Radars

When a DRFM equipped jammer is employed against a radar with a very high duty cycle, such as a pulse Doppler radar in high pulse repetition frequency (PRF), mode, the DRFM can collect data from a second pulse before retransmitting an earlier pulse (see Figure 8.40). This pipe-lining mode allows adequate time to generate the proper jamming pulse parameters. Note that high PRF radars normally operate on a single frequency to enhance fast Fourier transform (FFT) processing of received signals. Thus, one pulse looks like any other, and pipe-lining can be successfully employed.

Reference

[1] Andrews, Oliver, and Smit, "New Modelling Techniques for Real Time RCS and Radar Target Generation," *Proceedings of the 2014 EWCI Conference,* Bangalore, India, February 17–20, 2014.

9

Infrared Threats and Countermeasures

There have been significant developments in infrared (IR) weapons, sensors, and countermeasures in the last few years. In this chapter, we will talk about some principles, techniques, and current developments.

9.1 The Electromagnetic Spectrum

The purpose of electronic warfare (EW) is to deny an enemy the benefits of the use of the electromagnetic (EM) spectrum while preserving those benefits for friendly forces. That means the whole electromagnetic spectrum from just above dc to just above daylight. That said, most EW literature deals only with the radio frequency (RF) part of that spectrum. We will remedy this shortfall in this chapter.

Figure 9.1 is a much expanded view of the EM spectrum, with emphasis on the optical and infrared range. Note that the horizontal scale is in both frequency and wavelength. The relation between these two values is defined by the equation:

$$\lambda F = c$$

where λ is the wavelength in meters, F is the frequency in hertz, and c is the speed of light (3×10^8 m/s).

In the RF portion of the spectrum, we normally use frequency for convenience; however, the frequencies in the optical and infrared portion are inconveniently large, so we usually talk about these signals in terms of their wavelengths. The units used are micrometers (μm). Note that μ meters are also

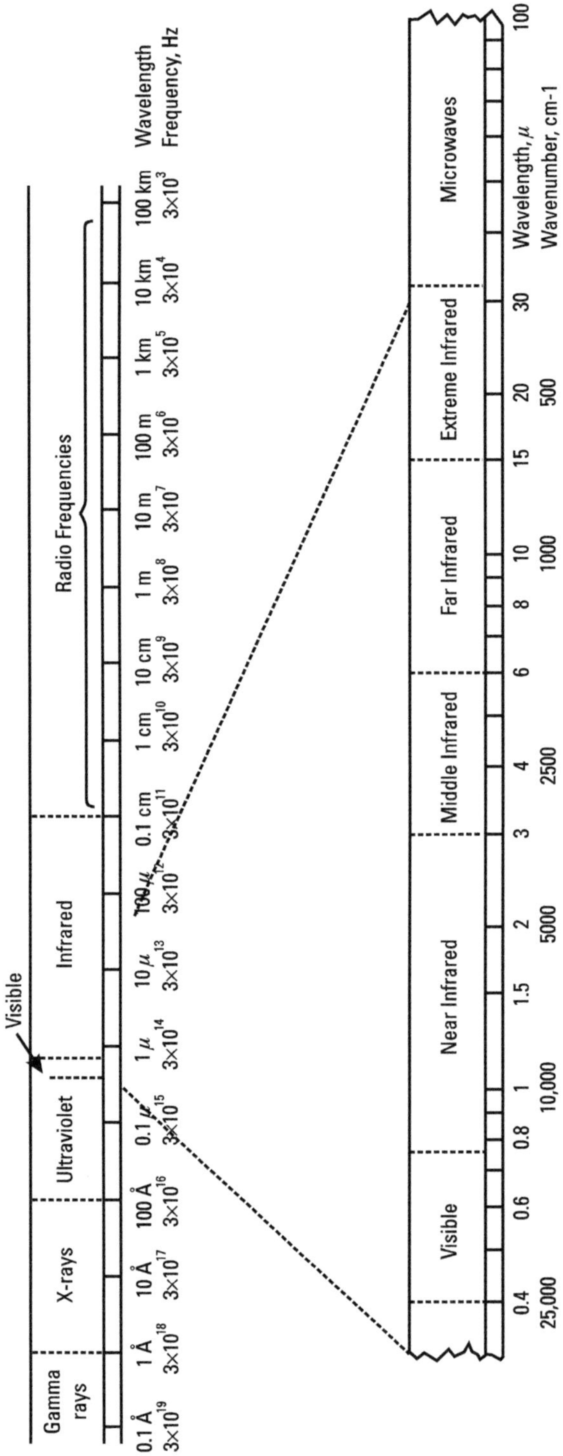

Figure 9.1 The electromagnetic spectrum includes much more than the RF frequency range.

called microns. There are three parts of the IR spectrum important to us in EW: near IR (0.78 to 3 μm), mid-IR (3 to 50 μm), and far IR (50 to 1,000 μm).

There are other bands and other band-edge wavelengths defined in literature, but we will use these definitions in this chapter.

In general, the near-IR signals are associated with high temperatures, the mid-IR signals are associated with lower temperatures, and the far-IR signals are associated with much lower temperatures like those in which humans can survive. This will be explained and expanded later in this chapter during our black-body theory discussion.

9.2 IR Propagation

9.2.1 Propagation Loss

In Chapter 6, we discussed line-of-sight attenuation for RF signals. In that discussion, it was stated that the formula comes from optics. We converted the units and the assumptions to make a convenient formula for RF applications: [Loss = 32 + 20log(F) + 20 log(d).] In the IR frequency range, we use the optics basics. Figure 9.2 shows the applicable geometry. The transmitter is located

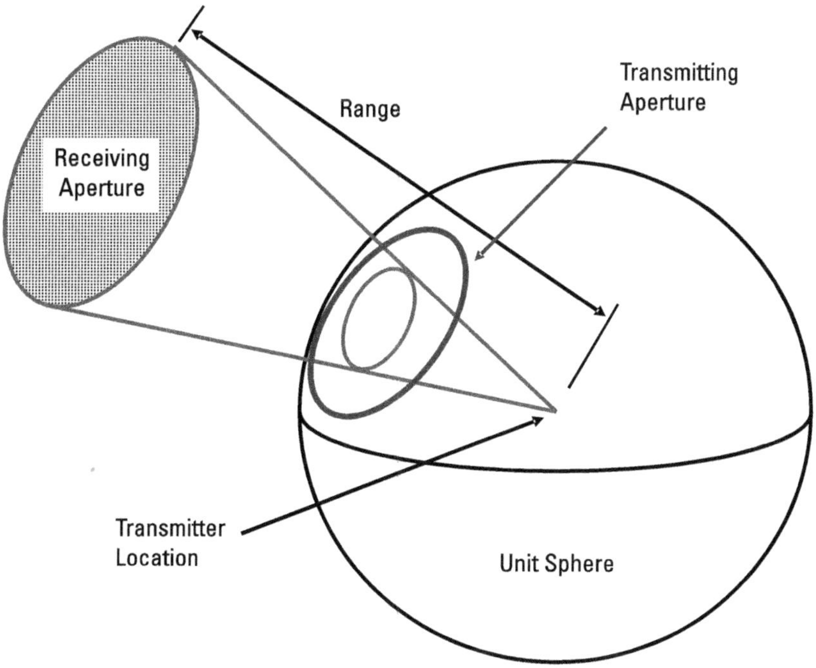

Figure 9.2 IR propagation attenuation is a function of the ratio of the transmitting and receiving apertures projected onto a unit sphere centered on the transmitter.

at the center of a unit sphere. The transmitting aperture is projected onto the surface of the sphere. The receiving aperture is projected back on the same unit sphere. The ratio of the receiving to transmitting areas on the unit sphere is the propagation loss factor. The longer the range, the smaller the receiving aperture will be on the unit sphere, so the greater the propagation loss.

9.2.2 Atmospheric Attenuation

Chapter 6 has a diagram of atmospheric *attenuation* per kilometer for the RF frequency ranges. The attenuation increases with frequency, but also has two attenuation peaks caused by atmospheric gases. One is for water vapor at 22 GHz and the other is for oxygen (O_2) at 60 GHz. Figure 9.3 deals with the IR frequency/wavelength range. It shows the *transmittance* percentage (as opposed to the attenuation) of IR signals through the atmosphere as a function of wavelength. Note that there are wavelength areas of high loss (i.e., low transmittance) caused by several atmospheric gases. The importance of this chart is that it shows propagation windows (i.e., areas of high transmittance) through which IR signals can be transmitted. Any system that depends on transmitting or receiving IR signals for communication, detection, tracking, homing, or imagery must generally operate at a bandwidth within one of these windows. If transmission or reception is attempted in one of the low transmittance (i.e.,

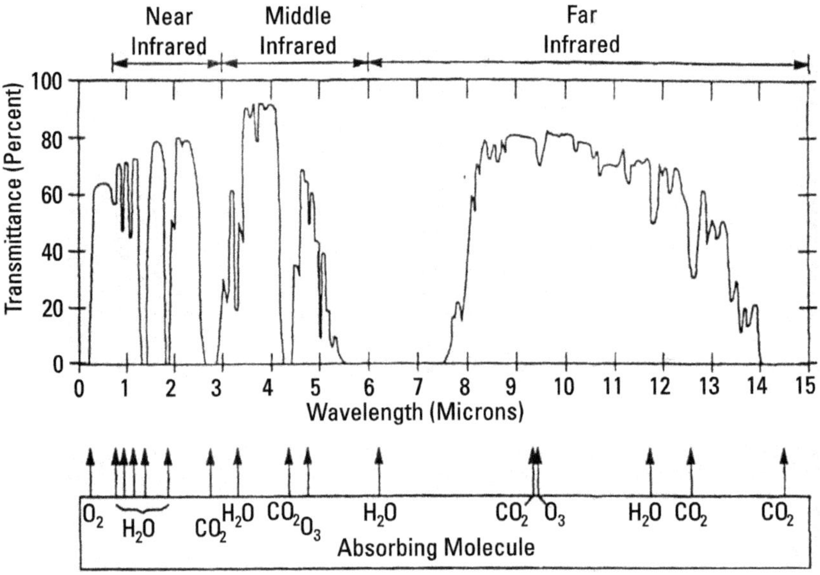

Figure 9.3 Atmospheric transmittance at IR wavelengths has transmission windows and drop-outs.

high loss) bands (for example, between 6 and 7 microns), there will be very little received power.

9.3 Black-Body Theory

A black body is an object that does not reflect any energy. In the lab, a black body is approximated by a pure carbon block with certain dimensions and characteristics. The black body is both a perfect absorber and a perfect emitter, with a well-defined profile of emission energy versus wavelength. Figure 9.4 shows the black-body radiation versus wavelength when the black body is heated to some specific temperature. The temperature is stated in degrees Kelvin (which is the centigrade scale indexed to absolute zero). Each curve is the emission versus wavelength for an object at a single temperature. Note that the peaks of the curves move to the left as the temperature of the object increases. Also note that the amount of energy emitted at any wavelength is greater if the temperature is higher.

As a note of interest, the Sun is a black body. Its surface temperature is 5,900K, causing its radiation peak to occur in the optical wavelength region.

Figure 9.5 shows the same emission power versus wavelength curve for lower temperatures. The point of these two figures is that measurement and analysis of the shape of the power versus wavelength in a received IR signal can determine the temperature of the object from which the signal is emitted. As you will see, this can be very significant to anyone attempting to counter IR-guided weapons.

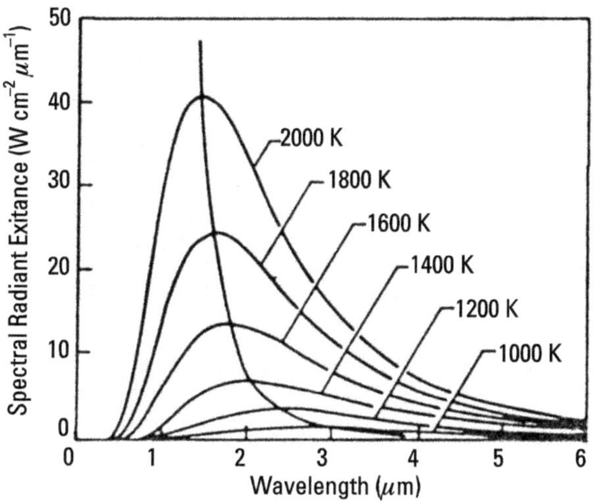

Figure 9.4 Black body radiation from any object varies with wavelength. The peak moves left as the temperature increases. This figure is for high temperatures.

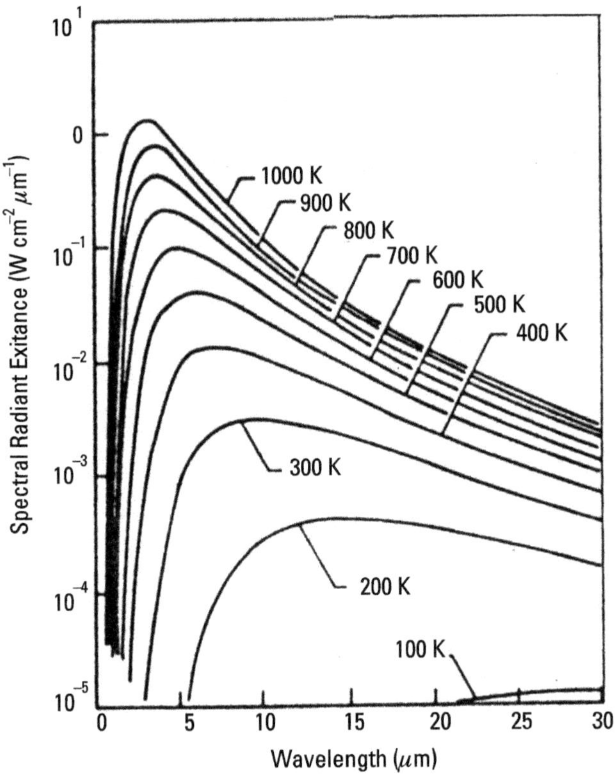

Figure 9.5 Black body radiation curves for lower temperatures show the continuing sift of peak energy wavelength with temperature.

9.4 Infrared-Guided Missiles

IR-guided missiles are significant threats to aircraft because the hot aircraft make easily distinguishable thermal targets against the cold sky. These can be air-to-air or ground-to-air missiles, including shoulder fired Man Portable Air Defense Systems (MANPADS). Some open source literature states that up to 90% of aircraft losses are caused by IR missiles.

IR missiles passively home on emitted IR energy from a target. As discussed in Section 9.3, the wavelength of the energy emitted by an object depends on its temperature. The hotter the object, the shorter the wavelength at which its IR emission peaks. IR missile sensor material is chosen for maximum response in the wavelength of peak emission at the temperature of the chosen target of the missile.

Early IR missiles operated in the near-IR region, requiring very hot targets. Their sensors needed to see the hot internal parts of engines, so the missiles were restricted to attack from the rear of a jet airplane. Later missiles use sensors that can operate against cooler targets, such as the engine plume or the aerodynamically heated leading edges of wings. Thus, they can attack from any aspect.

9.4.1 IR Missile Components

Figure 9.6 is a diagram of a heat-seeking missile. On the nose, there is a lens that is transparent at IR wavelengths. Behind the lens is an IR seeker that generates signals from which the guidance and control circuitry can determine the direction to the target. The guidance and control group controls steering surfaces, such as rollerons, which control the direction of flight. Then there is a fuse and warhead. Because the missile homes on the target, it will actually hit the target and can therefore use a contact fuse in many cases. Finally, there are a solid-state rocket motor and stabilizing tail fins.

9.4.2 IR Seeker

As shown in Figure 9.7, the seeker receives radiated IR energy from the target through the IR lens and focuses it onto an IR sensing cell using multiple shaped mirrors. The IR signals are filtered and passed through a reticle to the IR sensing cell that generates a current proportional to the power of the received IR signal. Note that the seeker is oriented along an optical axis that is offset from the missile's thrust axis. As shown in Figure 9.8, the missile uses proportional guidance so that it will approach the target at a gentle angle. If the missile were aimed directly at the target, it would be required to make a "high-g" turn near impact.

9.4.3 Reticles

There are several types of reticles with different characteristics. Figure 9.9 shows a rising sun reticle that was used in early IR missiles. This reticle has 50% transmittance over half of its surface and the other half has alternating clear and opaque wedges. This causes the IR energy into the sensing cell to receive IR energy from the target with the energy versus time characteristic shown in

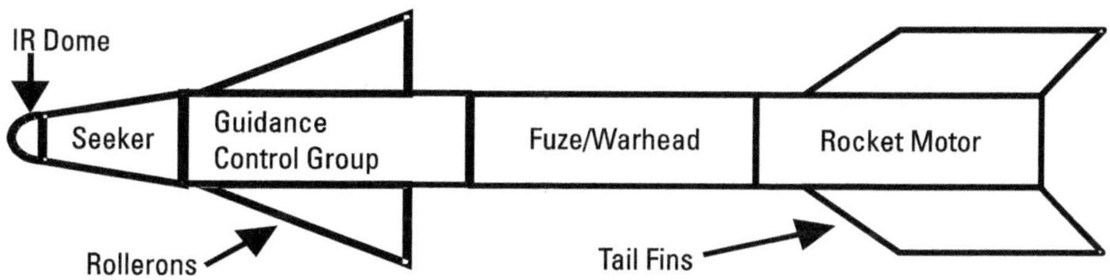

Figure 9.6 A heat seeking missile is guided by inputs from an IR sensor.

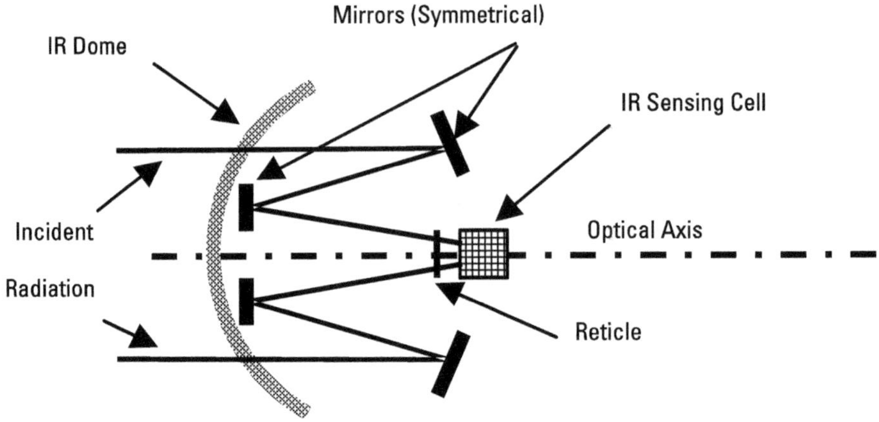

Figure 9.7 The IR seeker focuses received IR energy onto a sensing cell through a reticle.

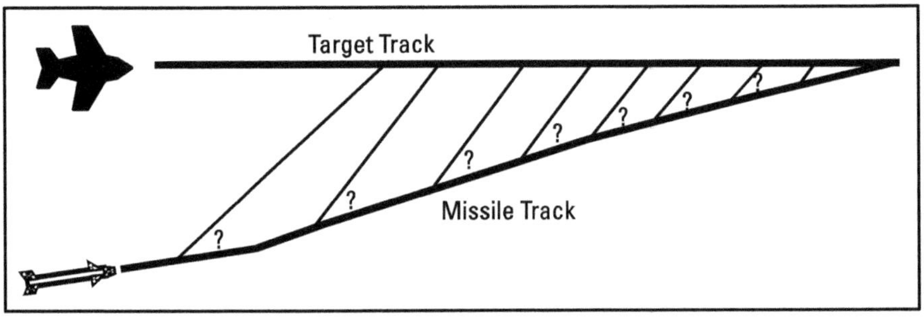

Figure 9.8 IR missiles use proportional guidance to avoid the requirement for a high g turn as they approach their targets.

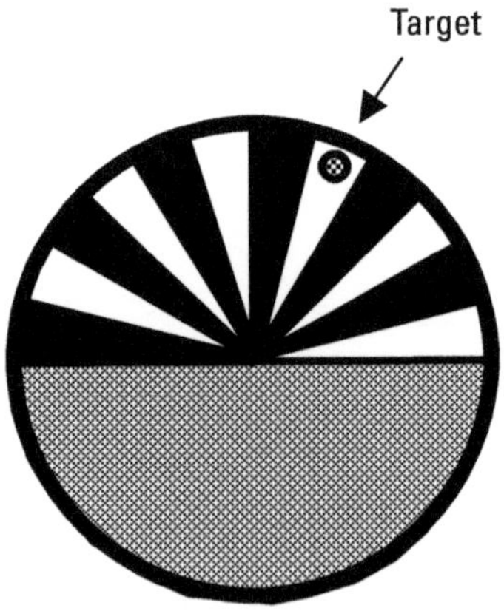

Figure 9.9 A rotating "rising sun" reticle has alternating clear and opaque areas over half of its area.

Figure 9.10. The square wave portion of the pattern starts as soon as the vector from the seeker toward the target enters the alternating portion of the reticle. This energy versus time pattern causes the sensing cell to output a current to the guidance and control group, which has the same pattern. As the direction to the target changes, the time at which the square wave portion of the waveform starts will be appropriately shifted in time. Thus, the guidance and control group can generate the proper steering command to center the optical axis of the seeker on the target. As the IR target direction approaches the center of the reticle, its energy is reduced by the narrowing of the clear wedges (i.e., part of the target is blanked by the opaque wedges). Thus, the error signal varies with the steering error angle as shown in Figure 9.11. One problem that this causes is that the greater signal energy from signals at the outer edge of the reticle will dominate over energy from a target at the center of the reticle. Thus, when the missile is tracking a target near the center of its reticle, a flare at the outer edge of the reticle would generate a larger signal, making it easier for the missile to be decoyed toward the flare. A second problem is that the ultimate aiming point occurs at the minimum received signal in the sensor cell. Later we discuss several other types of reticles that overcome these and other problems.

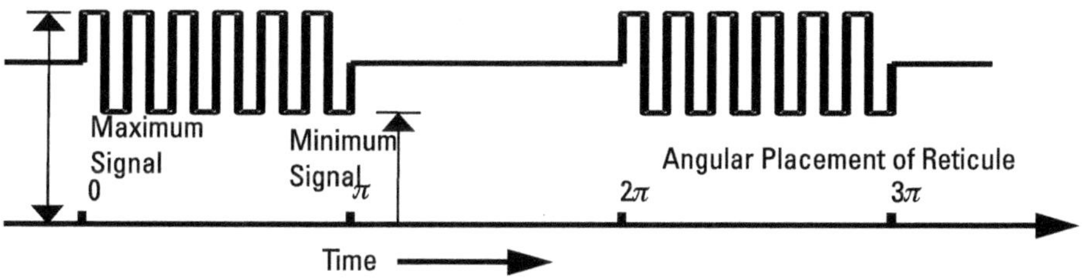

Figure 9.10 The IR energy into the sensing cell has a square wave pattern with a 50% duty cycle as the alternating part of the reticle passes the IR target.

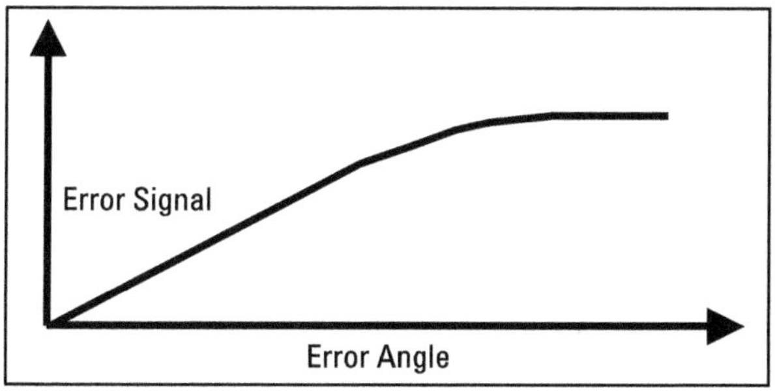

Figure 9.11 The amplitude of the signal into the sensing cell varies with the angle between the target and the optical axis of the seeker.

9.4.4 IR Sensors

Early sensors were made of lead sulfide (PbS), which operates in the 2 to 2.5 μm band (in the near-IR region). PbS sensors can operate without cooling, which simplifies the missile. Later missiles cooled PbS sensors to 77K for greater sensitivity and lower required target temperature, but these sensors still require a rear aspect attack on a target. Note that cooling to 77K can be done with expanding gas.

Later, all-aspect missiles used sensors made from several other chemicals, including lead selenide (PbSe), operating in the 3- to 4-μm band (in the mid-IR region) and mercury cadmium telluride (HgCdTe) operating around 10 μm (in the far-IR region). These sensors must be cooled to about 77K. In the atmospheric transmittance chart in Figure 9.3, note that each of these operating bands falls into one of the transmittance windows so that the IR energy from targets can be efficiently received by the missile's IR sensor.

9.5 Additional Tracking Reticles

In Section 9.4.3, we looked at the various components of heat-seeking missiles, including early tracking reticles. Now we will consider some more modern tracking reticles. These are chosen to illustrate various features and certainly do not include all of the available reticle designs. In each of these discussions, keep in mind that the objective is to determine the angular position of the target in the tracker's field of view so that the missile carrying the tracker can be steered to place the target at the optical axis.

9.5.1 Wagon Wheel Reticle

The wagon wheel reticle is not rotated, but rather nutated to move it in a conical scanning pattern. This causes a target to move through the tracking window in a circular pattern. As shown in Figure 9.12, the energy to the sensing cell has a number of nonuniform pulses when the target is off axis. To center the target in the tracker, the tracker's optical axis must be moved in the direction opposite to the narrowest pulses. Note that when the target is centered on the optical axis of the tracker, the clear and opaque segments of the reticle would cause a constant square wave pattern of energy to the sensor as shown in Figure 9.13. The rising sun reticle shown in Figure 9.9 causes the amount of energy in each pulse to the sensing cell to reduce as the target moves toward the optical axis of the tracker, causing a zero signal when the tracker is aimed directly at the target. The wagon wheel reticle has the advantage of a strong signal when the target is centered.

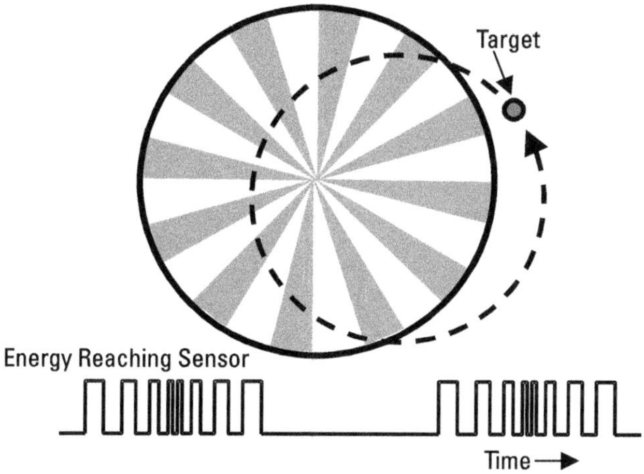

Figure 9.12 The wagon wheel reticle does not rotate. It is offset from the optical axis and moves in a conical pattern.

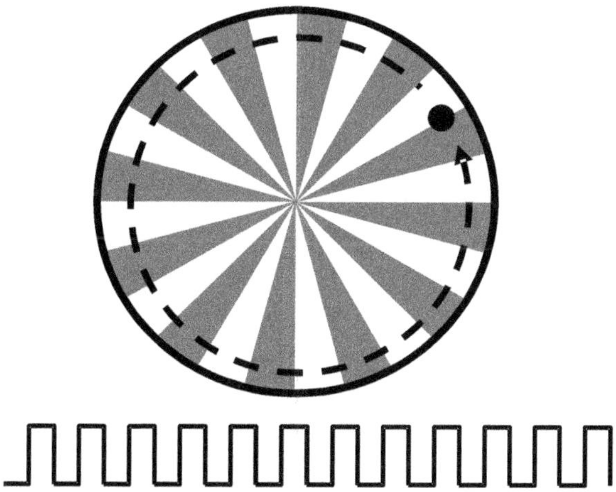

Figure 9.13 When the target is centered in the tracker (i.e., at the optical axis), the wagon wheel reticle produces a constant square wave of energy into the sensing cell.

9.5.2 Multiple Frequency Reticle

Note that the reticle shown in Figure 9.14 causes a series of energy pulses into the sensor half of the time just like the rising sun reticle. However, the number of pulses to the sensor as the target passes through the clear/opaque area of the reticle has differing numbers of pulses depending on the angle between the target direction and the optical axis of the tracker. The tracker is only tracking a single target, but the figure shows two targets to illustrate the different energy patterns. The target shown at the top of the diagram is farther away from the

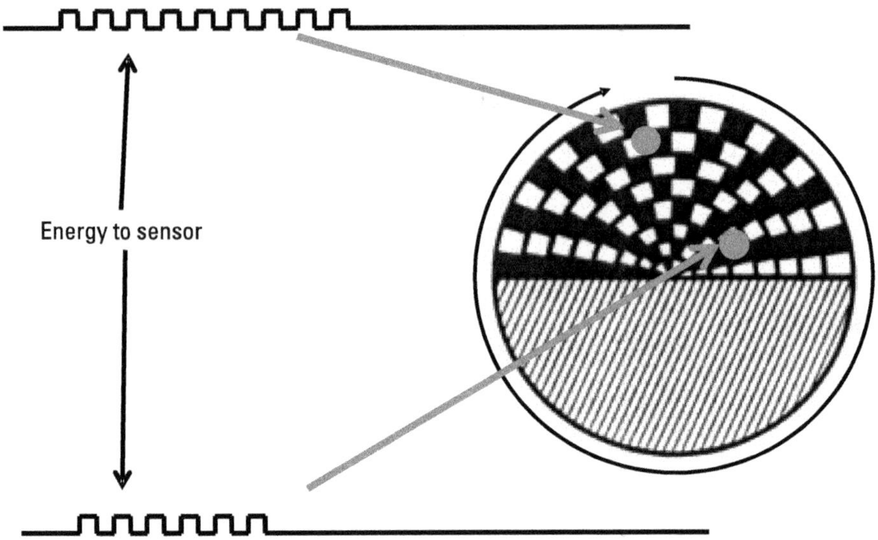

Figure 9.14 The multiple frequency reticle produces an energy pattern in which the number of pulses varies with the off axis angle of the target.

optical axis than the target shown near the center of the diagram. Note that the upper target causes a pulse pattern with nine pulses, and the lower target causes only a six-pulse pattern. This allows the tracking logic to determine the angular tracking error magnitude, so the correct steering correction can be made. Just as in the rising sun tracker, the direction the missile must turn to center the target in the tracker is derived from the time at which the pulse pattern starts.

9.5.3 Curved Spoke Reticle

The reticle shown in Figure 9.15 has curved spokes and has a large, functionally shaped opaque area. It is rotated around the optical axis of the tracker. The curved spokes are designed to discriminate against straight line optical interference. The horizon has a bright line, and reflections from various objects would reach the tracker as straight, bright lines that can interfere with the tracking processing.

Note that the shape of the opaque area causes a difference in the number of spokes through which a target passes as a function of the angle between the target and the optical axis. If the target is near the outer edge of the reticle, there will be seven pulses of energy covering half of the time. As the target moves toward the optical axis, the number of energy pulses increases, as does the percentage of time that the pulses are present. When the target is very near the optical axis, there are 11 pulses of energy and the pulses occupy nearly 100%

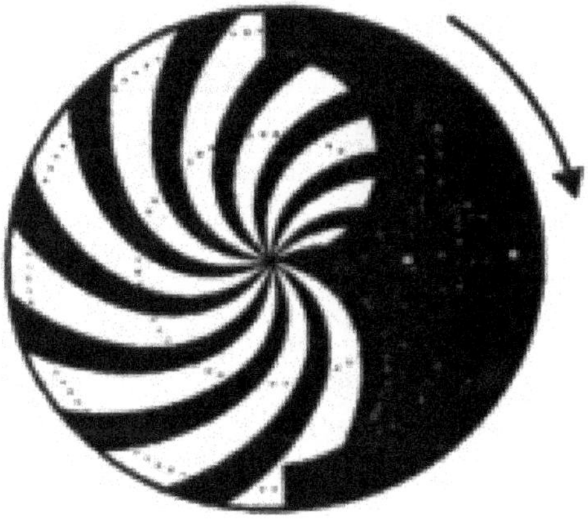

Figure 9.15 The curved spoke reticle discriminates against straight line extranious inputs (like the horizon). It also inputs an energy pattern with a number of pulses proportional to the off axis angle of the target.

of the time of a reticle rotation. This allows for proportional guidance just as in the multiple frequency reticle.

9.5.4 Rosette Tracker

The rosette tracker shown in Figure 9.16 moves the focal point of the sensor in the pattern shown. This movement is accomplished by two counter-rotating

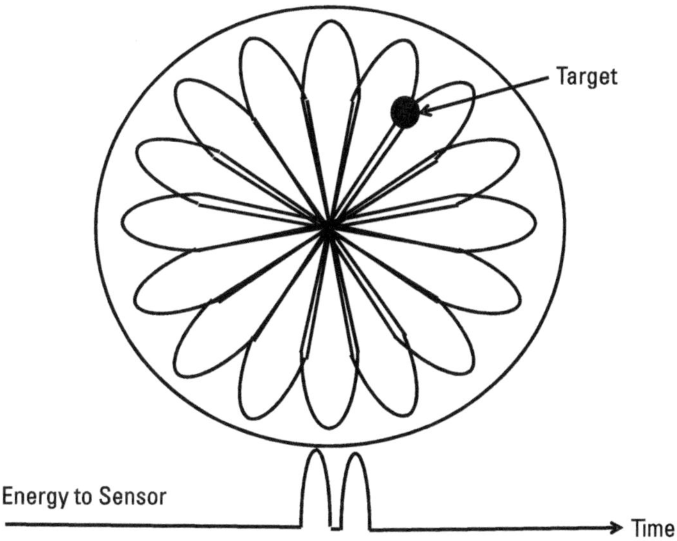

Figure 9.16 The timing of the energy bursts into the sensor following a rosette pattern determines the angular position of the target.

optical elements, and the rosette can have any number of petals. As the sensor is moved through the target, a pulse of energy reaches the sensor. In the figure, the target is shown in a location where it is covered by two petals. Thus, there are two response pulses. The location of the target relative to the optical axis is determined from the timing of the energy pulses.

9.5.5 Crossed Linear Array Tracker

The crossed linear array shown in Figure 9.17 has four linear sensors. The array is nutated to move it in a conical scan. As the target passes through each of the four sensors, an energy pulse is generated. The location of the target relative to the optical axis of the tracker is determined from the timing of the energy pulse in each sensor.

9.5.6 Imaging Tracker

The imaging tracker creates an optical image of the target. As shown in Figure 9.18, the tracker can have a two dimensional array of sensors or can move a single sensor in a raster scan pattern as is done in a commercial television camera. Each location creates a pixel from which the processor can create a representation of the size and shape of the target and its angular location relative to the optical axis.

As in all optical devices, the number of pixels determines the resolution that can be achieved. In general, the imaging tracker is usually thought of as a terminal guidance device because it will have relatively few pixels. Thus, to have enough pixels on the target to identify it as the tracked target, the missile

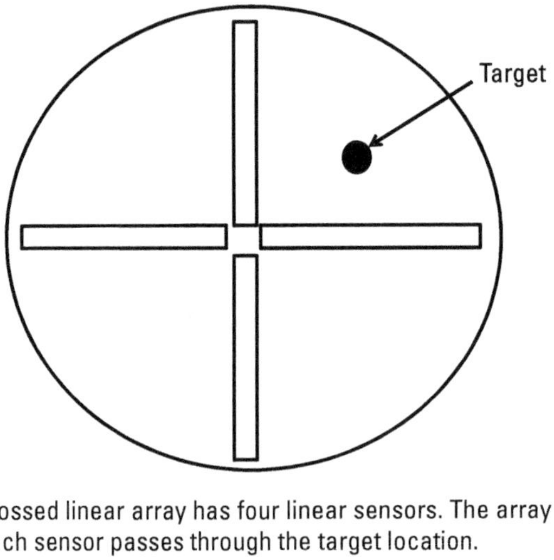

Figure 9.17 The crossed linear array has four linear sensors. The array is nutated, and outputs a pulse from each sensor passes through the target location.

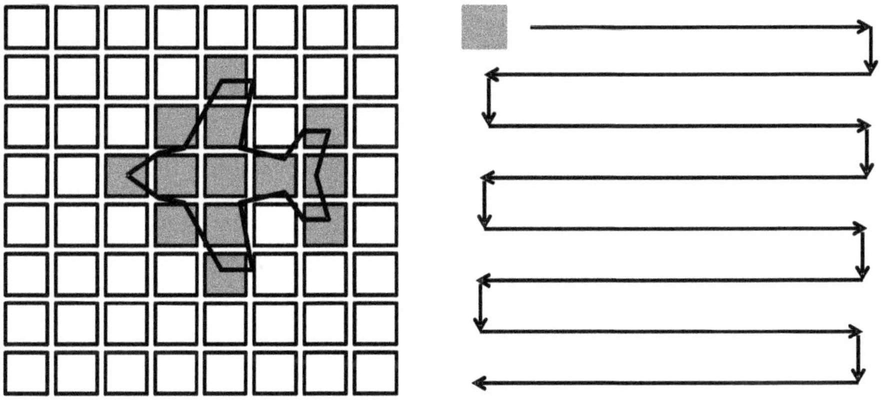

Figure 9.18 An imaging tracker has either a number of sensors in a two dimensional array or a single sensor which is moved over an angular area in a raster pattern. It creates an image of the target.

(carrying the tracker) must be relatively close. Some literature has given approximately 20 as the number of pixels that can receive target energy at the acquisition range. There will be more detailed discussion of this point later.

In the figure, the pixels on the target are shown in gray. This does not make a very clear picture of an airplane, but it looks radically different from a thermal decoy. The decoy would likely occupy only a single pixel, allowing the processor to reject the decoy in favor of the target aircraft.

9.6 IR Sensors

We have been talking about heat-seeking missiles, and Section 9.5 focused on various types of reticles. Now we will take a closer look at the actual IR sensors. The sensor generates a signal from received IR energy. Each type of sensor material responds to a spectral range, which determines the target temperature against which it is most effective.

9.6.1 Aircraft Temperature Characteristics

Figure 9.19 shows the approximate temperature ranges of the parts of a jet aircraft that can be targeted by heat-seeking missiles.

The compressor blades inside the engine are the hottest areas, and the external engine tail pipe parts are slightly cooler. Both are in the range of 1,000K to 2,000K, which means that their energy peaks in the 1- to 2.5-micron (μm) wavelength range. The plume from the engine is in the 700- to 1,000-K range, so it peaks in the 3- to 5-μm wavelength range. Aerodynamically heated aircraft skin, for example, the leading edges of wings, can be expected to be between

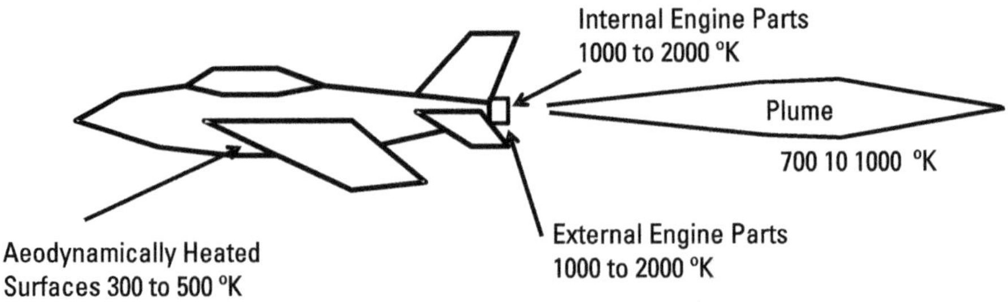

Figure 9.19 A jet aircraft can be attacked by a heat seeking missile which tracks the hot internal parts of the engine, the tail pipe, the plume, or aerodynamically heated skin surfaces.

300K and 500K so the energy from these areas will peak in the 8- to 13-μm range. Refer to Section 9.3 where the peak temperature versus wavelength relationship is discussed.

9.7 Atmospheric Windows

Another important issue, which also follows from Section 9.2.2, is the transmittance of the atmosphere. Figure 9.20 shows the four major windows through which infrared energy propagates well. The two lower windows are at 1.5 to 1.8 μm and 2 to 2.5 μm. These are in the near-IR region. The mid-IR region has

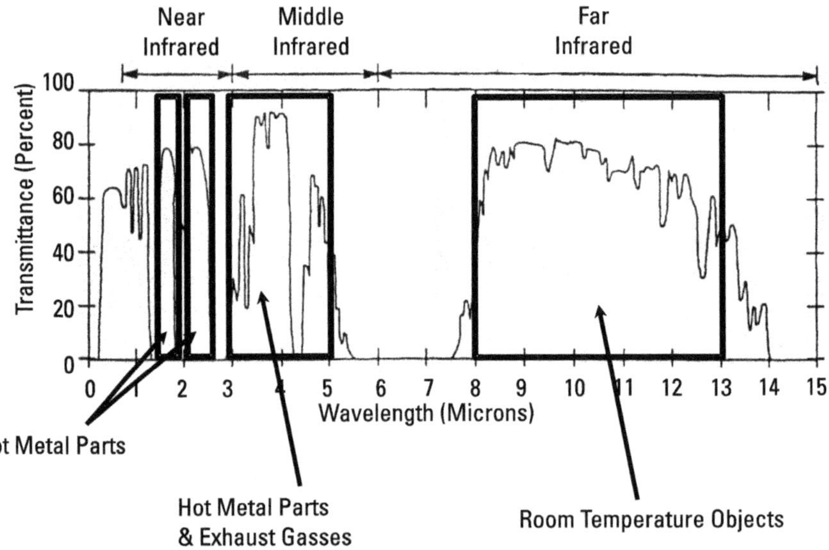

Figure 9.20 Atmospheric transmittance at IR wavelengths has transmission windows and drop-outs at well defined wavelength ranges.

two windows in the 3- to 5-μm region. The far-IR region has a large window from 8 to 13 μm.

Hot targets like the tail pipe or internal engine parts are tracked in the near-IR region, the plume is tracked in the mid-IR region, and heated skin targets are tracked in the far-IR region. In general, heat-seeking missiles like to target on the hotter targets.

9.8 Sensor Materials

Table 9.1 shows the peak response wavelengths of various important sensor materials and their typical applications.

All of the sensor materials except lead sulfide are cooled to 77K (which is the boiling point of nitrogen at one atmosphere) to increase sensitivity and signal-to-noise ratio and to discriminate against solar energy. Lead sulfide was used in the first heat-seeking missiles, which homed on the hottest part of the aircraft, the internal engine parts. For effective tracking, it was necessary for the missiles to approach the aircraft from the rear to achieve a clear view of the tracking point. These early sensors did not require cooling but were restricted in sensitivity.

With cooled lead selenide or indium antimonide sensors, it was possible to track on the aircraft's plume. Because the plume can be seen from the front or the side of the aircraft, the missiles could track from any angle, making these all-aspect missiles.

With mercury cadmium telluride sensors, missiles can track on aerodynamically heated aircraft skin, which allows all-aspect tracking. This material can also be used to make focal plane arrays, which allow image tracking, as discussed next.

Table 9.1
Properties of Sensor Materials

		Peak Response Wavelength (μm)		
Symbol	**Material**	**At 300K**	**At 77K**	**Typical Application**
PbS	Lead sulfide	2.4	3.1	Used at room temperature against hot targets
PbSe	Lead selenide*	4.5	5	Tracking plume
HgCdTe	Mercury cadmium telluride		10	Must be cooled to 77K. Used in missile trackers and FLIRs. Useful in both mid and long wavelength windows.
InSb	Indium antimonide	3.5	3	Tracking plume

9.9 One-Color Versus Two-Color Sensors

One of the issues faced by heat-seeking missiles is to discriminate against flares, the Sun, and other high-temperature distractions from the target. Conventional distractors are much hotter than the targeted parts of the target aircraft. Magnesium flares are 2,200K to 2,400K and the Sun is 5,900K. This causes the distractor to emit much higher energy than the target. Note that the black-body emission curves in Figure 9.21 (as explained in Section 9.3) show increased energy with increased temperature at any wavelength. Therefore, a very hot magnesium flare will capture a missile's tracker and lead it away from the target.

However, if the missile detects its target at two wavelengths, it can, in effect, calculate the temperature of the targeted object. This allows the missile to track a target at a chosen temperature, or at least to discriminate against false targets that are much hotter than the real target. Figure 9.3 deals with two hot objects, a distractor at 2,000K and a target at 1,600K, at two separate wavelengths (2 μm and 4 μm). Note that these temperatures and wavelengths are chosen to illustrate the effect rather than to represent the values of specific friendly or enemy sensors. The 2,000K flare emits 5.3 times the energy at 2 μm that it does at 4 μm. Now consider the 1,600K target. It emits only 3.1 times as much energy at 2 μm as it does at 4 μm. If only the tracking waveform for objects with the proper energy ratio range are input to the missile's processor, the missile will ignore the flare at the wrong temperature and track the target at the right temperature.

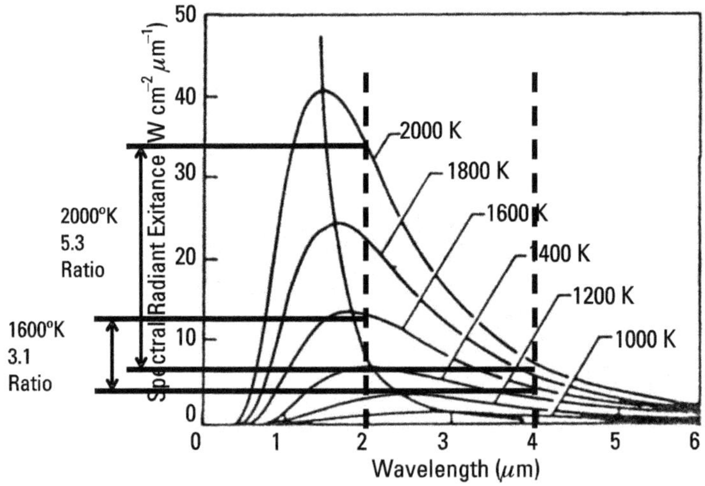

Figure 9.21 Black body radiation from any object varies with wavelength. The peak moves left as the temperature increases. When objects at different temperatures are sensed at two wavelengths, the ratio of the energy at each wavelength varies strongly with temperature.

The two selected wavelengths must be within atmospheric windows and can be selected to create significant ratio differences between flares and the targeted part of the target aircraft.

9.10 Flares

An important way aircraft are protected against heat seeking missiles is by the use of flares, which operate in three different roles. These roles (or tactical objectives) are seduction, distraction, and dilution.

9.10.1 Seduction

The seduction role involves deploying a flare into the physical area and wavelength range viewed by the tracker of a heat seeking missile. The flare must provide a stronger signal (in the missile's tracker) than that of the target being tracked. Unless the missile tracker has protective features to allow it to distinguish flares, the missile will transfer its attention from the target to the flare. The missile tracker then steers the missile toward the flare rather than the target. As the flare moves away from the target aircraft, the missile follows, as shown in Figure 9.22.

9.10.2 Distraction

In the distraction role, flares are deployed before the heat-seeking missile starts to track the target aircraft, and the flare is placed so that the missile tracker will see the flare before it sees the target. In this role, the flare does not need to produce a larger signal than that from the target, but it does need to be close enough so that the missile tracker will accept it as a valid target. If distraction is successful, the missile will track the flare and never actually see the target as shown in Figure 9.23. Note that this technique is also used to defend ships against heat-seeking anti-ship missiles; however, multiple flares (or thermal decoys) will probably be required to maximize the probability of capturing the missile tracker before it sees the target ship.

9.10.3 Dilution

The dilution tactic is used against threats that have imaging or track-while-scan capability. That is, the missile tracker can deal with multiple potential targets. In this defensive tactic, the objective is to cause the enemy to choose among many credible targets as shown in Figure 9.24. The flares (or thermal decoys) must look enough like real targets to avoid being rejected by the missile tracker.

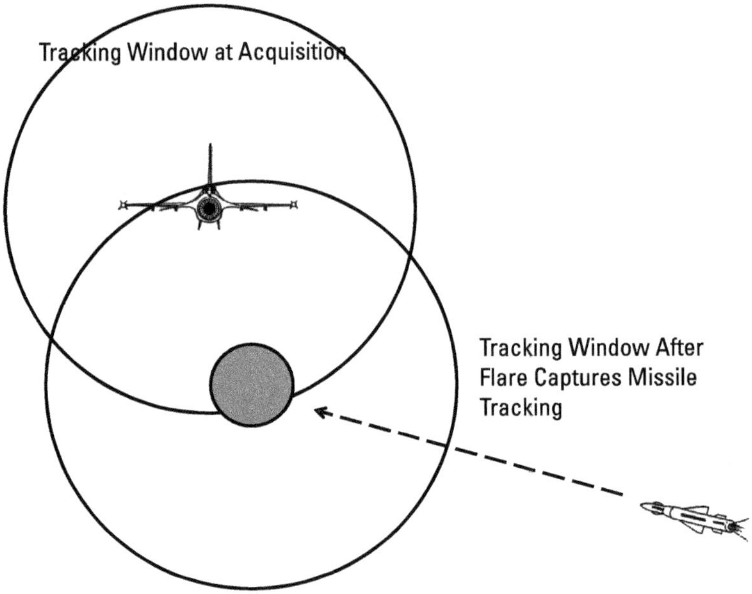

Figure 9.22 A flare used in a seduction mode captures the tracker in a threat missile and guides the missile away from the target aircraft.

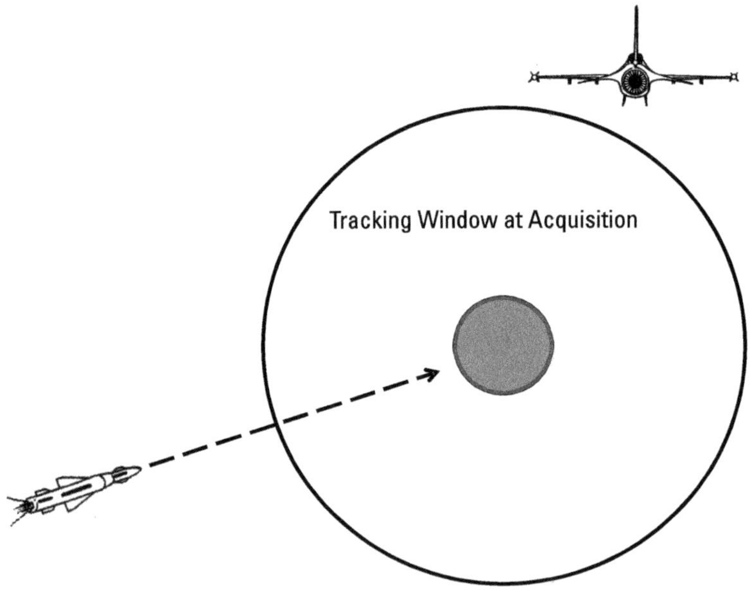

Figure 9.23 When used in the distraction mode, a flare captures the tracker of a threat missile before it acquires the target aircraft.

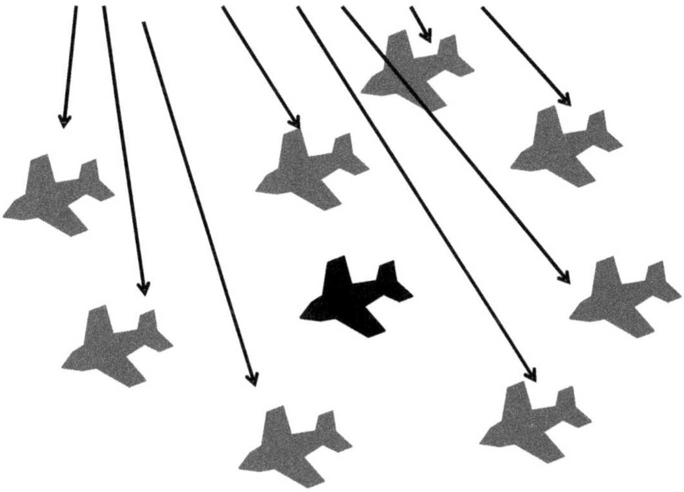

Figure 9.24 Flares used in a dilution mode cause a threat missile to chose among many false targets to attack the real target.

Naturally, the difficulty of this approach depends on the sophistication of the attacking missiles. Note that this approach is less desirable than either seduction or distraction because the missile may well choose the true target rather than one of the decoys deployed. The survival probability in this case is n/(n + 1) when n decoys are used against a single threat to protect a single target.

9.10.4 Timing Issues

This discussion is based on the seduction technique, but applies with qualifications to the other techniques as well.

Flares must come up to an effective energy level while they are within the tracking area of the attacking missile as shown in Figure 9.25. Depending on the design of the flare and the speed of the target aircraft, the aerodynamic deceleration of the flare may be as much as 300 m/s^2. Because the diameter of threat field of view is typically less than 200m at the time the flare is deployed. This calculates to a little over a half-second for the flare energy to exceed the target energy by enough to assure that the missile transfers its tracking from the target to the flare. Note that this energy level must be achieved in this time period at all of the wavelengths at which the threat missile may be tracking.

Figure 9.26 shows separation of a typical flare from the aircraft dispensing it from an altitude of 3 km. This figure shows vertical and horizontal separation as a function of the airspeed of the aircraft.

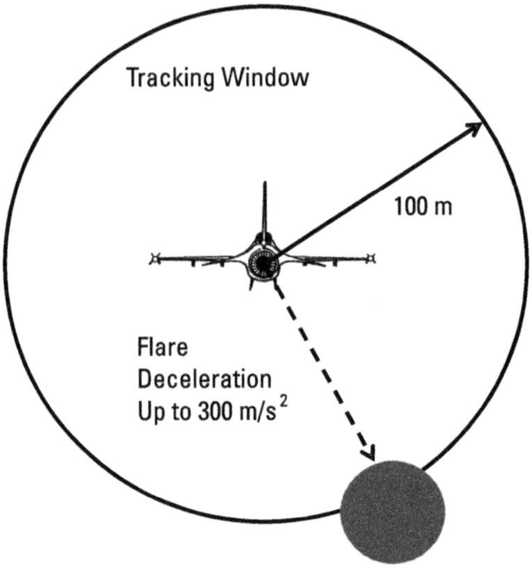

Figure 9.25 Because a flare can decelerate at 300 m/s^2 and the threat missile's tracking window is only about 200 meters in diameter at acquisition, the flare must reach an adequate energy level in about one half second in order to successfully seduce the missile away from the target.

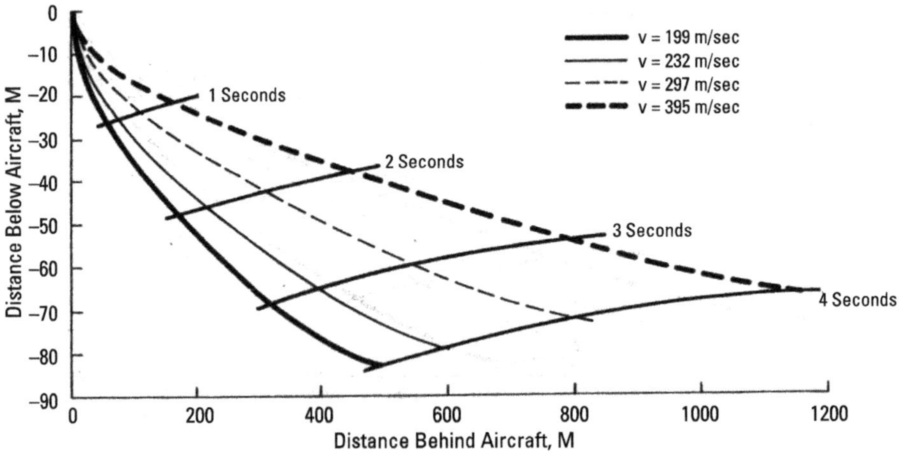

Figure 9.26 A deployed flare will fall behind and below the aircraft that deploys it by values which vary with the altitude and the velocity of the deploying aircraft.

The decoy must continue to provide adequate energy to overcome the energy of the target in the missile's tracker until the target is no longer in the missile's tracking volume so that the missile cannot reacquire the target. It is most desirable that the flare provide this level of protection until the missile has passed the target or can no longer maneuver to hit the target.

9.10.5 Spectrum and Temperature Issues

To be effective, a flare must radiate at the wavelength at which the sensor in the threat missile performs its tracking. Refer to the discussed black-body radiation and atmospheric transmittance in Sections 9.1 and 9.9. The missile tracker must operate in one of the atmospheric windows, but that covers a great deal of spectrum. To provide protection, the flare must provide adequate energy in the actual wavelength range used by the missile tracker.

In general, the fuel and binder materials used in flares radiate according to the black-body radiation energy characteristic. Thus, the temperature determines the spectral distribution of the energy. However, at any wavelength, the radiated energy from a black body increases with temperature. To create enough energy to capture the missile tracker in the small size of a flare, it is desirable to use very hot burning material (like magnesium powder with binders which enhance the burning). If the flare is significantly higher in temperature than the target, it will generate a significantly greater signal level in the missile tracker to capture the tracking function. As discussed in Section 9.9, two-color sensors can determine the temperature of the flare. This is one of the techniques that a missile tracker might use to discriminate against the flare.

Flares are a very effective countermeasure against heat-seeking missiles because a flare puts more energy into the tracker that the target does. Thus, the flare captures the seeker and leads the missile away from its target. However, there are various techniques that can be employed to allow the missile tracker to discriminate the flare from the target. If successful, they allow the tracker to ignore the flare and continue toward its intended target. Some of these techniques are chemical, some are temporal, and some are geometrical.

9.10.6 Temperature-Sensing Trackers

We have discussed the two-color sensor that can determine the actual temperature of the target and of the flare. The tracker then tracks only a target at the correct temperature. As discussed in Section 9.10.9, the tracker senses at two different wavelengths. If the sensed energy at the two wavelengths has the proper ratio, the tracker will conclude that it is tracking a valid target. If the flare is at a higher temperature, it will be rejected by the tracker and thus cannot lead the missile away from its intended target.

For a flare to be effective against a two-color sensor, it must create the correct energy ratio. This can be done by emitting at the proper temperature or by emitting at a higher temperature, but with the correct ratio between the energy at the two wavelengths sensed by the tracker.

As shown in Figure 9.27, a low-temperature flare can emit at the correct temperature, but then it must create higher energy by filling a larger volume than that of the missile's intended target. This can be accomplished by ejecting

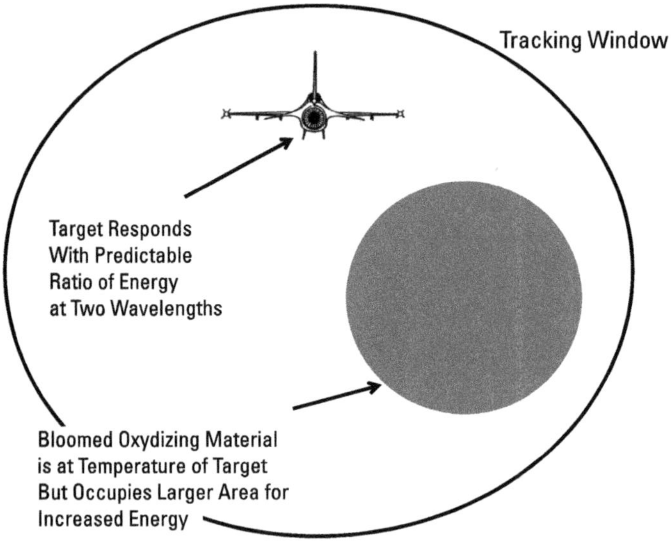

Figure 9.27 A low temperature flare blooms smoldering material over a large area to cause a high energy response in the missile sensor, seducing the tracker away from its intended target.

a cloud of small pieces of material coated with a rapidly oxidizing chemical that will spontaneously burn at the correct temperature. Igniting a cloud of flammable vapor can create the same effect. Low-temperature flares have the advantage of being less visible and of not setting fires when they hit the ground in forests or cities.

A second approach is to manufacture flares with two chemicals that burn at high temperatures but emit with the correct energy ratio at the two wavelengths sensed by the missile tracker. The energy ratio will cause the tracker to accept the flare as a valid target, and the high temperature will create an attractive response in the tracker so that it can be lured away from its intended target. These are called two-color flares and are shown in Figure 9.28.

9.10.7 Rise Time-Related Defense

As discussed in Section 9.10.4, a flare can decelerate at 300 m/s^2 and the tracking window is only about 200m wide at acquisition. The flare must thus reach its maximum energy in about half a second. This requires that the chemicals chosen for flare construction must build up their energy extremely quickly. This creates a much higher rate of energy rise than that from the afterburner of a jet engine. Thus, if the rate of energy increase from an object in the tracking window over a preset time interval is above a certain threshold, the tracker could stop tracking. Then, when the energy in the tracker drops to its previous level (as the flare leaves the tracking window), the tracker can start tracking again (see

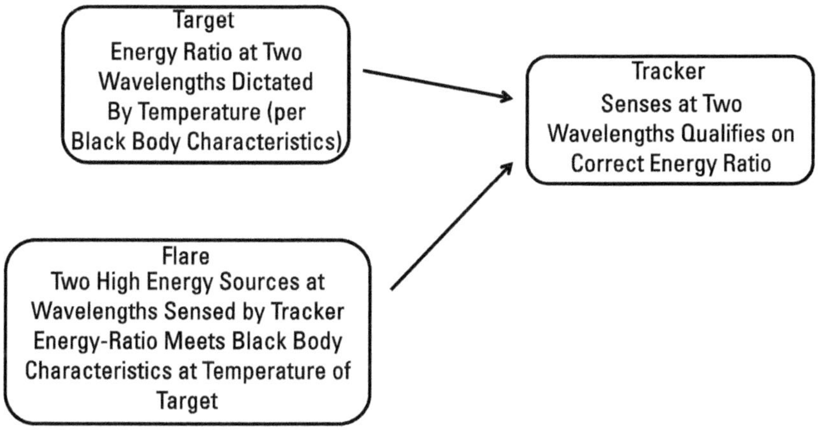

Figure 9.28 A two color flare emits with the correct energy ratio to match that of the target.

Figure 9.29). This flare countermeasure could be overcome by activating flares in anticipation of a missile attack rather than in response to a detected missile approach.

9.10.8 Geometric Defenses

If a missile is attacking from abeam an aircraft, the rate of change of angle to the flare will be much greater than that of the target. By sensing the rate of change

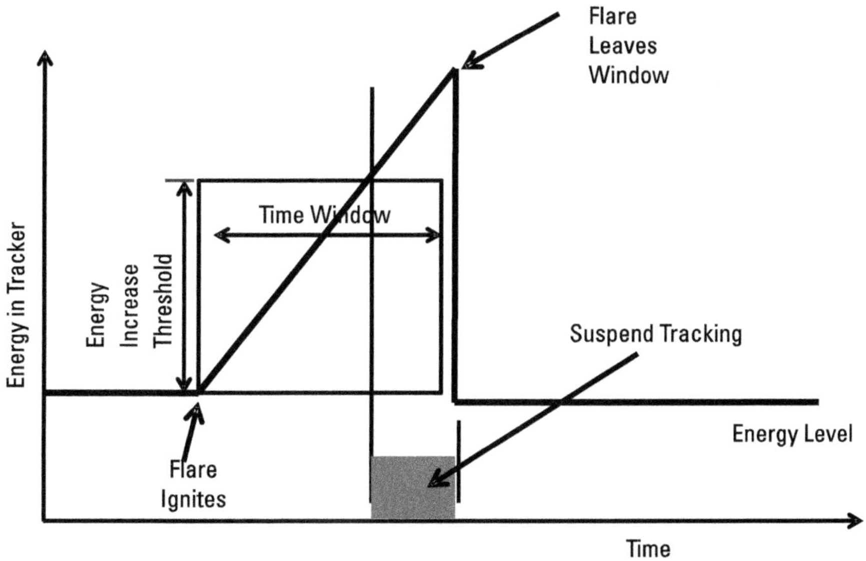

Figure 9.29 When a flare causes the energy in the tracker to rise more than a fixed amount during a fixed time interval, the tracking can be stopped until the flare leaves the window.

of aspect angle, the missile tracker can detect the presence of a flare and stop its tracking until the flare has left the tracking window. Note that the angular separation will be seen as much smaller during an attack from the front or rear of the target aircraft, so this defense will be much less effective.

Also in an abeam attack, the missile seeker will see two targets because a flare decelerates relative to the launching aircraft as shown in Figure 9.30. In this case, if the tracker focuses on the leading target, it will discriminate against the flare.

Because a launched flare will fall below the launching aircraft, as shown in Figure 9.31, the tracker could place a filter over the lower part of the tracking window or in the lower rear quadrant if tracking from abeam. This will reduce energy received from the flare and thus allow the tracker to see the intended target as the more attractive target.

Note that these geometric defenses are defeated if the flare has forward thrust or lift.

9.10.9 Operational Safety Issues for Flares

IR flares generate a great deal of energy and generate heat very quickly. Because of this, there are some serious safety issues that must be considered in their application. In this section, we will discuss the different types of flares used to

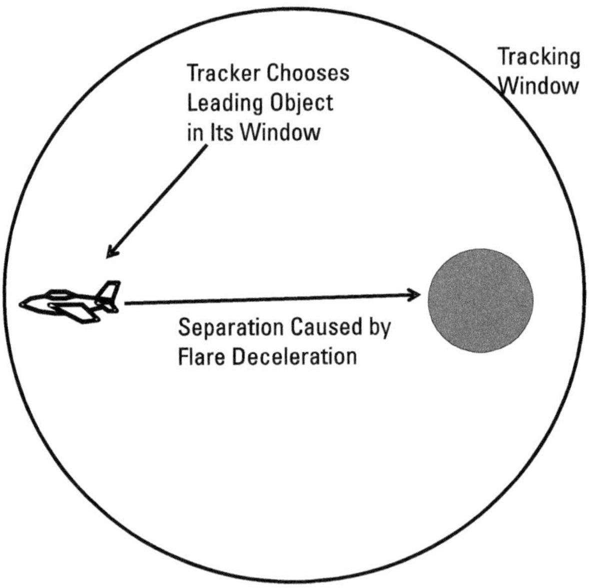

Figure 9.30 In an attack from an abeam, the tracker can chose the leading object in its window to discriminate against flare.

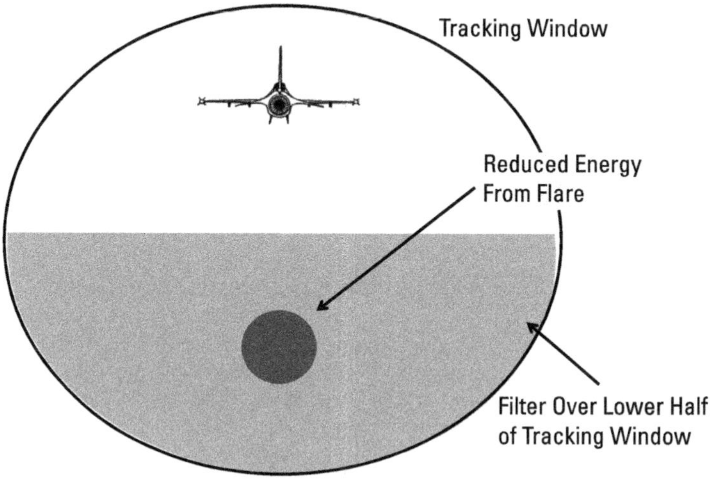

Figure 9.31 A filter over the lower half of the tracking window will cause the tracker to lock onto the target because the energy from the flare is reduced.

protect aircraft against heat seeking missiles and their related safety issues. We will also discuss some of the required testing and safety features that are used.

Flares are placed into tubs on aircraft. They are aluminum and hold fiberglass magazines in which the actual flares are held and from which they are fired. The U.S. Navy uses round flares which must all be the same size (36 mm in diameter and 148 mm long). The U.S. Air Force and Army use flares that are 1 × 1 inch or 1 × 2 inches by 8 inches long, and the Air Force also uses some that are 2 × 2 × 8 inches. These are the NATO standard sizes. The larger flares are to provide greater energy to overcome the IR profiles of larger aircraft engines. There are a number of additional sizes and shapes of flares used by various countries and aircraft types. All types of flares contain materials that are launched from the aircraft and produce hot targets to lure heat-seeking missiles away from the protected target. They can be either pyrotechnic or pyrophoric.

9.10.9.1 Pyrotechnic Flares

A pyrotechnic flare is launched from the aircraft by an electrically initiated ejection charge. Figure 9.32 is a sketch of a pyrotechnic flare. The flare payload is a pyrotechnic pellet that must be ignited, either by the same charge that launches the flare or by a secondary charge that is ignited by the launching charge. The pellets in the earliest flare types are magnesium-Teflon (MT) with various binding materials that provide mechanical integrity and enhance its performance. MT flares burn at very high temperatures to cause the energy differential required to capture the missile seeker. These are still used, but as we have discussed, there are now also flares that work against two-color sensors

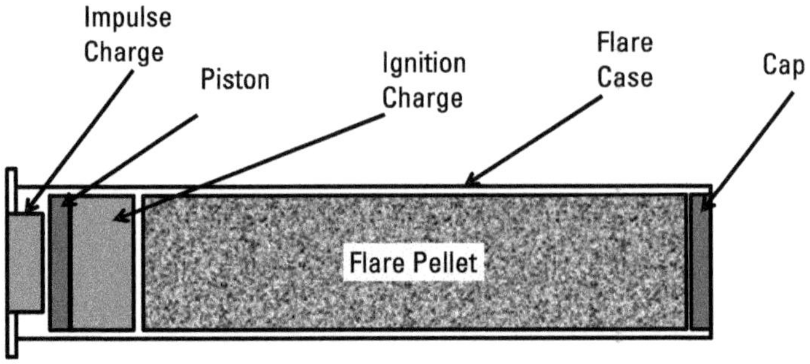

Figure 9.32 A pyrotechnic flare has a payload that must ignited. It can be compressed magnesium-Teflon or other compounds. Sometimes the launching charge lights the payload and sometimes it lights a secondary charge that lights the payload.

designed not to react to the spectrum of burning magnesium. These are called spectrally matched flares.

Spectrally matched flares typically burn pyrotechnic materials that produce a more correct ratio of energy in the low and mid-IR bands to satisfy the selection criteria of two-color missile seekers, even though the actual combustion temperatures may be much higher. In general, the safety issues have been with these more recently developed types of flares.

9.10.9.2 Pyrophoric Decoy Devices

Pyrophoric decoy devices are sometimes called cool flares, but are not properly called flares because they do not burn. They actually oxidize very rapidly to create IR radiation not visible to the naked eye, to look like a target to a missile seeker. Figure 9.33 is a sketch of a pyrophoric decoy device.

Earlier, some pyrophoric decoy devices used liquids, but these were found to be dangerous and difficult to use, so the pyrophoric foil payloads are typically used now. The basic approach is to manufacture a thin metal foil with highly porous surface that can then oxidize very rapidly when exposed to air. These devices do not burn; they smolder, producing a dull red glow. Thus, they are not visible by day or night at operational ranges, except for the flash caused by the charge which ejects the payload. Small pieces of 1- to 2-mil-thick treated foil cut to fit the round or square decoy body are ejected and bloom to make a large cross section that provides an attractive target for the missile seeker. If a decoy is at the correct temperature, its energy versus wavelength profile will match the black-body radiation characteristic shown in Figure 9.4. These are sometimes called black-body flares; however, because they do not have the perfect emissivity of a true black body, they are more correctly called gray body decoys.

Pyrophoric decoys can come up to temperature in less than a half second, just as required of pyrotechnic flares.

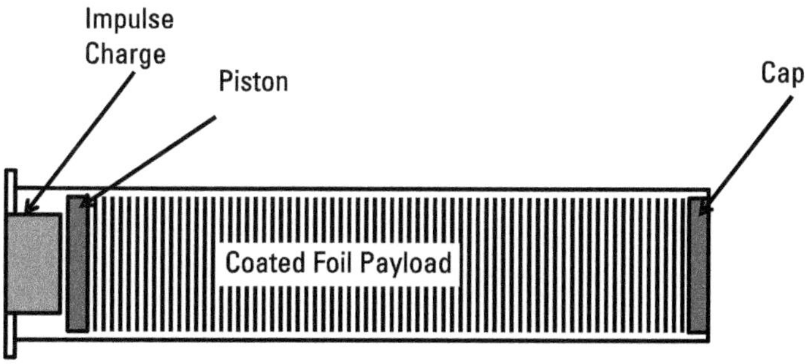

Figure 9.33 The payload of a pyrophoritic decoy is a large number of coated thin iron foil pieces that bloom when deployed and oxidize very rapidly to create a thermal target for heat seeking missiles.

9.10.9.3 Safety Issues

In addition to bullet impact resistance to firing, there are also standards for resistance to electromagnetic radiated power. The concern is that the squib that ignites the launching charge could be initiated by radar signals; the ignition of the actual charge from RF power is not considered a danger, but rather that energy will be coupled into the squib bridge wire. This is not a big issue to Air Force applications, but is very significant on aircraft carriers that have powerful radars in close proximity to launch-ready aircraft. There are reports of accidents (on aircraft carriers) caused by the firing of flares from radar energy. This category of hazards is known as HERO (Hazard of Electromagnetic Radiation to Ordnance). There are no reports of flare ignition from radars in the air.

As a minimal safety requirement, most squibs are required to withstand a current of 1 amp without functioning. Because they also have 1-ohm resistance, 1W produces 1 amp. The no-fire criterion is typically 1 amp and the all-fire specification is usually 4 to 5 amps. There are also HERO-safe squibs that have lowpass filters to reduce the energy from RF sources such as radar, but they are not in universal use.

9.10.9.4 Confined Function Test

Concern about inadvertent flare ignition and ejection failures have led to requirements for confined function testing. In these tests, the tube is blocked and the flare is fired and allowed to complete its burn. The passing criterion may be slightly different for the different agencies, but is generally that there should be no damage beyond the flare dispenser.

9.10.9.5 Bore Safety

Pyrotechnic flares can have bore-safety devices called sliders (see Figure 9.34). These devices are intended to prevent ignition until the flare has left the magazine. They are not universally used, because often flares are ignited within the

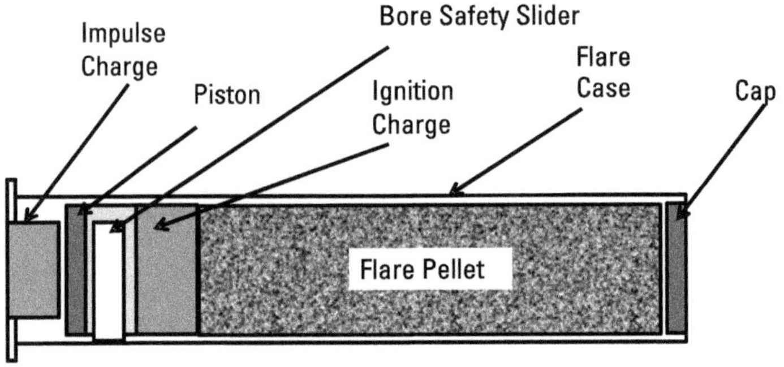

Figure 9.34 A bore safety device is used in pyrotechnic flares to prevent ignition of the flare pellet until it has cleared the flare case.

magazine. The combustion gasses produced will immediately eject the flare with little no damage to the flare dispenser or the aircraft.

9.10.10 Flare Cocktails

Flares are normally launched in combinations of two or three types to effectively counter missile attacks. Figure 9.35 shows a typical flare combination. The mix and the order of deployment of the flares is chosen to optimize response to the expected threats

9.11 Imaging Trackers

Imaging infrared trackers for anti-aircraft missiles have received great benefit from commercial developments in focal plane arrays and are becoming operationally important. An imaging tracker homes on a target that looks like an airplane. This allows it to discriminate against IR decoys.

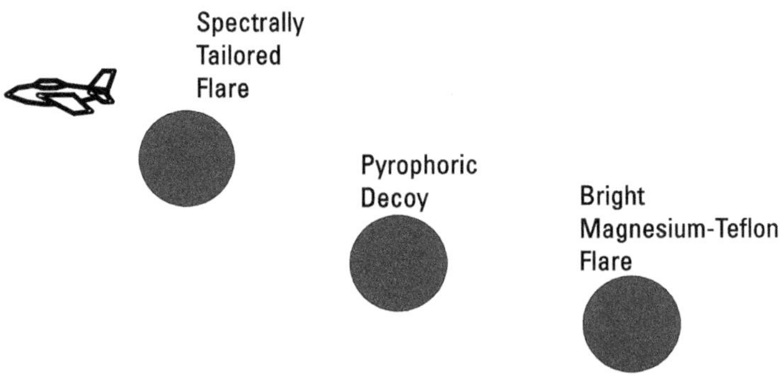

Figure 9.35 A combination of multiple types of flares is normally launched to provide optimum defense of an aircraft against the various types of IR missile trackers.

We have discussed IR decoys that capture a missile's tracker by creating a larger thermal profile and the types of missile trackers that are vulnerable to these types of decoys. Now we consider a new type of missile tracker that can discriminate against decoys on the basis of the shape of the energy pattern detected by the missile tracker. Early imagery trackers used scanned linear arrays to capture the scene in the tracker's field of view. These trackers were large and heavy (on the order of 40 pounds). Developments in weaponized focal plane arrays (FPA) now allow staring arrays with up to 256 × 256 pixels in trackers that weigh only about 8 pounds (including a recycling refrigerator that allows continuous operation for long periods).

The number of pixels that cover a target aircraft as a function of range is shown in Figure 9.36. At the typical acquisition range of 10 km, the target is in 1 or 2 pixels. At 5 km the target covers 4 × 4 pixels, at 1 km it covers 20 × 20 pixels, at 500m it covers 40 × 40 pixels, and at 250m it covers 80 × 80 pixels.

Imaging trackers operate in the atmospheric window at about 3 microns, which is where the plume signature peaks. These FPAs use indium antimonide (InSb) sensors cooled to 77K.

9.11.1 Imaging Tracker Engagement

As shown in Figure 9.37, there are three phases in the engagement: acquisition mid-course, and end game. Each phase has its own challenges.

9.11.2 Acquisition

At acquisition, the target looks like a white spot against a gray background. The big challenge is the thermal signal-to-noise ratio. Aerodynamic heating of the tracker dome is a significant source of thermal noise, and much development

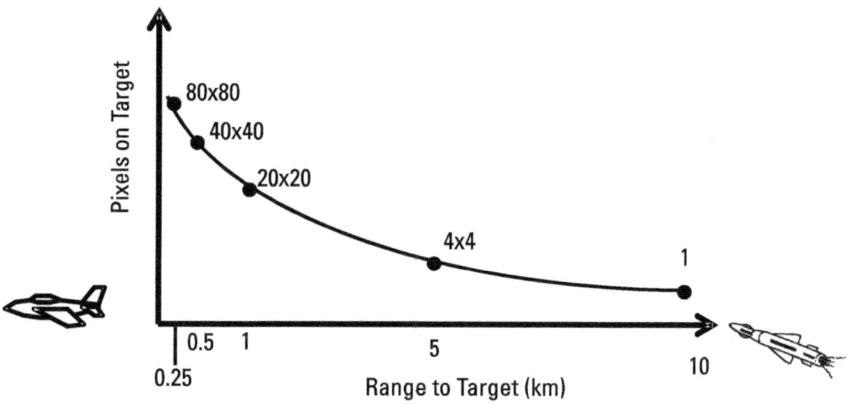

Figure 9.36 As the missile approaches the target the image resolution increases dramatically.

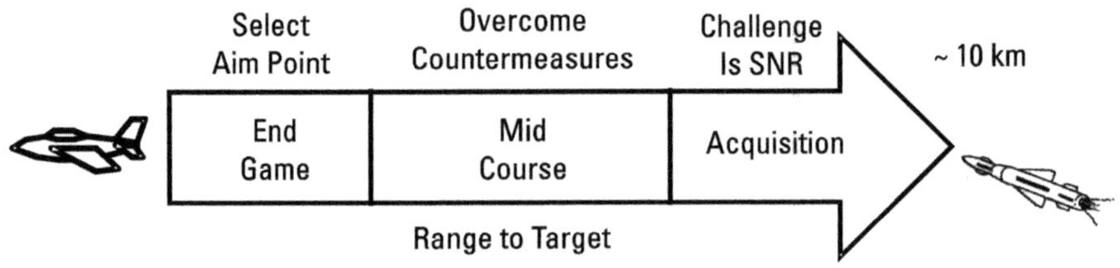

Figure 9.37 The engagement has three distinct phases, acquisition, mid course, and end game.

effort has been focused on optimum dome materials. The dome material has to be physically tough against rain impact and yet of high optical quality in the spectral region of interest. The current prevalent material is synthetic sapphire, which is cut into a flat lens mounted at an angle to the path of flight as shown in Figure 9.38.

9.11.3 Mid-Course

During the mid-course engagement phase, UV, IR, and, in some cases, radar missile warning systems (MWS) detect the approaching missile and initiate countermeasures. The primary mid-course challenge is rejection of these countermeasures. The countermeasures are either decoys to draw the missile away from the target aircraft or jammers to interfere with the missile tracker's operation. To continue tracking the target, the tracker must distinguish decoys from the target and reject them. As discussed earlier, decoys capture the attention of trackers by presenting greater energy at the tracking wavelength and in the tracking window than that of the target aircraft. There are sophistications in decoys to overcome two-color trackers and angular or rise-time decoy discrimination capabilities of trackers. However, imaging trackers present a new challenge

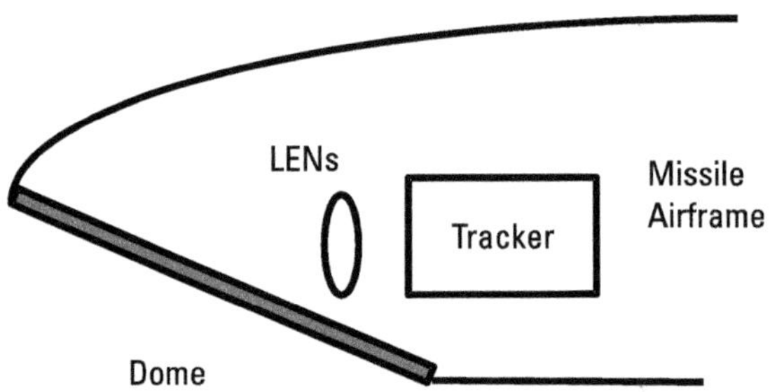

Figure 9.38 The dome can be a flat plate of material mounted at an angle to the missile airframe to minimize aerodynamic drag and heating, enhancing the thermal SNR. The lens is mounted on a gimbal behind the dome to give a large field of regard (FOR).

because they discriminate decoys from the intended target by physical size and shape.

When an imaging tracker is tracking a target and a decoy is deployed from the target, the tracker uses sophisticated software to perform correlation tracking. In one implementation, the tracker asks if a new energy source has the same shape as the shape which has been recently tracked; if not, the new energy source is rejected and the missile continues to track the original energy source.

During mid-course, the tracking FPA will typically have 7 × 7 or 9 × 9 pixels on the target aircraft. Figure 9.39 shows a 7 × 7 pixel array looking at a target, a hot flare, and a gray-body decoy. Note that the target presents a complex pattern of pixels receiving energy. The hot flare is physically small and therefore puts (lots of) energy into a single pixel. The gray-body decoy puts the amount of energy equivalent to a valid target into multiple pixels. Remember that this type of decoys uses rapidly oxidizing foil pieces that bloom to fill a large volume. However, the shape of the energy pattern is changed from the spatial energy distribution of the target. The key is that the shape does not have to look like an a priori stored image of what an aircraft should be. Rather, the tracker can reject the decoy because it does not correlate with the energy distribution seen a short time before.

Laser jammers present a significant challenge to imagery trackers because they can put significant energy into the tracker's FPA to saturate or even damage the array, thereby preventing target tracking. It is interesting to consider that IR missiles have been dealing with various types of decoys for 40 to 50 years, during which many tracking sophistications have been developed and deployed. However, laser-based jammers have only been used for one decade.

Look for significant hardware and software development to improve tracker performance in the presence of such countermeasures. This will be accompanied by improved jamming tactics.

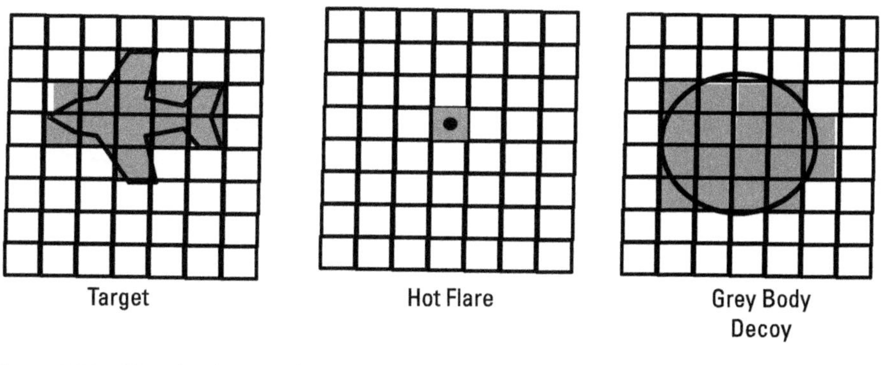

Figure 9.39 The distribution of energy in FPA pixels for a target, a hot flare, or a grey body decoy supports correlation tracking.

9.11.4 End Game

During the end game, the missile tracker has plenty of energy and lots of pixels on the target. Its challenge during this phase (the last second of flight) is to pick the optimum point at which to impact the target for maximum lethality. As shown in Figure 9.40, these high lethality aim points would include the cockpit, an engine, or an aircraft fuel tank. If the sensor energy level in each element of an FPA is quantized to 10 bits, the FPA dynamic range would be about 30 dB, plenty to distinguish the cockpit and other important elements of vulnerability for hit point selection.

9.12 IR Jammers

Platforms which must dwell within the lethal range of IR-guided threats would require an extremely large number of flares to provide adequate protection. Therefore, the use of IR jammers is the best solution.

As shown in Figure 9.41, an IR missile tracks on the IR energy from some part of the targeted aircraft. The jammer on the target aircraft directs modulated IR energy at the attacking missile. Received IR energy is processed to determine the direction in which the missile must be steered to home on the target. Figure 9.42 shows the missile tracker components. IR energy passes through a lens and a reticle to a sensing cell that generates a video signal from which a processor generates missile guidance commands.

The IR jammer creates a modulated IR emission that is transmitted toward the attacking missile and inputs modulated energy to the sensing cell. This energy causes the processor to output improper tracking information, either breaking the lock onto the target or steering the missile away from its intended target.

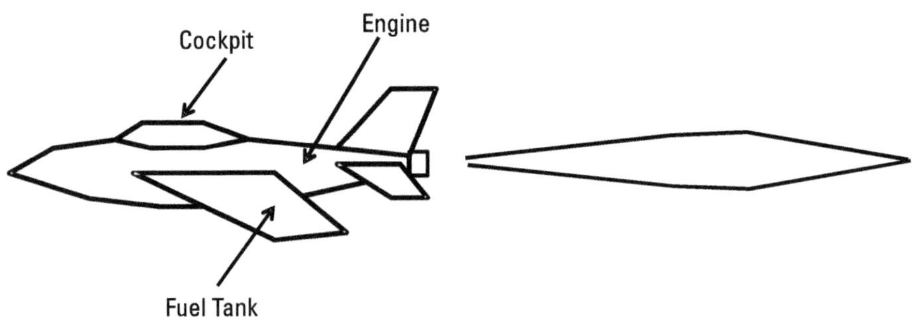

Figure 9.40 During the end game, the missile tracker can target the most vulnerable parts of an aircraft, such as the cockpit, engine or fuel tanks.

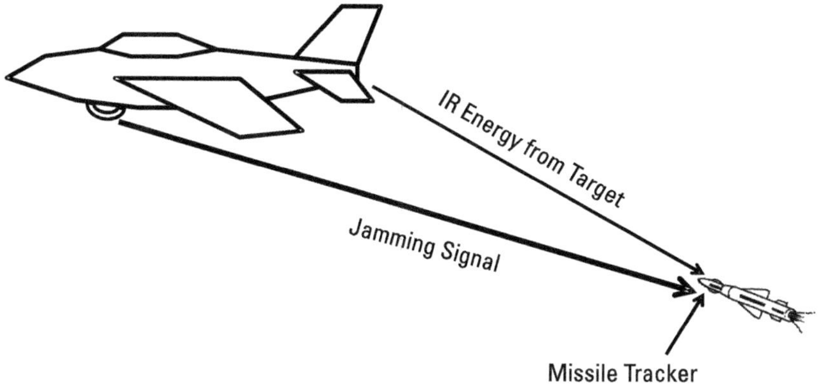

Figure 9.41 The IR jammer directs IR energy into the missile tracker with a waveform that causes the missile to fail to recognize a valid target or guide away from the target.

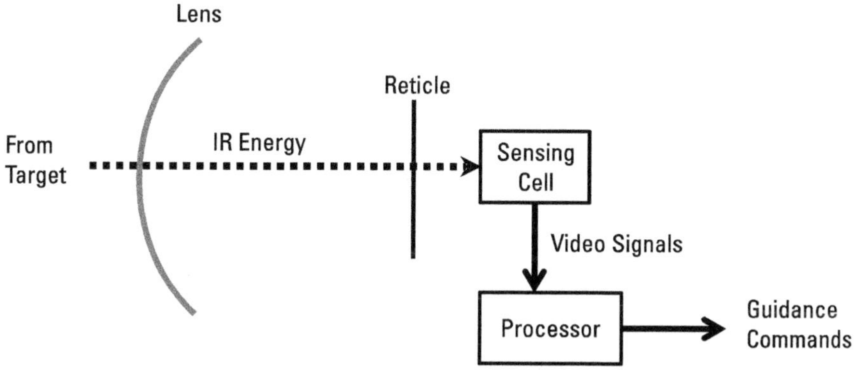

Figure 9.42 The tracker in the missile passes IR energy from the target through a reticle to a sensing cell which generates signals to its processor from which guidance commands are generated.

9.12.1 Hot-Brick Jammers

The earliest IR jammers had heated silicon/carbide blocks that emitted a high level of IR energy. As shown in Figure 9.43, these blocks are mounted in cylindrical housings with lenses over their vertical surfaces. Each lens has a mechanical shutter that is opened and closed to create an energy waveform like that which is created by the operation of the reticle in the missile's tracker. Thus, the jamming signal is accepted by the missile tracker's processor as a valid IR target. This type of jammer, sometimes called a hot-brick jammer, outputs jamming signals over a wide angular area, so it does not require accurate information about the location of the attacking missile and can jam multiple attacking missiles.

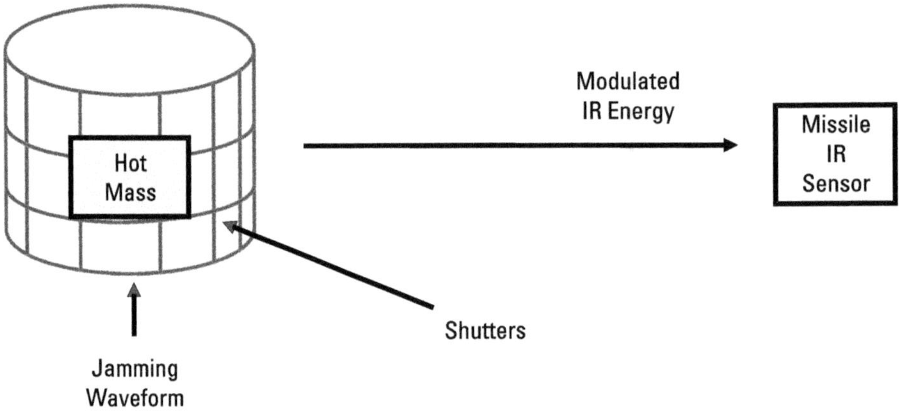

Figure 9.43 An early IR jammer had a heated mass inside a housing with mechanical shudders around 360 degrees. It emitted bursts of IR energy that look like the pulses reaching the tracker's sensor through a reticle.

9.12.2 Effect of Jammer on Tracker

Figure 9.44 shows some of the reticle types described earlier along with the IR energy patterns that they output to their sensing cells. The processor uses the timing or width of the video pulses from the sensing cell to determine the direction in which the missile must be steered to home on the target aircraft. In some missiles, the amplitude of the pulses or the number of pulses in each burst determines the magnitude of the angular offset of the tracker's optical axis from the target direction.

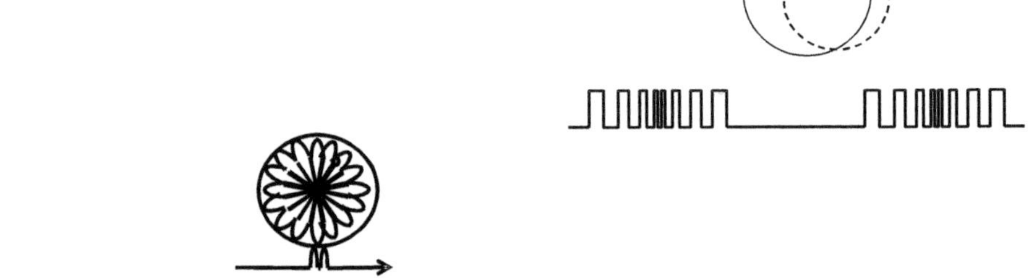

Figure 9.44 Each type of reticle causes the tracker sensor to see a different modulation waveform on the received IR energy.

Figure 9.45 shows the video generated from the target's IR energy (after passing through the reticle) for one type of missile. Also shown is a jamming signal. Both IR energy patterns enter the sensing cell, and their combined energy patterns cause a complex video signal pattern into the processor. Note how this combined pattern prevents the processor from accurately determining the number of pulses in a burst, the timing of the bursts, or the amplitude of the video pulses. Also note that the video signals from the jammer are much larger than those from the target. This is from the jamming-to-signal ratio (J/S), in this case, the ratio of the received jamming energy to the received target signal energy.

Section 9.11 discussed imaging trackers, for which the energy patterns are much different. This complicates the jamming approaches available, and this subject will be discussed later.

9.12.3 Laser Jammers

Figure 9.46 shows another type of IR jammer, capable of generating very high J/S. In this type of jammer, an IR laser generates the required jamming energy pattern and it is directed at the attacking missile with a steered telescope. These jammers are called directed infrared countermeasures (DIRCM) systems. There are several current programs using this technique, including: the common IRCM (CIRCM), the large aircraft IRCM (LAIRCM) and others. The telescope allows a very high level of IR energy to be placed into the missile's tracker (i.e., high J/S), but causes two significant requirements on the jamming system. First, the laser must generate signals at the correct wavelength to be accepted by the missile tracker. This requires multiple wavelength operation. Second, the system must know where the missile is located in order for the telescope to be properly oriented. Thus, the system must incorporate a missile tracking capability. This function can be performed by a radar, but the jamming

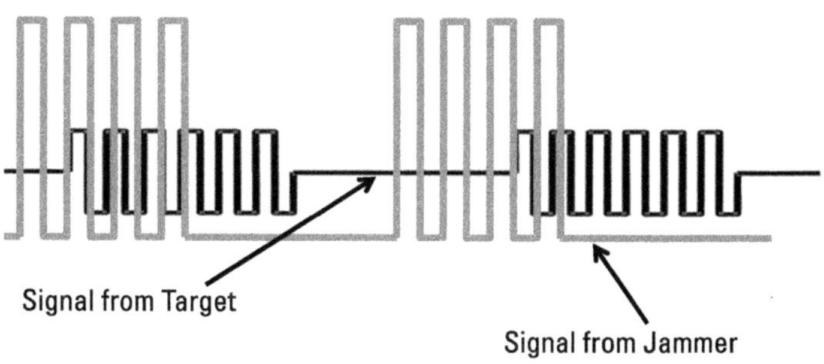

Figure 9.45 The processor in the attacking missile receives superimposed video waveforms from the target's IR energy and from the jammer energy. The presence of the jammer video prevents the tracker from determining the relative location of the target aircraft.

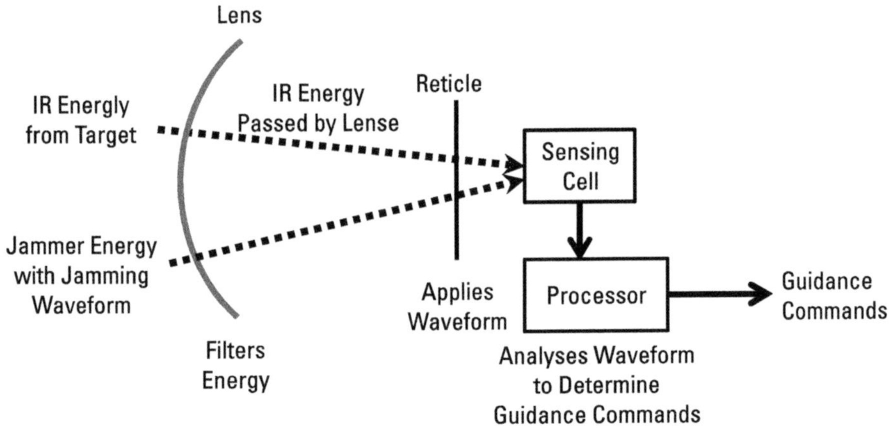

Figure 9.46 A laser based jammer detects and locates the missile. The laser is modulated with the proper jamming waveform and the telescope directs the laser jamming signal at the missile.

system most often locates and tracks the missile through the ultraviolet (UV) energy of its plume or the IR signature of the missile from aerodynamic heating. Whatever the technique, the missile must be located accurately enough for the jammer's telescope to get enough IR energy into the missile tracker to create the required J/S level.

9.12.4 Laser Jammer Operational Issues

Now we examine some specifics about jammers which use lasers. Because it is directed at the missile tracker, a laser can generate a significant energy level into the missile tracker's sensing cell, thus creating a significant J/S. However, as missile trackers become more sophisticated, the jamming patterns also need sophistication. The object is to cause the missile's logic to send the missile away from the target or to convince the missile's processor that there is no valid target, so the missile is blocked from being fired.

Target trackers in missiles have dealt with challenges posed by flares for decades, and many counters to these countermeasures have been developed and deployed. However, IR jammers are relatively new and bring new challenges. IR missile trackers and IR jammers are in a competitive cycle that will see a continuing series of measures and countermeasures on both sides for the next few years.

As mentioned above, a laser-based jammer must detect and locate hostile missiles so its telescope can be aimed at that missile to direct energy to the missile's tracker. As shown in Figure 9.47, the tracker's lens provides filtering that allows only signals in the operating band to pass into the tracker. The shorter wavelength bands track hotter targets like internal jet engine parts, but longer

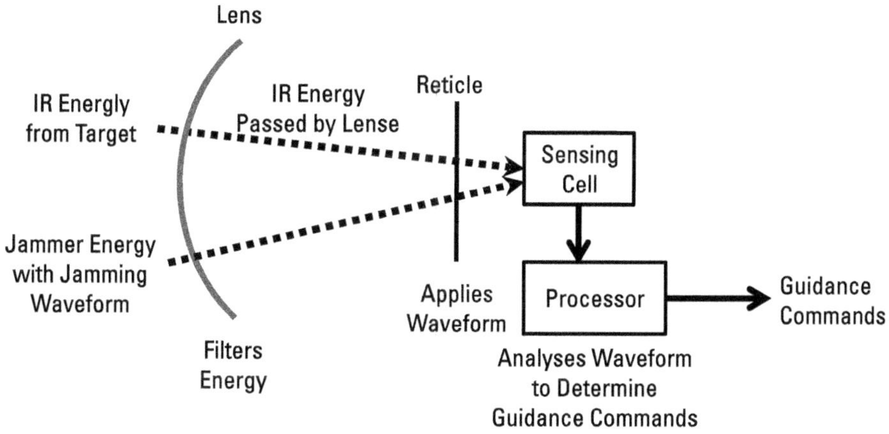

Figure 9.47 The missile tracker lens filters the energy to the wavelength at which the tracker is designed to operate.

wavelengths are required to track lower temperature targets such as plumes and aerodynamically heated airframe surfaces. Imagery tracking also requires long wavelengths. These longer wavelength trackers need to be cooled, typically to 77K. Because missiles operate only for a few seconds, they can usually use expanding gas for cooling, however longer engagements require longer-term refrigeration-type cooling. This longer-term cooling is also required in the missile detector portions of laser-based IR jammers. This is particularly important when a jammer is operating in a preemptive mode, keeping a missile from acquiring the target. To reduce the time to bring trackers to the proper temperature, work has been done on higher temperature sensors at about 100K. Simplifying the cooling system reduces the complexity of the trackers, improving system reliability.

9.12.5 Jamming Waveforms

A sophisticated jammer will have a library of jam codes that can be tried very quickly. The subsystem that tracks the attacking missile then must look for erratic missile movement to determine that the correct jamming code has been applied. The correct jamming code will look like the waveform created by the specific missile's reticle, but will interfere with the tracker's operation. First, let's consider the types of rotating and nutated reticles. The jamming waveform must be accepted by the tracker and cause the tracker to steer away from the target. Here are two examples.

9.12.5.1 Nutated Tracker Reticle

The waveforms from the nutated tracker are as shown in Figure 9.48. At the left, the target is centered in the reticle because the missile is locked onto its tar-

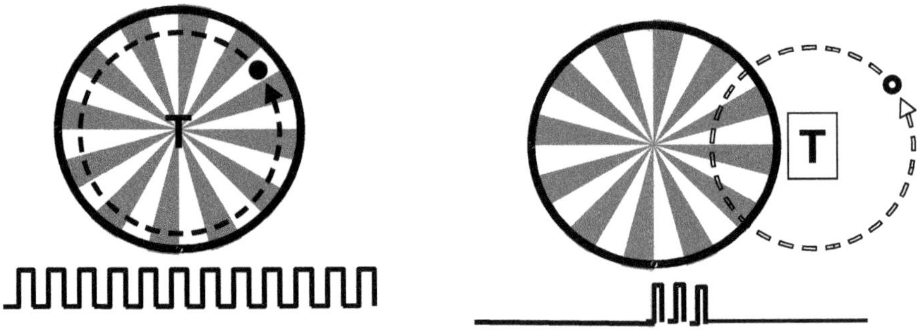

Figure 9.48 To jam a tracker with a nutated reticle, energy must be input in a pattern that will move the target out of the reticle.

get. This produces a square wave energy pattern to the sensing cell. On the right of the figure, the target is outside the reticle and the energy pattern is much different. If a jammer applies a strong signal with this energy pattern, the tracker will move to the right to try to center the target in the reticle. This will cause the missile's aiming point to move to the right, away from the intended target.

9.12.5.2 Proportional Guidance Reticle

Figure 9.49 shows a rotating reticle that has different numbers of clear and opaque segments as a function of the angle of the target from the center of the reticle. At the left of the figure, the target is centered in the reticle because the missile is locked onto the target. Thus, the energy pattern to the reticle is zeroed. At the right of the figure, the target is at the edge of the reticle and the

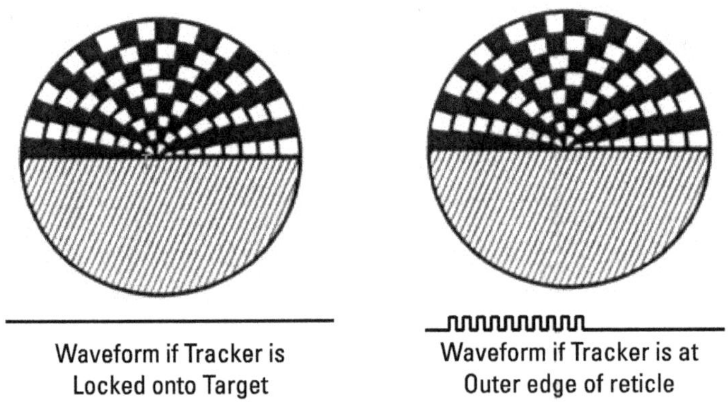

Figure 9.49 For a tracker with a multiple frequency reticle, the jamming signal must cause the tracker's processor to conclude that the tracking point should move in a way that will cause it to steer away from the target.

energy waveform to the sensing cell has 10 pulses for each rotation of the reticle. If a strong jamming signal with 10 pulses is transmitted into the sensing cell, the tracker will move in the direction it thinks is required to center the target (and thus cause the energy waveform to zero), so the tracker will move away from the actual target location.

9.12.5.3 Imagery Tracking

Imagery tracking requires an FPA of IR sensors as shown in Figure 9.50. The trend is toward larger numbers of pixels in arrays because this allows more accurate images for better target discrimination. We have talked about pattern tracking. The location of the thermal image of the target in the FPA determines the direction the missile needs to move to lock onto the target. When a flare is used to lure the tracker away from the aircraft, a sophisticated tracker will compare the image to the image it was tracking a second or so before and reject the larger flare signature. This presents a tough problem to counter-IR missile defense systems. The IR image of a tracked aircraft is constantly changing as the aircraft maneuvers, so generating a standard pattern that could be moved away from the center of the FPA looks very difficult. One promising approach (in discussions with people in that business) seems to be to put a very strong signal into the FPA to saturate it, causing the display to bloom and thus fail to detect the aircraft. Another approach discussed is to use even more energy to burn out pixels in the FPA. It is important to note that it takes about three orders of magnitude more power to damage circuitry than that required to temporarily disable it.

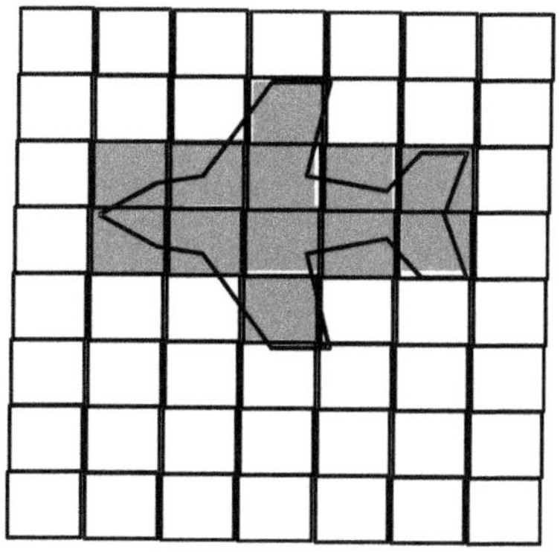

Figure 9.50 The FPA in an imagery tracker generates a digital signal capturing the pattern of pixels illuminated by energy from the target. The shape of the image changes as the aircraft maneuvers.

10

Radar Decoys

10.1 Introduction

The purpose of any decoy is to make a sensor believe it is seeing something real. This, of course depends on the way that the sensor receives its information. If the sensor is optical or thermal, the decoy must create the proper optical image, including size, shape, and color (or wavelength). For example, before the Normandy invasion in World War II, fake facilities were built in locations that would make enemy airborne photo reconnaissance conclude that the invasion was to take place in Calais.

A radar identifies potential targets by analysis of the signals reflected from objects illuminated by its transmitter. Thus, radar decoys must generate false returns that the radar will decide are real targets.

In this series, we will discuss decoys in terms of their operational missions, the way they generate false targets in the radars they counter, and how they are deployed.

10.1.1 Missions of Decoys

There are three basic missions of radar decoys as shown in Table 10.1: saturation, seduction, and detection.

Saturation decoys, as shown in Figure 10.1, create many false targets that look enough like real targets to force the radar to expend time and processing resources to distinguish real from false targets. Ideally, the radar will be unable to make this differentiation and must therefore expend many weapons to destroy the few targets among the many false targets. Even if the decoy cannot completely fool the sensor, it should be difficult enough to differentiate that

Table 10.1
Missions and Platforms Versus Types of Decoys

Decoy Type	Mission	Platform Protected
Expendable	Seduction and saturation	Aircraft and ships
Towed	Seduction	Aircraft
Independent maneuver	Detection	Aircraft and ships

Figure 10.1 Saturation decoys create many false targets to overload the capability of a target sensor or of the weapons it controls.

the detection process is significantly slowed. The mission of the decoys is thus to saturate the enemy's information throughput so that there will not be time during an engagement to defend against an attack. In this case, the decoys must look enough like real targets to fool the radar to some level. The radar's analysis capabilities dictate the necessary decoy features; the more sophisticated the radar processing, the more complex the decoys must be.

Seduction decoys are placed within a radar's resolution cell along with a real target that is being defended, as shown in Figure 10.2. The resolution cell is the volume in which the radar cannot determine whether a single target or multiple targets are present. The cell is shown in two dimensions for simplicity, but is actually a three-dimensional volume. In this mission, the decoy must look more like the target than the target does. To be successful, the seduction decoy must "seduce" the radar's tracking circuits away from the real target to itself. Now, the radar is tracking the decoy rather than its intended target, and the radar will center its resolution cell on the decoy. As the decoy moves away from

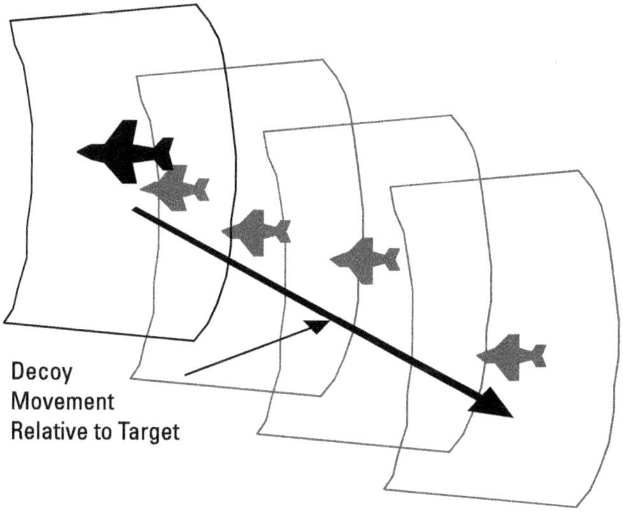

Figure 10.2 A saturation decoy "seduces" the tracking of a radar away from its intended target and leads it to another location.

the protected target, it takes the resolution cell with it. When the resolution cell no longer contains the real target, the weapon being guided by the radar will be guided to the decoy.

Detection decoys look enough like real targets to cause the radar to acquire and track them. When a radar is searching for targets, a false target can cause the radar to perform its design function. If the radar is a dedicated acquisition radar, the target will be passed to a tracking radar. As discussed in Chapter 4, the operating philosophy in a defensive network is now normally hide, shoot, and scoot. That is, radars remain off the air as long as possible before launching weapons and then move away from their launching locations as quickly as possible. If a radar decoy looks like a credible target, an enemy will be forced to bring up its tracking radars. As shown in Figure 10.3, these tracking radars can be attacked by antiradiation missiles.

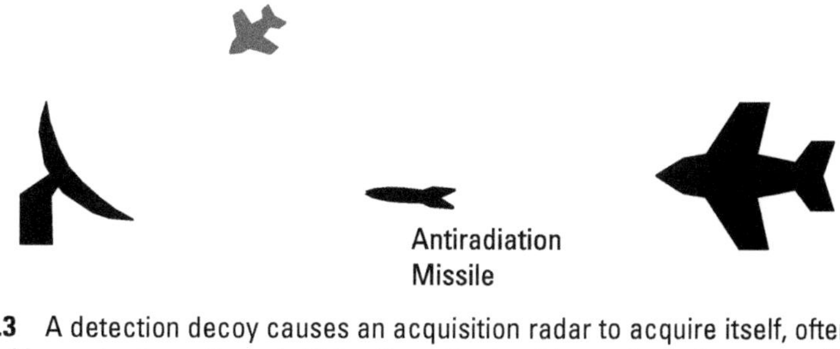

Figure 10.3 A detection decoy causes an acquisition radar to acquire itself, often requiring that a tracking radar be activated. This allows the tracking radar to be targeted by an antiradiation missile.

Later in this chapter, we will deal with sophistications in modern radars that require the creation of very detailed radar cross sections to make decoys look like credible potential targets.

10.1.2 Passive and Active Radar Decoys

A *passive* decoy creates a radar cross section physically. Obviously, if the decoy is the same size, shape, and material as the real target that it is simulating, it will have the same radar cross section. However, there are ways that a decoy can make itself look larger. One common technique is to incorporate a pattern of corner reflectors. A corner reflector produces a radar cross section significantly larger than its actual size. The formula for the radar cross section of a corner reflector with circular edges as shown in Figure 10.4 is:

$$\sigma = \left(15.59L^4\right)/\lambda^2$$

where σ is the radar cross section in square meters, L is the length of a side, and λ is the wavelength of the illuminating signal.

If the side is one-half of a meter and the illuminating signal is at 10 GHz (i.e., the wavelength is 3 cm), the radar cross section is 1,083 m^2.

Chaff, which comprises a large number of half-wavelength pieces of aluminum foil or plated strands of fiber glass, can be deployed into a cloud that has a very large radar cross section and can thus act as a decoy.

An *active* decoy, as shown in Figure 10.5, includes electronic gain to create a radar cross section. This can either be an amplifier or a primed oscillator that generates a powerful signal to simulate a radar return from an object much

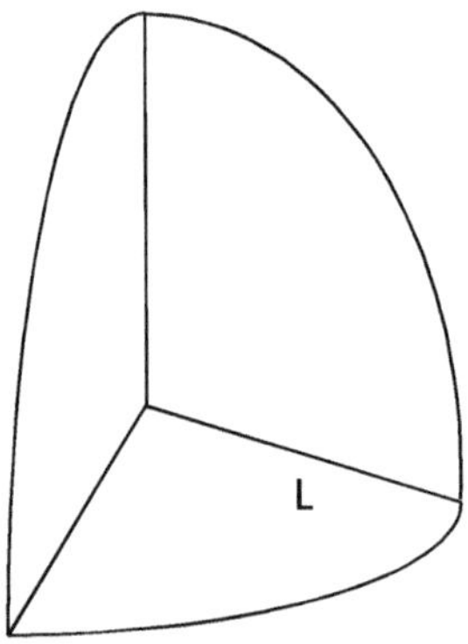

$$\sigma = \frac{15.59\ L^4}{\lambda^2}$$

Figure 10.4 A corner reflector can generate a radar cross section much larger than its physical size.

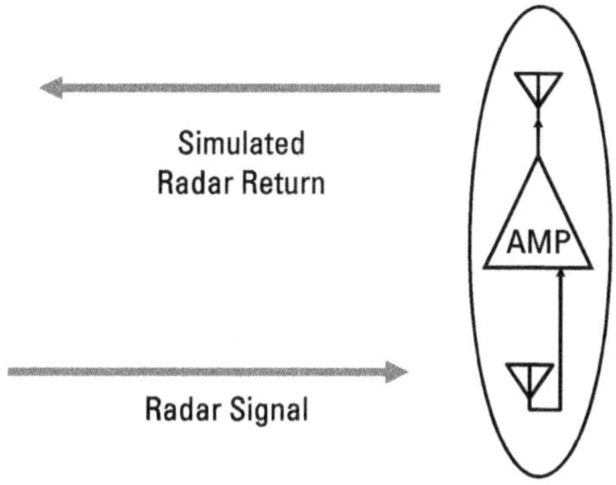

Figure 10.5 An active decoy creates a large radar cross section by amplifying and rebroadcasting signals received from a target radar.

larger than the decoy, but with the same frequency and modulation as the signal produced by the target radar. As we will see later in this series, the signal returned to the radar must sometimes have complex modulation to avoid being rejected as a false signal by the radar.

10.1.3 Deployment of Radar Decoys

Radar decoys must be physically separated from the platforms they protect against radar-controlled weapons. As shown in Figure 10.6, this separation can be achieved by expending decoys from the protected platform, by towing decoys behind the platform, or by independently maneuvering those decoys. There are important examples of decoys using each of these deployment techniques that will be discussed later. As you will see, the features and capabilities of modern radars have had great impact on the nature of each of these types of decoys.

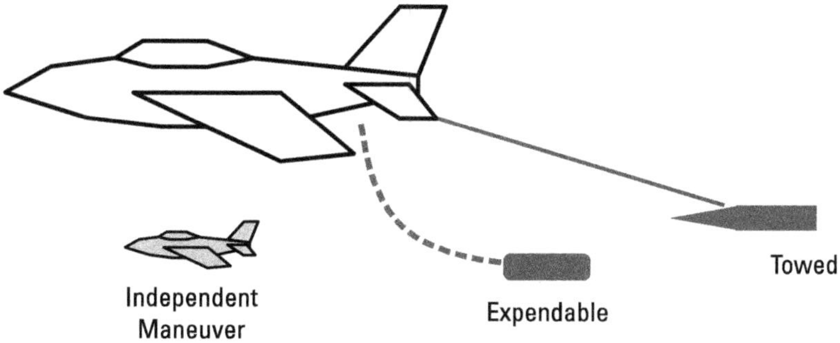

Figure 10.6 Decoys can be separated from the platforms they protect by being expended or towed, or by independently maneuvering.

10.2 Saturation Decoys

Saturation decoys protect friendly assets by providing false targets to hostile weapons. These decoys can be airborne, sea-based, or ground-based. In each case, the decoy must be perceived by the enemy weapon system's sensors as a credible target. If the hostile sensors are not sophisticated, the decoy may just have to create a radar cross section of the same order of magnitude as that of the protected asset. However, many modern weapon sensors have become more sophisticated and it appears that the level of sophistication will continue to increase for the foreseeable future.

In this chapter, we will not limit our discussion to existing systems, but will consider all of the weapon and decoy techniques that seem practical within the predicted state of the art. The philosophy here is: If it has not yet been developed, it will be pretty soon. Therefore, we should be thinking about what we will do about it.

10.2.1 Saturation Decoy Fidelity

To be useful, saturation decoys create credible false targets. Consider how a radar can differentiate a real target from a decoy. First, there is the size and shape of the target platform. A decoy is typically much smaller than the aircraft or ship it simulates; therefore its radar cross section (RCS) must be enhanced. This can be done mechanically by the addition of corner reflectors or some other highly reflective shape features. However, it is normally most practical to enhance the radar cross section electronically by providing gain to increase the signal propagated back to the illuminating radar. The RCS perceived by the hostile radar is given by the formula (in algebraic form):

$$\sigma = \lambda^2 G / 4\pi$$

where σ is the RCS produced by the decoy in square meters, λ is the wavelength of the radar signal in square meters, and G is the combined gain ratios of the decoy's receiving and transmitting antennas and its internal electronics as shown in Figure 10.7.

The following is the same formula, but in decibel form

$$\sigma = 38.6 - 20\log_{10}(F) + G$$

where σ is the RCS in dBsm, F is the radar frequency in megahertz, and G is the combined gains in the decoy in decibels.

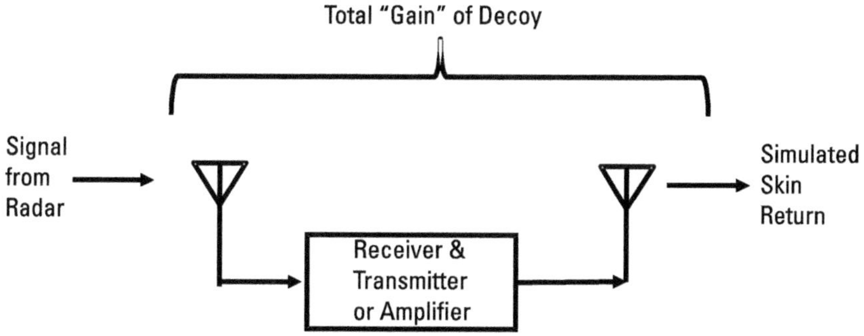

Figure 10.7 The sum of receiving antenna gain, transmitting antenna gain and processing gain of an active decoy determines the RCS it will simulate.

For example, if the radar signal is at 8 GHz, the decoy's receiving and transmitting antennas each have 0-dB gain and its internal electronics gain is 70 dB, the decoy will simulate a 1,148.2 m^2 RCS.

$$\sigma(\text{dBsm}) = 38.6 \text{ dB} - 20\log(8{,}000) + 70 \text{ dB} = 38.6 - 78 + 70 = 30.6 \text{ dBsm}$$
$$\text{antilog}(30.6/10) = 1{,}148.2 \text{ m}^2$$

10.2.2 Airborne Saturation Decoys

Figure 10.1 showed a large number of airborne targets, including one real target and a number of decoys. For a hostile radar to take the decoys seriously, they must look (to the radar) much like the real target. This means that they must have approximately the same RCS. However, there are also other considerations.

Chapter 4 discusses pulse-Doppler radars, which are widely represented among modern threat radars. The processing circuitry of a pulse-Doppler radar includes a time versus frequency matrix as shown in Figure 10.8 in which the time of arrival and received frequency of multiple targets is captured. For each target, the time of arrival represents the range to the target, and the received frequency is determined by the Doppler shift in the received signal. Because the Doppler shift is a function of the rate of change of range to the target, this chart can be considered a range versus velocity matrix. The frequency data comes from a bank of filters, usually implemented in software. Note that this bank of filters can also analyze the spectrum of a received signal.

Because aircraft have significant velocity, they present significant Doppler shifts. If decoys are expended from an aircraft, they will slow down quickly in response to atmospheric drag. This will make a significant change in Doppler shift for which the retransmitted radar signal must compensate. A hostile radar may be able to reject a decoy that returns a time-varying signal frequency with

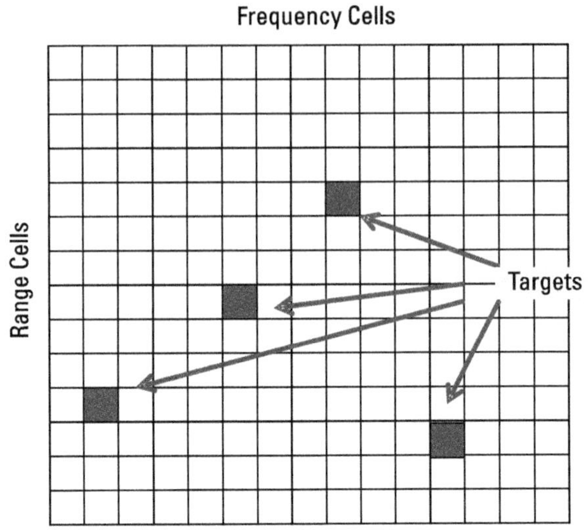

Figure 10.8 Pulse-Doppler radar processing includes a matrix of range vs. frequency cells that allows the determination of the frequency of each received return signal.

the characteristic shape of the atmospheric drag deceleration curve. This means that a decoy may need to return radar signals with an appropriate frequency shift to simulate the correct Doppler shift of a real target. Figure 10.9 shows the velocity versus time for an object (e.g., a decoy) expended from a moving

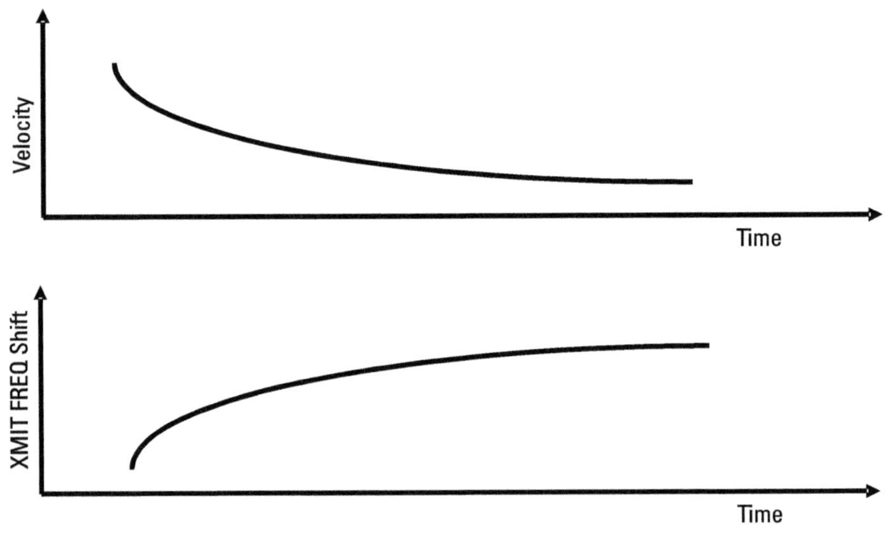

Figure 10.9 A decoy expended from an aircraft is slowed by atmospheric drag which reduces as the object slows. To simulate the velocity of the aircraft which expended the decoy, the transmitted frequency from that decoy must be increased by a time varying amount.

aircraft and the frequency shift required to make a radar believe that the expended object has the same velocity as the aircraft which expended it.

Jet engine modulation (JEM) is complex amplitude and phase modulation from the motion of the moving internal parts of a jet engine. It changes with aspect angle and can be detected in the skin return to a radar which is up to 60° off of the flight path of a jet powered aircraft. If a hostile radar can detect JEM modulation, it will notice that a decoy which does not have a jet engine, will not have this modulation feature and can easily discriminate against it. Thus, JEM modulation may need to be placed on simulated skin returns from decoys.

In Chapter 8, we discussed the way digital RF memories (DRFM) can simulate the complex RCS of a tactical aircraft. As shown in Figure 10.10, if a hostile radar has the capability of analyzing the frequency spectrum of a skin return, it will determine that a decoy return has a much simpler waveform than that returned by an aircraft. This would allow for the rapid rejection of a decoy as a potential target. To overcome this capability of a sophisticated radar, the decoy must modulate its output signal to create a complex, realistic RCS characteristic.

10.2.3 The Radar Resolution Cell

At this point, let us take a moment to discuss the radar resolution cell. This is the spatial volume in which the radar cannot determine whether there is a single target or multiple targets. Figure 10.11 shows this in two dimensions for simplicity, but it is actually three-dimensional, comprising a range slice of the conical volume within the antenna's beamwidth. The dimensions of the resolution cell are normally calculated as:

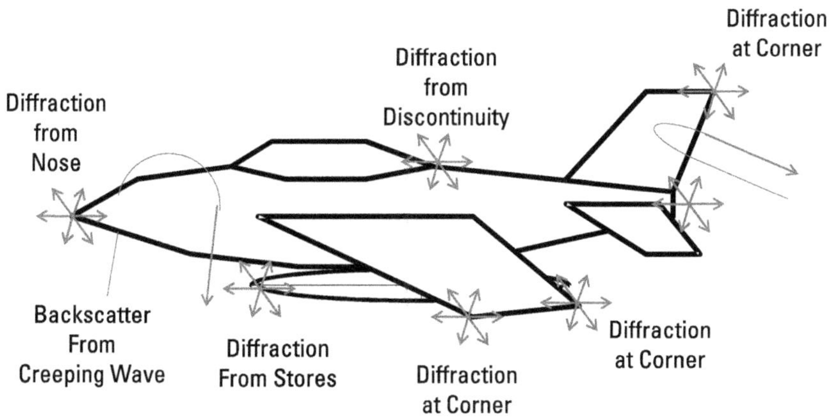

Figure 10.10 There are many contributing factors to the RCS of an aircraft. Together, they cause an RCS with complex amplitude and phase components.

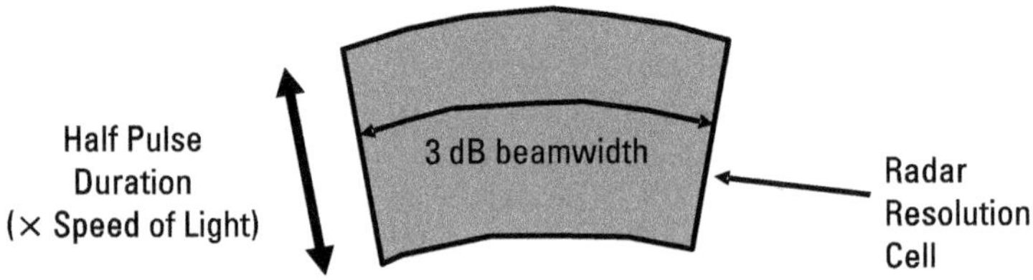

Figure 10.11 The radar resolution cell is the volume in which the radar cannot determine if there is only one target or multiple targets.

- Cross-range resolution = $R \times 2\cos(\mathrm{BW}/2)$, where R is the range to the target from the radar and BW is the radar antenna's 3-dB beamwidth.
- Down-range resolution = $c \times \mathrm{PW}/2$, where c is the speed of light and PW is the radar's pulse width.

For CW radars, the down-range resolution is calculated from the same formula, but with the radar's coherent processing interval replacing pulse width.

In Chapter 4, two techniques for improving the range resolution (chirp and Barker code) were discussed. At this point, note that these techniques along with some multipulse techniques can reduce the effective size of the resolution cell.

10.2.4 Shipboard Saturation Decoys

Active or passive decoys can be used to protect ships from anti-ship missiles. As shown in Figure 10.12, decoys with about the same radar cross section as the ship being protected can be placed in a pattern around the ship. When an anti-ship missile is fired at a ship from an aircraft, a ship, or a shore-based site, it will be inertially guided to the location at which the ship was detected. Then, when the missile comes within radar range, its on-board radar will acquire a target as shown in Figure 10.13. This acquisition range depends on the type of

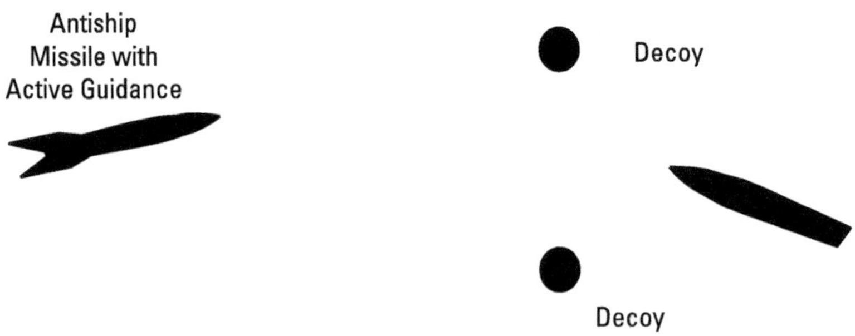

Figure 10.12 Distraction decoys create many false targets to overload the capability of a target sensor or of the weapons it controls.

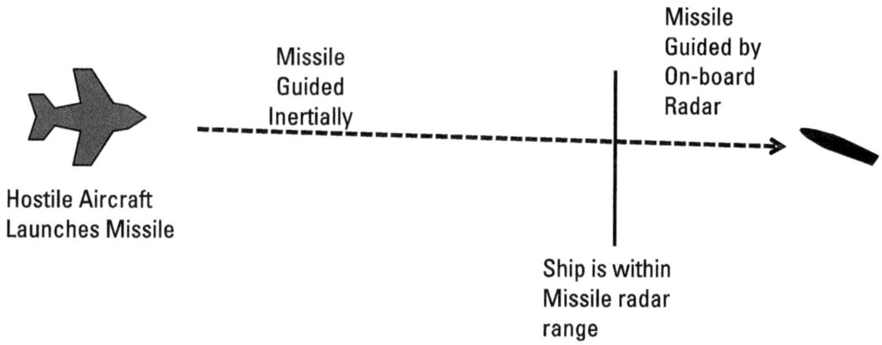

Figure 10.13 An anti-ship missile is launched from long range. It is guided to ship general location inertially—when within radar range, its on-board radar guides it to the target.

missile and the type of target, but will typically be 10 to 25 km. Ideally (from the missile's point of view), the missile's on-board radar will acquire its desired target and allow the missile to be guided to the center of that target. However, if the missile cannot distinguish the target from the decoys, it may acquire a decoy rather than the ship. If there are n decoys, the probability of acquiring the ship is reduced by the factor: n/n+1.

Like an aircraft saturation decoy, the ship protection saturation decoy must present an RCS roughly equivalent to that of the ship. Because a decoy is much smaller than the ship it protects, its RCS must be enhanced. This can be done by the incorporation of corner reflectors or the electronic generation of large signal returns. Note that, like an aircraft, a ship has a rather complex RCS. If the anti-ship missile can distinguish the features of a ship's RCS from those of a decoy, it can quickly reject the decoys as targets. Against such a missile, the decoys must present complex RCS patterns like those discussed above for aircraft protection decoys. Note that the generation of a complex, multifacet RCS requires significant processing power, typically provided by multiple digital radio frequency memories (DRFM) implement on locally programmable gate arrays (LPGA).

As shown in Figure 10.14, chaff clouds can also be used as distraction decoys. Each distraction chaff cloud has approximately the RCS of the protected ship and is placed near but outside of the radar's resolution cell. If the attacking missile sees a distraction chaff burst before it sees the ship and cannot distinguish it from the ship, the missile will home on the chaff cloud.

Note that the placement of distraction decoys or chaff clouds must not direct the missile toward another friendly ship as in Figure 10.15. The missile has a delayed contact fuse, so it will not be detonated by the decoy or chaff cloud. If it emerges from the chaff cloud (or passes the decoy), it will go back into its acquisition mode, and if it then acquires another ship as a target, there will be no time for the newly targeted ship to take effective countermeasures.

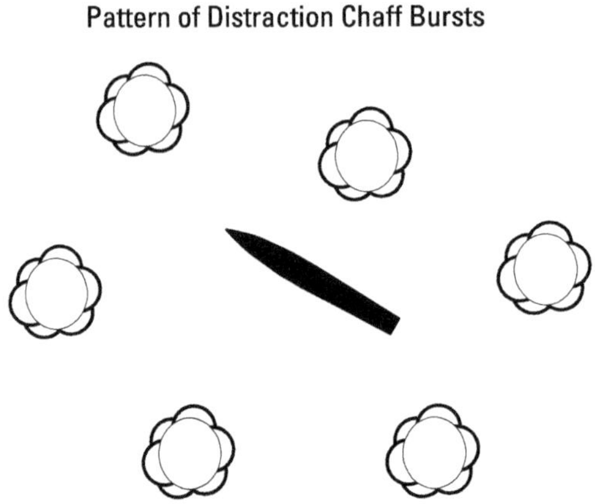

Figure 10.14 Saturation decoys create many false targets to overload the capability of a target sensor or of the weapons it controls.

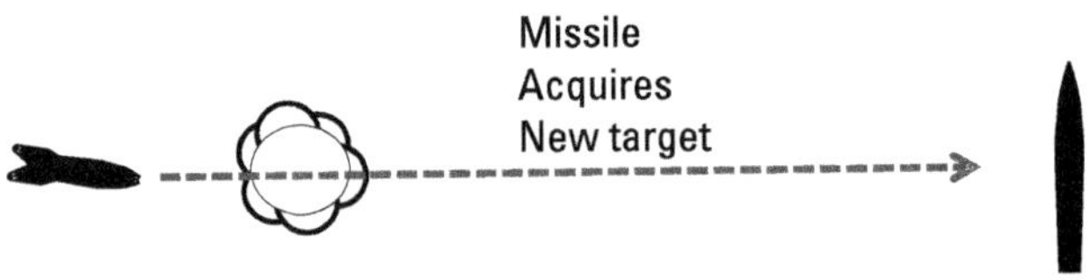

Figure 10.15 An anti-ship missile does not fuse on a chaff cloud, so it can acquire a new target when it passes the cloud.

10.2.5 Detection Decoys

What we are calling detection decoys are the same as the distraction decoys that we have been discussing; however, their purpose is different. In this case, the object is to cause an enemy to expose its electronic assets. For example, as in Figure 10.16, an enemy acquisition radar would acquire a decoy as a valid target and hand it off to a tracking radar. The tracking radar, which had been off the air (and thus nondetectable), would start to emit. This allows the tracking radar to be detected and located by a friendly asset. The enemy tracking radar can then be destroyed by a radar homing missile [like the high speed antiradiation missile (HARM)] or by other types of bombs or missiles.

10.3 Seduction Decoys

The mission of a seduction decoy is to capture the tracking function of a threat radar, causing the radar to lose track on its chosen target and acquire the decoy as a false target. This is done both for the protection of ships and aircraft. The decoy turns on within the radar's resolution cell as shown in Figure 10.17.

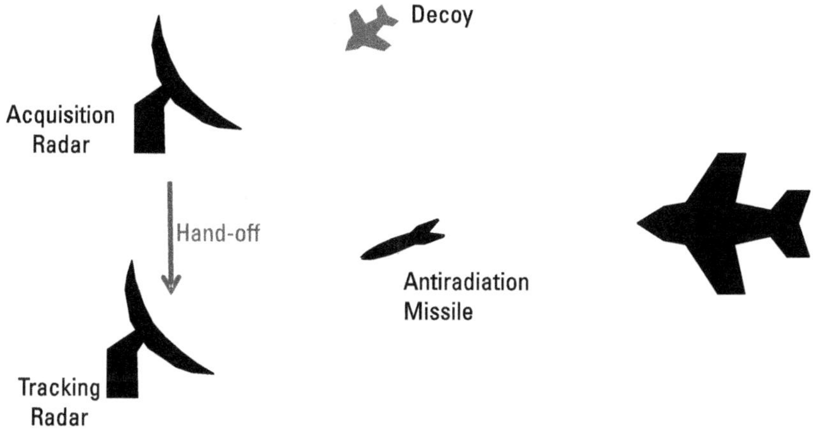

Figure 10.16 A detection decoy is detected by an acquisition radar—which hands off the target to a not yet active tracking radar. This causes the tracking radar to emit—so that it can be detected and attacked.

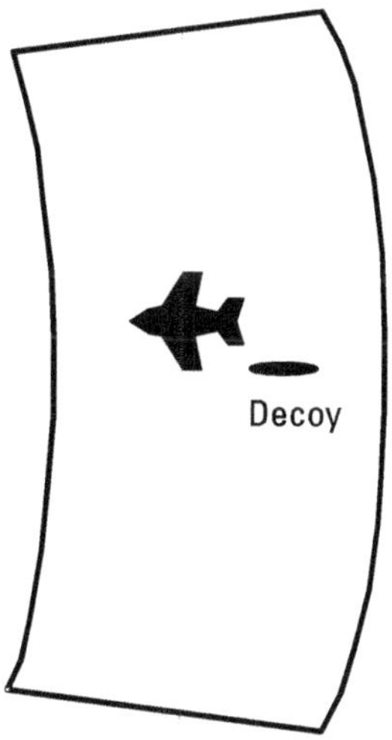

Figure 10.17 When tracking, the threat radar centers its resolution cell on the target. The seduction decoy turns on within the threat resolution cell, presenting an RCS significantly larger than that of the target.

Inside the resolution cell, the radar cannot detect the presence of a second target. It assumes there is only one target, located between the two targets in its cell. The assumed target location is proportionally closer to the target with the greater radar cross section (RCS) as shown in Figure 10.18. This means that the decoy must present a larger RCS. Double the RCS is highly desirable.

If the radar has pulse compression as discussed in Chapter 4, the decoy must start within this reduced volume.

If the radar is initially tracking the target, it will see the target RCS as shown in Figure 10.19. Then, when the decoy turns on, the radar will see the combined RCS of the decoy and the target. The decoy will move away from the target, so the target will ultimately leave the resolution cell. Then the radar will see only the RCS of the decoy.

It is natural to worry that the radar will detect these changes in RCS and reject the decoy. Understand that the actual measured RCS of either a ship or an aircraft typically looks like a fuzzy ball, with very quick changes of RCS over small angular changes. The data is smoothed (i.e., averaged over a small number of azimuth or elevation degrees) before it is plotted. Therefore, the observed RCS can change quite a bit as the target and/or radar platform maneuver, but with a much slower changing average RCS. In discussions of possible processing sophistication, I often say, "This is a rocket, not a rocket scientist." That said, it is a real possibility that future sophistication of radar processing will enable this counter-countermeasure tool. As a side note, you might want to review some of the processing tricks used by IR missiles in Chapter 9.

Figure 10.20 shows the location of the resolution cell after a short time, if the decoy is successful. This is very powerful countermeasure, because the radar cannot even see the target after it leaves the resolution cell.

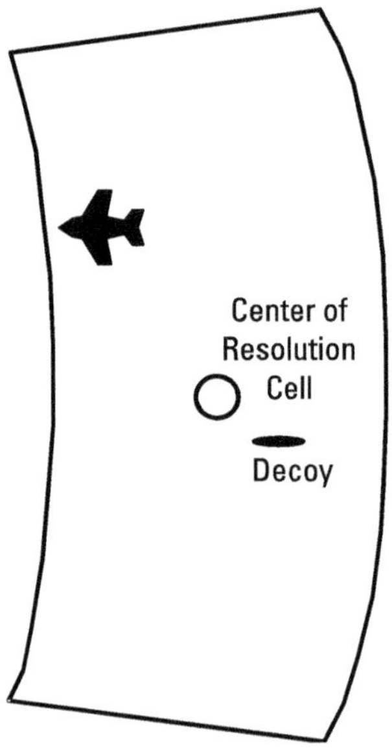

Figure.10.18 When the seduction decoy turns on within the threat resolution cell, presenting an RCS significantly larger than that of the target, the cell is centered on a point closer to the decoy by the ratio of the decoy RCS to the target RCS.

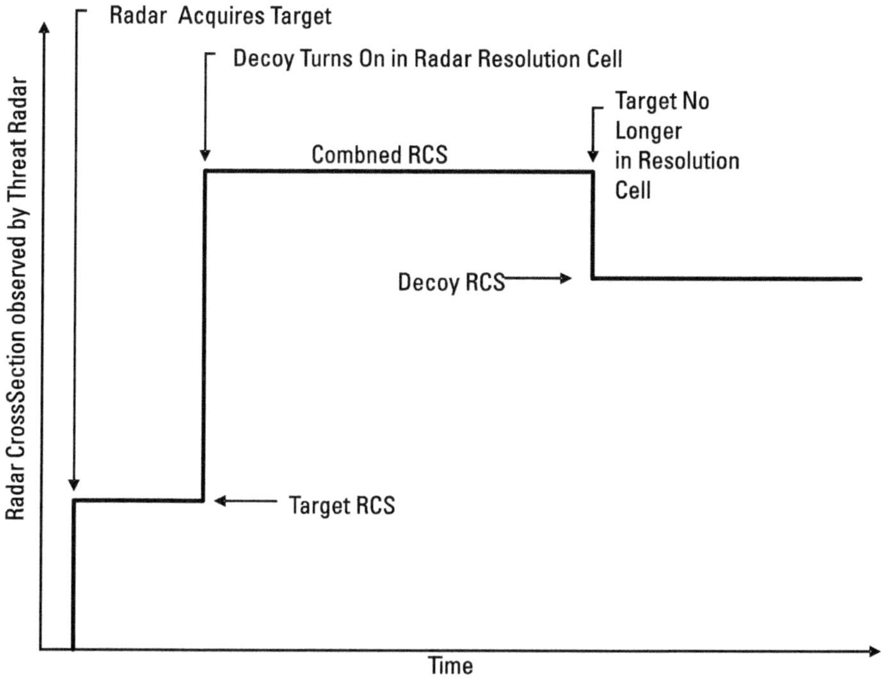

Figure 10.19 When the decoy turns on, the hostile radar sees a large increase in RCS. Then when the decoy leaves the resolution cell, the radar sees only the RCS of the decoy.

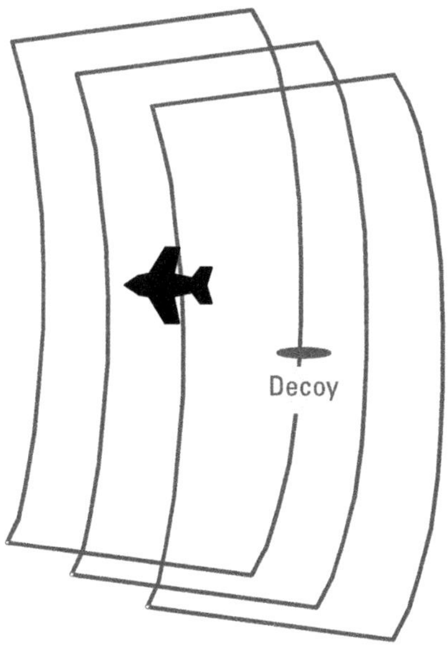

Figure 10.20 The greater RCS of the decoy causes the threat radar's resolution cell to track it as it moves away from the target.

Figures 10.17 and 10.20 show a radar tracking an aircraft. Figure 10.21 shows a ship being attacked by an anti-ship missile. The radar on the missile turns on when it is within radar range. The anti-ship missile actively guides itself to impact the ship with its on-board radar. When the radar is tracking the ship, the resolution cell of the missile's radar is centered on the targeted ship.

The ship launches a decoy, for example, a Nulka, which turns on within the resolution cell and then maneuvers away from the ship. The decoy has more RCS than the ship, so it steals the radar's tracking away from its target. As shown in Figure 10.22, the missile's radar resolution cell follows the decoy. Naturally, the decoy moves away from the ship in a direction that will not aim the missile at another friendly ship.

As with all decoys, the seduction decoy must present a credible radar return (with an appropriate RCS) to the missile's radar to be effective.

10.4 Expendable Decoys

Expendable decoys are used to protect both ships and aircraft. They can fulfill either a distraction or seduction role.

These are active decoys that are much smaller than the platforms they protect, so a decoy must enhance its apparent radar cross section (RCS) electronically by one of two approaches, a straight-thorough repeater as shown in Figure 10.23 or a primed oscillator as shown in Figure10.24. Note that the receiver does not need to be physically located on the decoy. In either case, the

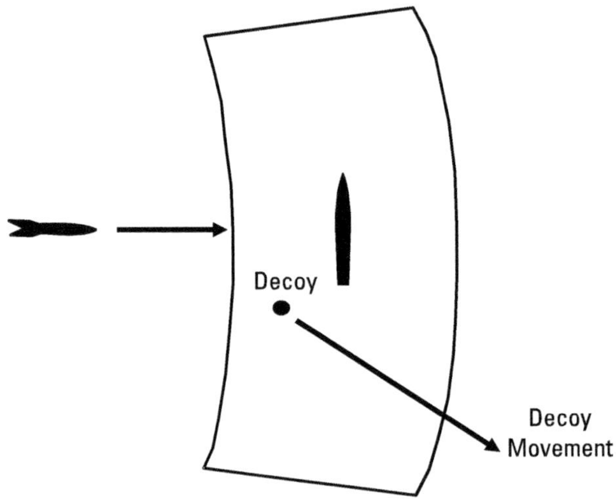

Figure 10.21 A ship protection seduction decoy starts within the resolution cell of the attacking missile's radar and moves away from the ship's location.

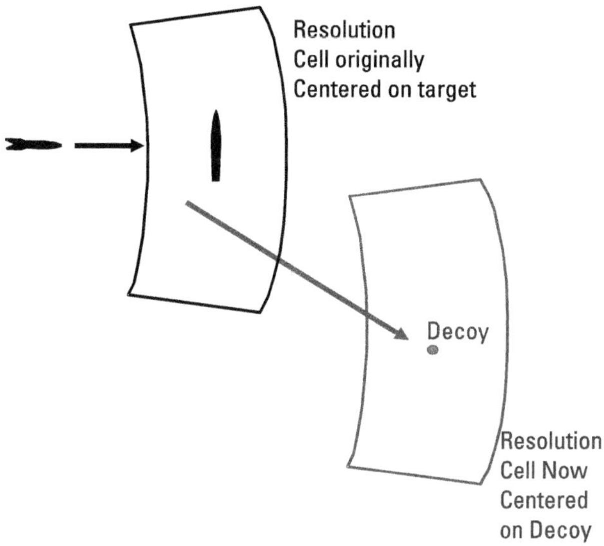

Figure 10.22 The missile radar resolution cell remains centered on the decoy as it leaves the ship's location.

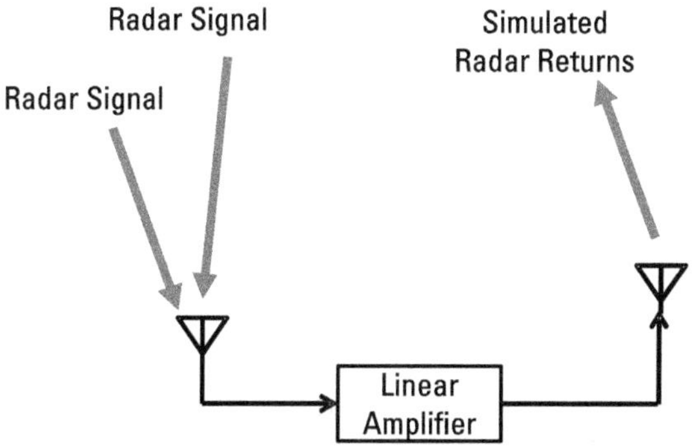

Figure 10.23 A straight through repeater decoy amplifies and rebroadcasts one or more radar signals.

effective radar cross section is calculated from the throughput decoy gain by the following formula (from Section 10.2.1):

$$\sigma = 38.6 - 20\log_{10}(F) + G$$

where σ is the RCS in dBsm, F is the radar frequency in megahertz, and G is the decoy throughput gain in decibels.

If the decoy is a repeater, G is the sum of the receiving antenna gain, the amplifier, and the transmitting antenna gain, less any losses.

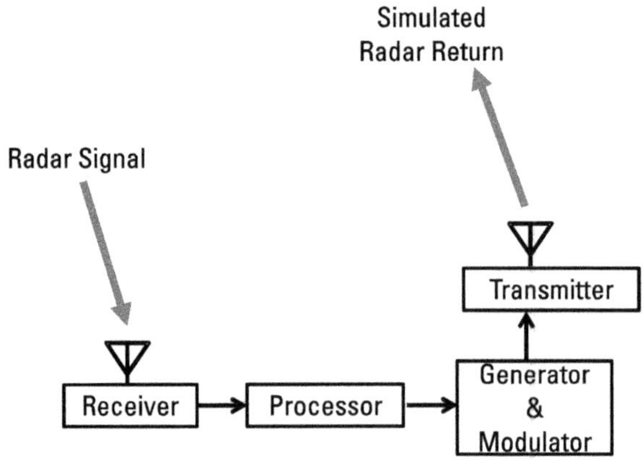

Figure 10.24 A primed oscillator decoy receives one radar signal and determines its frequency and modulation. Then it generates a matching return signal with large ERP to represent a large RCS.

If the decoy is a primed oscillator, G is the effective radiated power from the decoy's transmit antenna divided by (or subtracted from in decibels) the radar signal strength arriving at the decoy's receiving antenna. The arriving signal strength is determined from the formula:

$$P_A = ERP_R - L_P$$

where P_A is the signal strength arriving at the decoy receiving antenna (in dBm), ERP_R is the effective radiated power of the radar toward the decoy (in dBm), and L_P is the propagation loss from the radar to the decoy (in decibels).

The repeater can decoy more than one radar, and creates the same radar cross section for each. The primed oscillator has a constant effective radiated power (ERP), so weaker received signals have more gain and thus have more simulated RCS.

10.4.1 Aircraft Decoys

Expendable aircraft decoys are launched from the same dispensers that deploy chaff or flares. For U.S. Air Force and Army aircraft the flares have a 1 × 1 inch square shape factor with an 8-inch length as shown in Figure 10.25. For U.S. Navy aircraft they are cylindrical, 36 mm in diameter by 148 mm long as shown in Figure 10.26. In either case, the decoy is fired electrically to launch it into the slip stream. The decoy turns on as soon as it is launched.

Because of its small size, the aircraft decoy is expected to be powered by a thermal battery, which has a lifetime of a few seconds. This is plenty of time for the decoy to perform its mission.

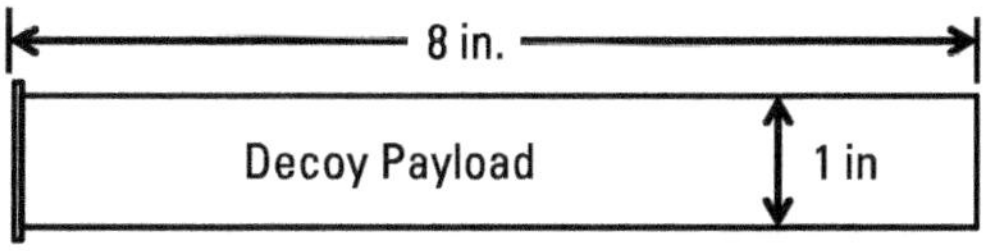

Figure 10.25 The USAF aircraft decoy is 1 inch square and 8 inches long. It is the same shape factor as USAF chaff cartridges and the smallest flare cartridges.

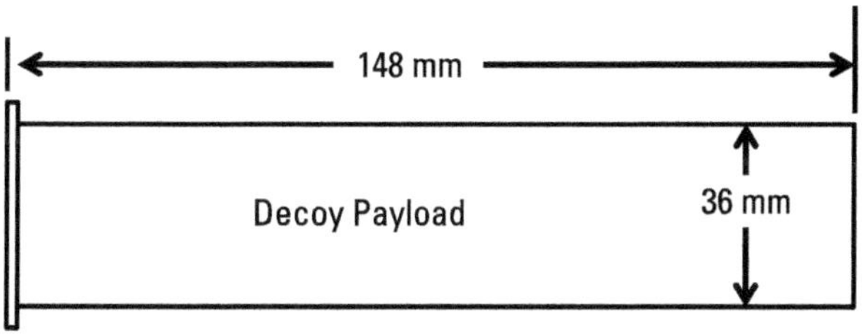

Figure 10.26 U.S. Navy expendable aircraft decoys are cylindrical, 36 mm in diameter and 148 mm long. They are expended from the same dispenser as naval airborne flares and chaff cartridges.

10.4.2 Antenna Isolation

If the decoy is a repeater, there must be adequate isolation between the receiving and transmitting antennas as shown in Figure 10.27. Without adequate isolation, the system will oscillate just like an audio system howls when a microphone is too close to an amplified speaker. Because of the small size of an

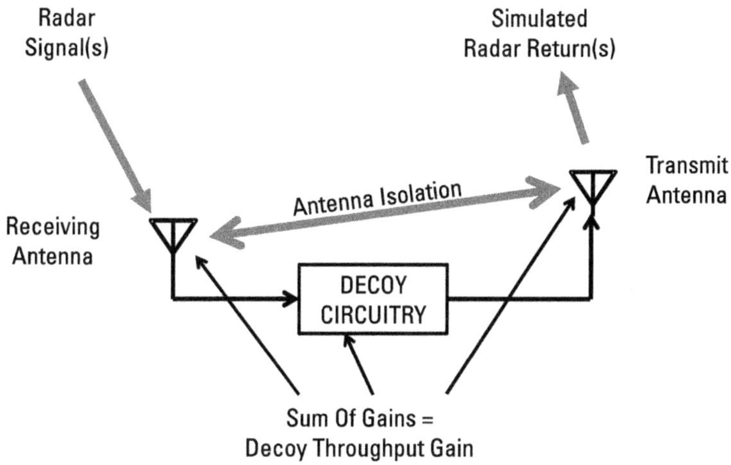

Figure 10.27 For proper decoy operation, the antenna isolation must at least equal the decoy throughput gain.

aircraft expendable decoy, this can be a significant challenge. The isolation must be greater than the decoy throughput gain.

10.4.3 Aircraft Distraction Decoys

If the decoy is successful in a distraction role, it will be acquired by an acquisition radar, which will hand it off to a tracking radar. The tracking radar will establish a track on the decoy as it falls away from the targeted aircraft and will therefore not acquire or track the aircraft. A distraction decoy must have approximately the RCS of the target aircraft and must present a realistic enough radar return that the threat radar processor cannot distinguish it from the target. Depending on the threat radar, this may require that the decoy present a complex RCS or such signal characteristics as jet engine modulation (JEM).

10.4.4 Aircraft Seduction Decoys

If used in a seduction role, the decoy will operate against a threat radar that is already tracking the aircraft. The resolution cell of the threat radar will be centered on the targeted aircraft. To fulfill its function, the decoy must be fully operational before it leaves the resolution cell. If the effective RCS of the decoy is twice that of the aircraft, the radar will set its resolution cell twice as far from the aircraft as it is from the decoy. Then, as the decoy moves away from the aircraft, it takes the radar's resolution cell with it, so if the threat fires a missile, it will fire at the decoy.

10.5 Ship-Protection Seduction Decoys

Like the aircraft protection seduction decoy, the ship-protection seduction decoy captures the tracking mechanism of the threat radar and leads it away from its intended target. The decoy must activate within the threat radar's resolution cell and simulate a larger RCS than the target ship. The threat radar is located on the anti-ship missile, and it observes the ship's RCS from its attacking aspect. Figure 10.28 shows the anti-ship missile seduction geometry.

10.5.1 Ship Seduction Decoy RCS

Like the airborne seduction decoy, a simulated RCS twice that of the target is desirable. Because of the large size of a ship, the RCS simulated by the decoy must be thousands of square meters.

The ship will generally have a larger RCS if attacked from a beam than if attacked from the bow or stern aspects. Figure 10.29 is a sketch of a typical

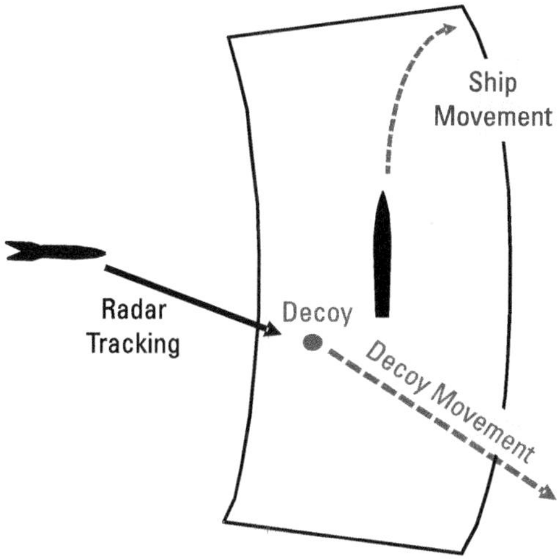

Figure 10.28 For successful seduction, the decoy captures the radar's tracking within the resolution cell. Then the ship and/or the decoy move to separate the decoy from the targeted ship.

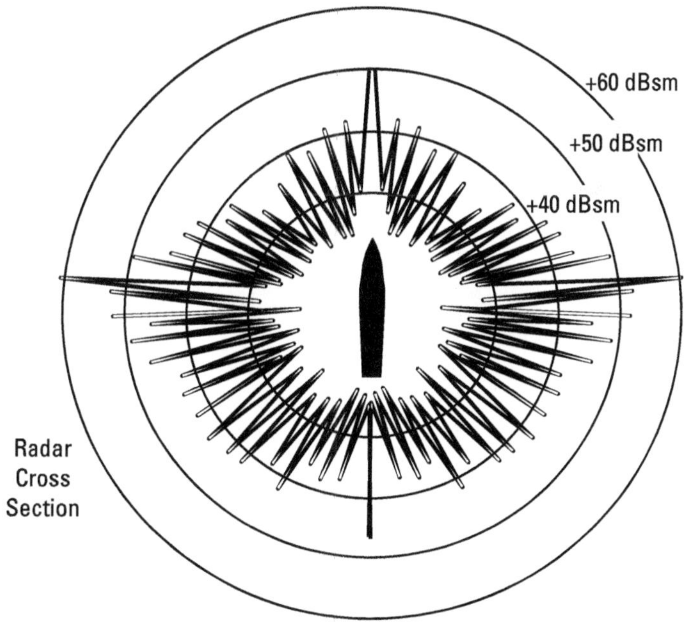

Figure 10.29 An older ship has many external features that become complex and efficient radar reflectors. This makes the ship RCS both complex and large.

RCS versus aspect angle for an older ship, while Figure 10.30 shows the RCS of a modern ship with external geometry designed to reduce radar reflection.

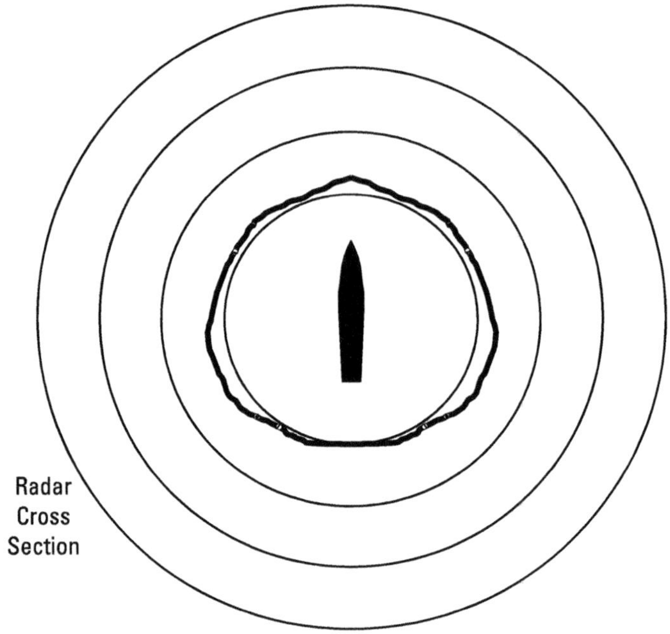

Figure.10.30 A new ship with external features designed to reduce radar reflections will have a much smaller and simpler radar cross section than an older ship.

10.5.2 Decoy Deployment

Decoys (along with chaff and Infrared decoy cartridges) can be fired from super rapid blooming offboard chaff (SRBOC) launchers or fired as rockets from stands on the ship. The SRBOC rounds are 130 mm in diameter. The SRBOC rounds or rockets are launched when a hostile tracking radar is detected by the ship's radar warning system.

The decoy can either be launched into the water, or independently maneuvered. If launched into the water, the decoy will stay in position while the ship cruises away. The ship can be maneuvered to minimize the RCS seen by the attacking anti-ship missile and to maximize the miss distance as shown in Figure 10.31.

If it is independently maneuvered, the decoy can hover above the water under a manned helicopter or in an unmanned flying platform. It can also be located in a small, powered watercraft. Either way, the decoy maneuvers along an optimum path to decoy the attacking missile away from the ship as shown in Figure 10.32.

As mentioned above, an anti-ship missile turns on a tracking radar when it is within radar range of the target ship. As the decoy presents more RCS than the ship, the missile will track to the decoy if the decoy is successful.

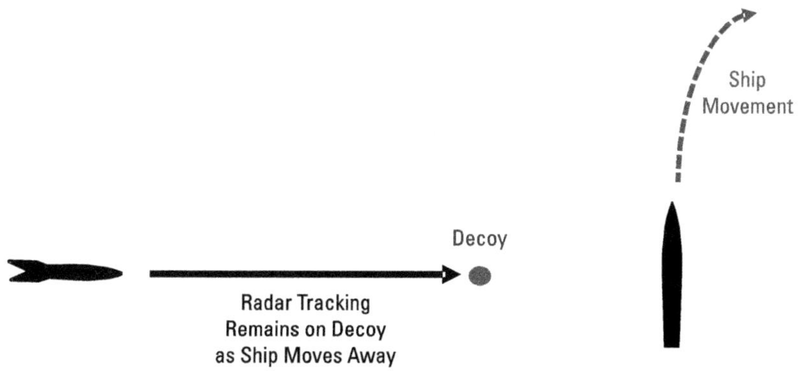

Figure 10.31 A floating decoy is launched to capture the attacking radar's tracking within the resolution cell. The radar continues to track the stationary decoy as the ship moves away.

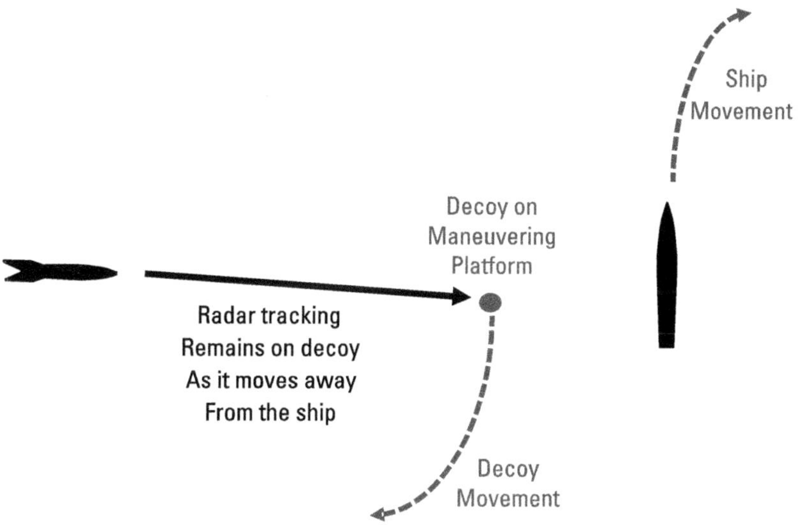

Figure 10.32 An independently maneuvering decoy can be in a hovering rocket, an unmanned helicopter, a manned helicopter, a ducted fan vehicle, or an unmanned small craft.

If the attacking missile radar processing performs waveform analysis of received signals, it may compare the details of the skin return from the ship with simulated skin returns from decoys. This would make it possible for the missile radar to reject simple returns from a decoy while accepting the more complex returns from the ship. The RCS of the ship can have many components from various physical features. To overcome this, it will be necessary for the decoy to have multiple digital RF memories to generate a complex waveform that will be accepted by the radar as a valid return. Note that this process is explained in Chapter 8.

10.5.3 Dump Mode

If a seduction decoy is placed outside the attacking radar's resolution cell as shown in Figure 10.33, a shipboard deceptive jammer could be used to move the radar's tracking centroid to the location of the decoy. The decoy would then capture the radar's tracking and hold it away from the targeted ship as shown in Figure 10.34. This technique is called a dump mode.

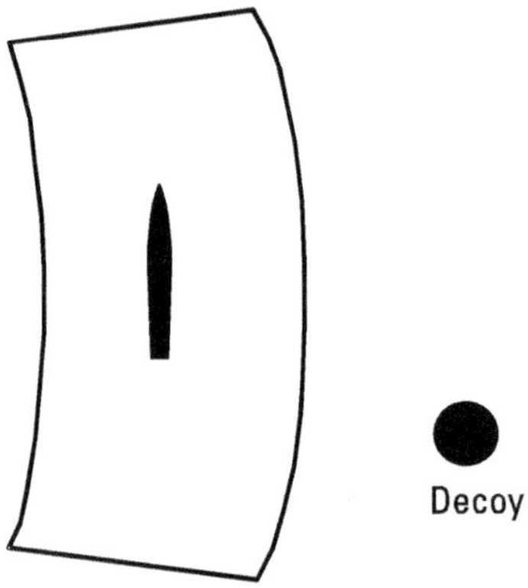

Figure 10.33 In "dump" mode, a decoy is placed outside the resolution cell, but reasonably close.

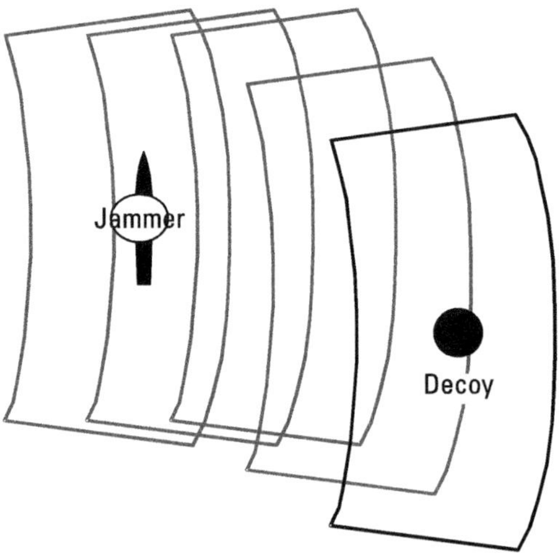

Figure 10.34 A deceptive jammer on the targeted ship moves the radar tracking center to the location of the decoy.

10.6 Towed Decoys

A towed decoy can provide terminal defense for an aircraft attacked by a radar guided missile. This is most importance when a threat missile has home-on-jam capability or when an aircraft must fly closer to a radar than the burn-through range allowed by available jamming support. The towed decoy is launched from an aircraft and goes to the end of a tow cable. When it reaches the end of the cable, it turns on.

The decoy generates the effect of an RCS significantly larger than that of the protected aircraft. This will cause a radar guided missile to track to the decoy rather than the aircraft. In a recent conflict, there were 10 towed decoys shot off of aircraft. Thus, the tow cable must be long enough that the aircraft will be outside the burst radius of the likely attacking missile.

The towed decoy has a seduction mission. This means that the decoy must be within the resolution cell of the attacking radar at the time of acquisition. The larger RCS of the decoy will cause the radar to track (and guide its missile) to the decoy rather than the targeted aircraft.

Some towed decoys are single use devices. When no longer needed, they are cut loose from the aircraft. Later decoys can be retrieved when no longer required. These retrievable decoys also have the feature of selectable spacing from the protected aircraft. This feature will allow optimum trade-off of close spacing for ease of capturing the threat radar's tracking against long spacing for greater distance from a decoy that is actually destroyed by a missile.

As shown in Figure 10.35, the towed decoy system includes a receiver and processer in the towing aircraft and the decoy itself. The receiver and processor determine the frequency and optimum modulation for the simulated radar return from the decoy and transmit the actual decoy signal down the tow cable (at a low power level). As shown in Figure 10.36, the decoy carries only an amplifier and antennas. Power for the amplifier is also passed to the decoy from the aircraft over the tow cable. The antennas are located at the front and back

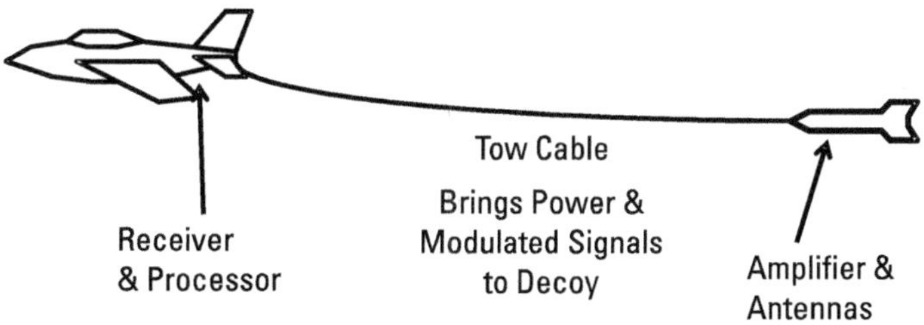

Figure 10.35 The towed decoy is attached to the towing aircraft by a tow cable, which also carries signals from a receiver/processor in the aircraft to an amplifier and antenna in the decoy.

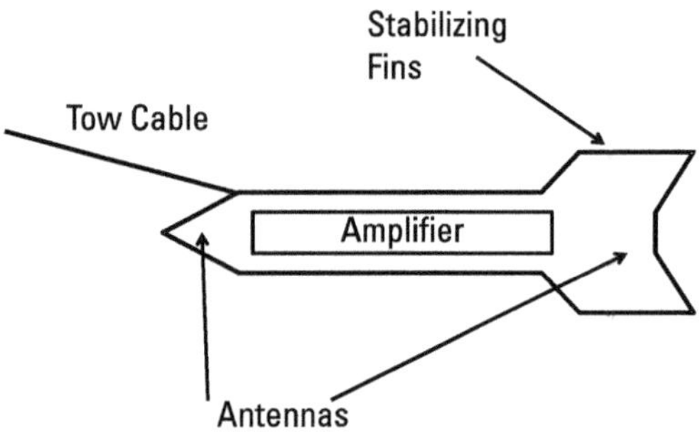

Figure 10.36 The decoy contains only an amplifier and fore and aft transmitting antennas.

of the decoy and have fairly broad beam width so the decoy can be oriented a few degrees away from the radar and still be effective.

Figure 10.37 shows an engagement with a threat radar. The aircraft and the decoy are treated like a single target by the attacking radar. The radar signal is received and analyzed in the aircraft, and a simulated skin return signal is broadcast from the decoy with enough power to create a much larger RCS than that of the aircraft. In using the formula

$$\sigma = 39 - 20\log_{10}(F) + G$$

With the constant rounded, to determine the effective RCS of the decoy, the gain term (G) is the difference (in decibels) between the effective radi-

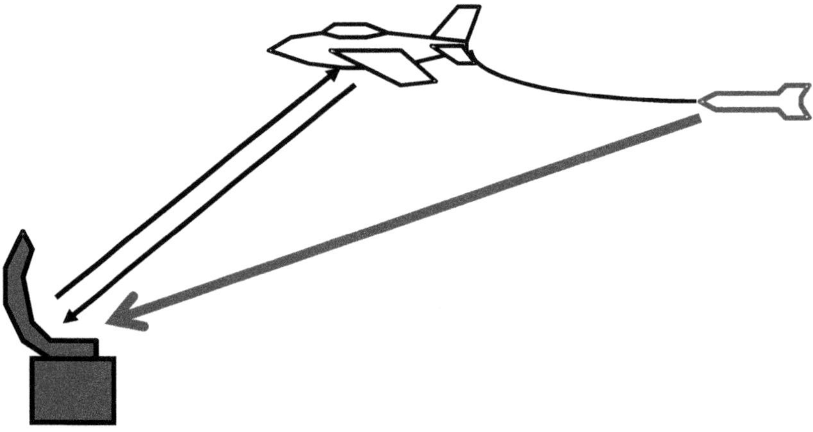

Figure 10.37 The radar signal is received in the aircraft and an amplified simulated skin return is rebroadcast from the decoy with any required extra modulation to make the decoy return credible.

ated power of the simulated skin return from the decoy and the signal strength arriving at the receiving antenna on the towing aircraft.

10.6.1 The Resolution Cell

Figure 10.38 shows the resolution cell of the attacking radar and the effective area of the resolution cell with chirp or Barker code pulse compression. The resolution cell and pulse compression are discussed in detail in Chapter 4. The point here is that both the towing aircraft and the decoy must be within the resolution cell (including compression if present) to be effective.

When a radar pulse is highly compressed, the resolution cell is much wider than it is deep as shown in Figure 10.39. This means that the radar may be able to detect both the aircraft and the decoy, and ignore the decoy. To prevent this, it is necessary to first capture the radar's tracking, after which the radar will only see the decoy in its shallow resolution cell. One tactic that could accomplish this is to notch the radar. That is, turn 90° to the radar so the aircraft and decoy are both in the shallow compressed cell. Then, when the aircraft turns back toward the radar, only the decoy will remain in the cell.

10.6.2 An Example

Consider the situation pictured in Figure 10.40. An aircraft with 10-m2 RCS is 10 km from an 8-GHz radar with 100-dBm ERP. The signal strength arriving at the aircraft receiving antenna is –30 dBm (using formulas found in Chapter 3). The effective radiated power of the decoy is 1 kw (which is +60 dBm). Thus, the gain of the decoy is 90 dB.

So the RCS simulated by the decoy is: 39 + 90 – 20 log(8,000) = 51 dBsm. This converts to 125,893 m^2 of simulated RCS created by the decoy.

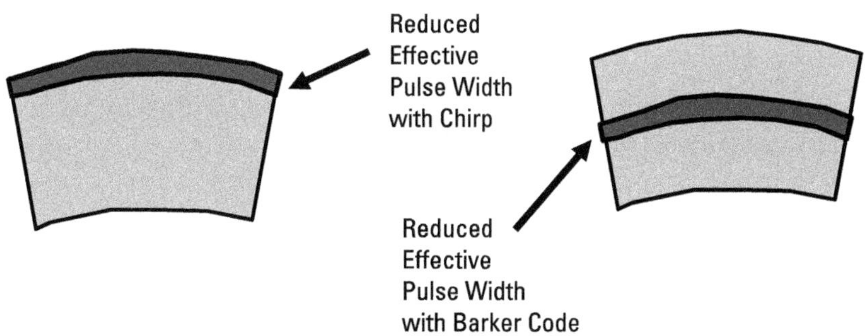

Figure 10.38 The resolution cell of an attacking radar can be compressed in range by chirp or Barker code techniques.

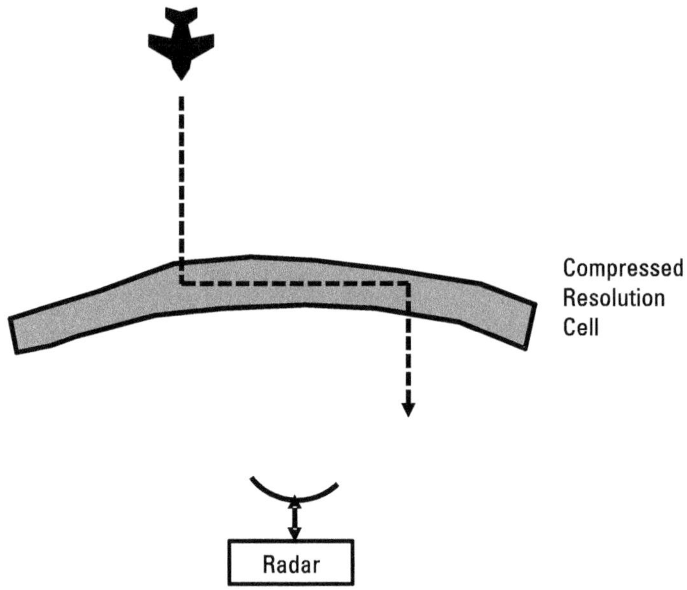

Figure 10.39 By flying at 90 degrees to the tracking radar, the towing aircraft could bring a towed decoy into the shallow range dimension of the radar's compressed resolution cell.

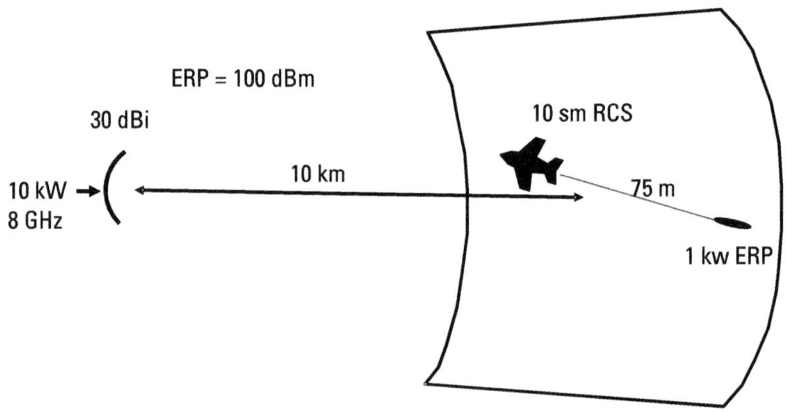

Figure 10.40 A towed decoy with 1 kw ERP that is 10 km from a 100 dBm 8 GHz radar will produce a 125,893 square meter effective RCS.

Comparing this to the 10-m^2 RCS of the aircraft shows the power of this towed decoy to protect the aircraft.

11

Electromagnetic Support Versus Signal Intelligence

11.1 Introduction

In this chapter, we will discuss the differences between electromagnetic support (ES) systems and signal intelligence (SIGINT) systems, both of which are designed to receive hostile signals. The differences between SIGINT and ES have to do with the reasons those signals are received, as summarized in Table 11.1. There are also some technical differences between the typical environments in which these systems work that dictate differences in system design approach and system hardware and software.

11.2 SIGINT

SIGINT is the development of militarily significant information from received signals. It is commonly divided into communications intelligence (COMINT) and electronic intelligence (ELINT) as shown in Figure 11.1. Each of these subfields is somewhat related to ES as shown in Figure 11.2. ES is commonly divided into communications ES and radar ES as shown in Figure 11.2. The nature of communication and radar signals dictates differences in mission between these two subfields. The following sections will focus on systems handling each type of signal, differentiating the intelligence and ES roles.

Table 11.1
SIGINT Versus ES

	SIGINT Systems	ES Systems
Mission	COMINT: Intercept enemy communications and determine enemy capabilities and intentions from information carried on signals. ELINT: Find and identify new threat types.	Communications ES: Identify and locate enemy communications emitters to allow development of electronic order of battle and to support communications jamming. Radar ES: Identify and locate enemy radars to allow threat warning and to support radar countermeasures.
Timing	Timeliness of outputs is not too critical.	Timeliness of information is central to mission.
Data collected	Gather all possible data on received signals to support detailed analysis.	Gather only enough data to determine threat type, operating mode, and location.

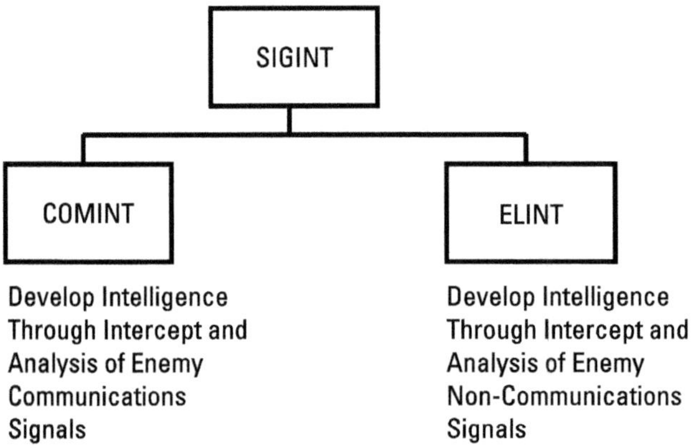

Figure 11.1 SIGINT comprises COMINT and ELINT to develop intelligence from enemy communications and noncommunications signals.

11.2.1 COMINT and Communications ES

Figure 11.3 is a flow diagram showing the relationship between COMINT and communications ES systems.

The dictionary definition of COMINT is "gathering of intelligence by intercept of wire or radio communications." Basically, this is listening to what an enemy says to determine their capability, their force structure, and their intentions. This implies that a COMINT system deals with the internals (i.e., the information carried in the modulation) of transmitted enemy signals. Because of the nature of military communication, important signals can be expected to be encrypted, and of course in the enemy's language. Decryption and translation of signals can be expected to delay the availability of the information recovered. Thus, COMINT can be considered more valuable to strategic and high

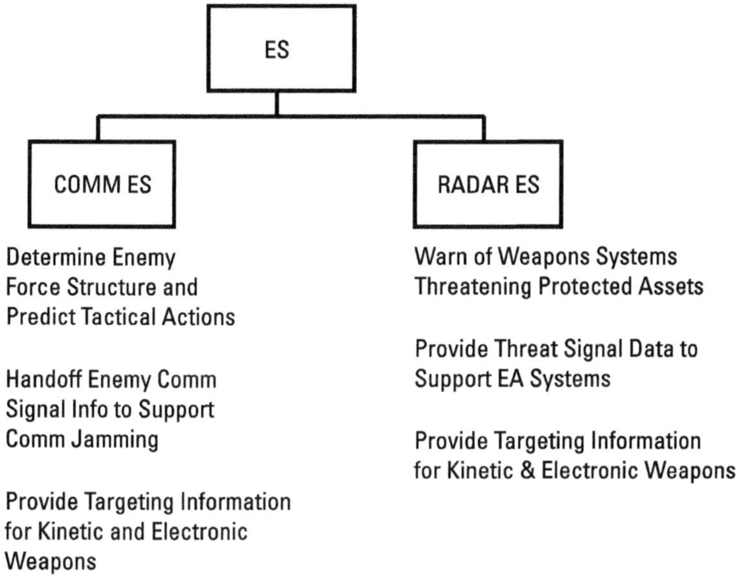

Figure 11.2 ES comprises Comm ES and Radar ES. Both provide information about enemy emitters currently operating in support of EA and weapon engagement.

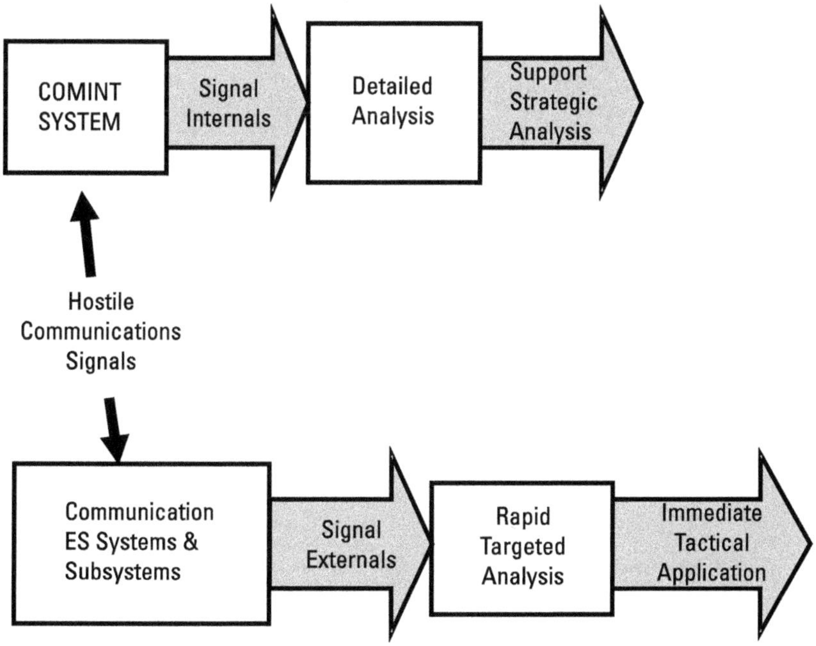

Figure 11.3 COMINT classically deals with signal internals to support strategic actions; Comm ES deals with signal externals to support immediate tactical decision making.

level tactical considerations than to determination of appropriate immediate tactical response.

Communications ES focuses on the externals of communications signals: the type and level of modulation and the location of the transmitters. It supports tactical responses to current situations by determining the types and locations of enemy emitters. By modeling all of the types of emitters against the types of emitters used by various enemy organizations, it allows estimates of the enemy force structure to be made. The location and location history of the observed emitters can be used to indicate the location and movement of the enemy's forces. The total laydown of transmitters is called the electronic order of battle (EOB), and can be analyzed to determine the enemy's capabilities and even their intentions.

In summary, COMINT determines the enemy's capabilities and intentions by listening to what is said (i.e., signal internals), while communications ES determines the enemy's capabilities and intentions by analysis of signal externals.

11.2.2 ELINT and Radar ES

ELINT involves the interception and analysis of noncommunications signals, primarily from radars. The purpose of ELINT is to determine the capabilities and vulnerabilities of newly encountered enemy radars. As shown in Figure 11.4, the ELINT system gathers enough data to support detailed analysis. The first task when a new radar signal type is received is to determine whether the received signal is, in fact, a new threat. Two other possibilities exist: it may be

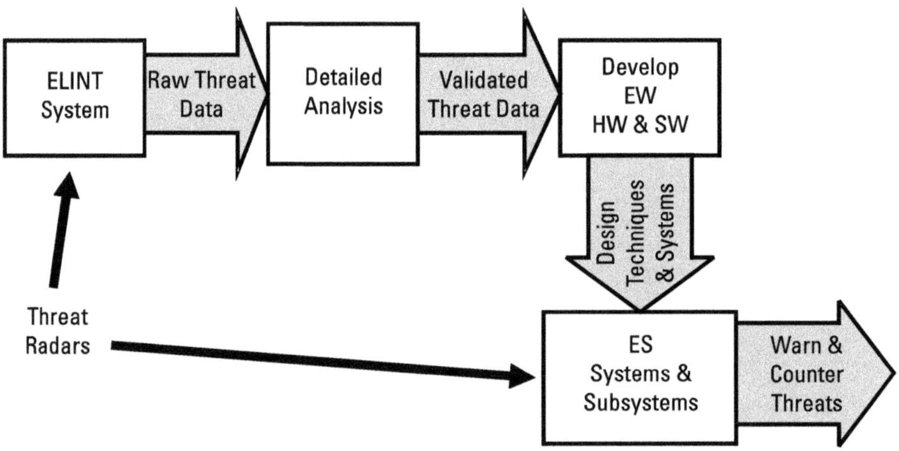

Figure 11.4 ELINT systems gather threat data to support the development of ES systems and subsystems for threat warning and countermeasure selection.

an old threat radar that is malfunctioning or there may have been something wrong with the intercept system. If the received signal is a new type of radar or a new operating mode, the detailed analysis will allow modification of ES systems so that they will be able to recognize this new threat type.

Radar ES systems also receive hostile radar signals, but their purpose is to quickly determine which of the enemy's known weapons is being deployed against a target at the moment. After threat type and mode identification is complete, this information is displayed to operators along with the location of the threat emitter and/or passed to other EW systems or subsystems to support countermeasure initiation. If a signal of an unfamiliar type is received, it is considered an unknown. In some ES systems, the operator is merely notified that an unknown threat has been received. However, in other systems, an attempt is made to guess the threat type. In some ES systems, unknown threats are recorded for later analysis.

In summary, ELINT determines what capabilities the enemy has, while radar ES determines which of the enemy's radars is being used at the moment and where the emitter (hence the weapon it controls) is located.

11.3 Antenna and Range Considerations

There are some technical differences between ES and SIGINT systems dictated by mission and environment considerations. These differences have to do with the anticipated intercept geometry, the different types of information taken from intercepted hostile signals, and time criticality of intercepts.

11.4 Antenna Issues

Antennas can be characterized as directional or nondirectional. This is a great oversimplification. Antennas like whips and dipoles are sometimes (incorrectly) described as omni-directional. This is not true, as both antenna types have nulls in their coverage. However, both types, if vertically oriented, provide 360° of azimuthal coverage. There are also circular arrays of directional antennas that provide full azimuthal coverage. Directional antennas (including but not limited to parabolic dishes, phased arrays, or log periodic antennas) restrict their coverage to a reduced angular sector.

Angular coverage has a significant impact on the probability of intercepting a hostile signal at an unknown direction of arrival. As shown in Figure 11.5, a 360° coverage antenna (or array of antennas) "looks" in all directions all of the time, so it will input any new signal to a receiver as soon as it occurs. The directional antenna must be scanned to the direction of arrival of a new signal before it can be received. If a hostile signal is present for a limited time, the

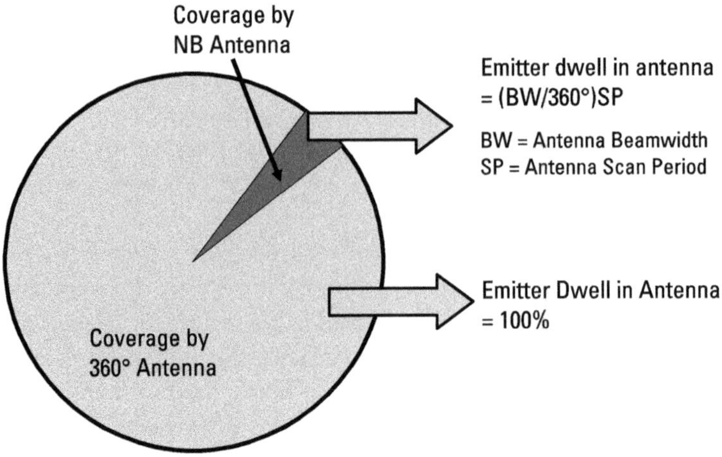

Figure 11.5 A 360 degree antenna, like a dipole or whip provides 100% coverage of all azimuths of arrival, while a narrow beam antenna must be scanned to the correct direction of arrival.

probability of intercept is a function of the antenna beam width and the scan rate of the antenna. For an intercept to occur, the antenna must be moved to place the signal's direction of arrival into the antenna beam coverage area.

As shown in Figure 11.6, the beamwidth determines the percentage of possible angles of arrival covered by the antenna. To use this part of the figure, draw a line straight up from the beamwidth to the solid line and then draw right to the right side ordinate value. This considers only one search dimension (e.g., azimuthal search); a two-dimensional search is significantly more difficult. In the same figure, the amount of time that a scanning antenna will dwell on the signal's angle of arrival (also in azimuth only) is shown as a function of beamwidth for various circular scan periods. To use this part of the figure, draw straight up from the beamwidth to the dashed line for the selected scan period and then draw left to the left side ordinate value. It should be noted that a frequency search must be made during the time the antenna is pointed at each possible angle of arrival. The narrower the antenna beam, the slower the receiving antenna must be scanned to allow for frequency search. Thus, the longer it will take to find a signal of interest at unknown frequency and angle of arrival. Frequency search will be discussed in the context of receiver types in Section 11.5.

Normally, SIGINT intercepts are less time-critical than ES intercepts. Thus, a delay in intercept caused by scanning a narrow beam antenna is likely to be acceptable. However, because ES systems must typically intercept a hostile signal within a small number of seconds, a wide coverage antenna or array of antennas is usually required.

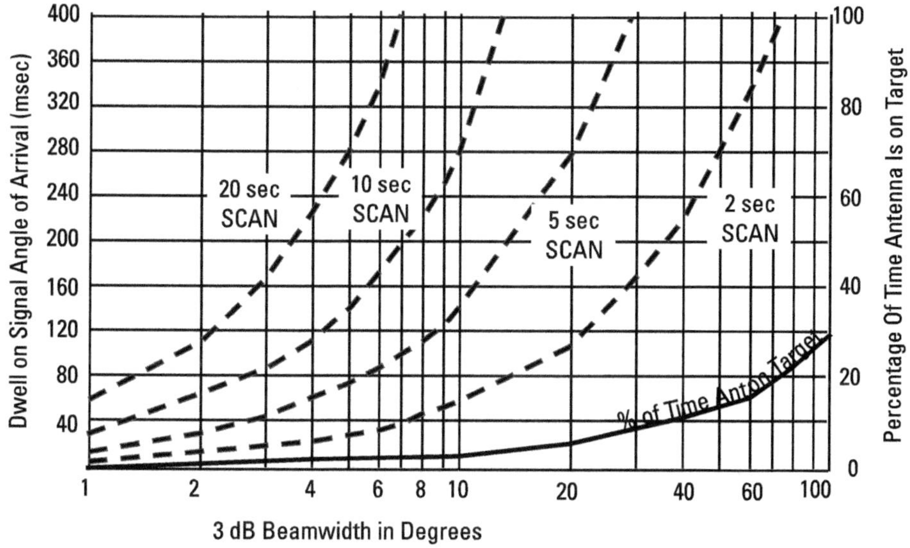

Figure 11.6 The percentage of angular space within the antenna beam varies inversely with the beamwidth, as does the dwell time at the signal's angle of arrival.

As shown in Figure 11.7, there is a trade-off of the half-power (3-dB) beamwidth of an antenna and the antenna gain. This figure is for a 55% efficient parabolic dish antenna, but this trade-off applies to all types of narrow

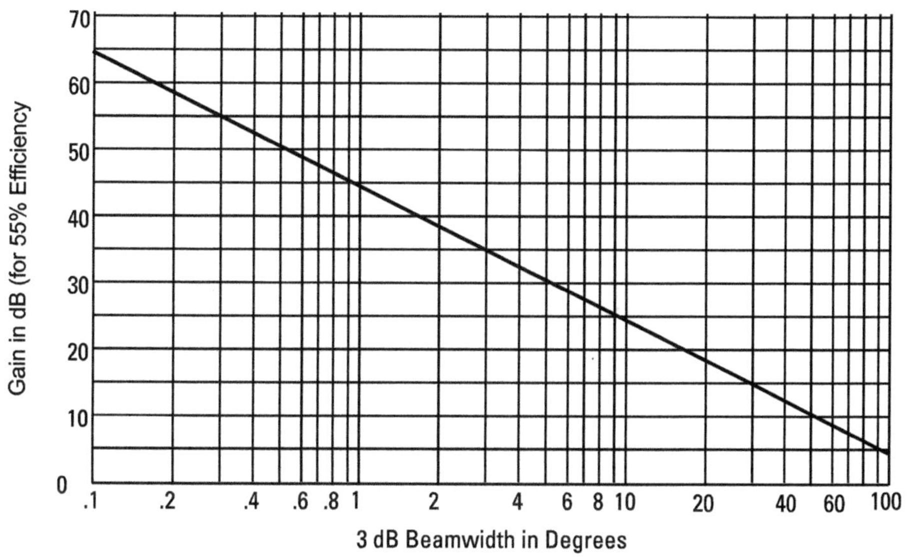

Figure 11.7 The gain of a narrow beam antenna varies inversely with its beamwidth.

beam antennas. The receiving antenna gain is an important consideration in the range at which a hostile signal can be intercepted, as discussed next.

This means that wide coverage (hence low gain) antennas are almost always required for ES systems, while narrow beam (hence high gain) antennas may be the best solution for SIGINT systems.

11.5 Intercept Range Considerations

Figure 11.8 shows an intercept situation for either an ES or SIGINT system. Note that the range at which a receiving system can intercept a hostile signal depends on the effective radiated power of the target signal, the applicable propagation mode, the receiving antenna gain in the direction of the emitter, and the sensitivity of the receiving system. Propagation modes are discussed in detail in Chapter 5.

Radar and data link signals typically propagate in the line-of-sight mode. In this mode, the intercept range is given by the formula:

$$R_I = \text{antilog}\left\{\left[ERP_T - 32 - 20\log(F) + G_R - S\right]/20\right\}$$

where R_I is the intercept range in kilometers, ERP_T is the effective radiated power of the target emitter in dBm, F is the frequency of the transmitted signal, G_R is the gain of the receiving antenna in the direction of the target emitter, and S is the sensitivity of the receiver system in dBm.

Communication signals will propagate in the line of sight or two-ray modes, depending on the link distance, antenna heights and frequency. If propagation is in the two-ray mode, the intercept range is given by the formula:

$$R_I = \text{antilog}\left\{\left[ERP_T - 120 - 20\log(h_T) + 20\log(h_R) + G_R - S\right]/40\right\}$$

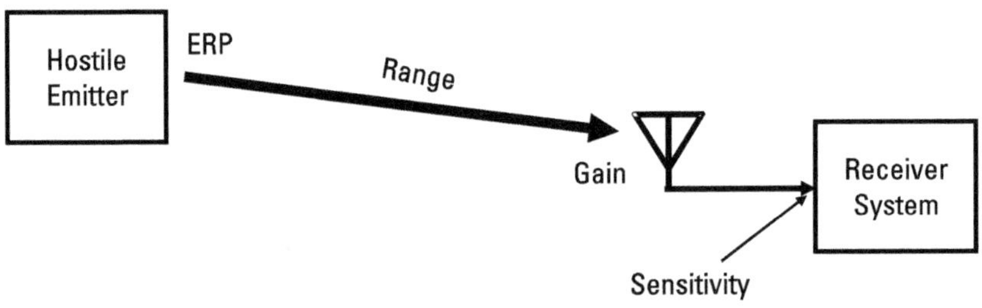

Figure 11.8 The range at which a receiving system can intercept a hostile emitter signal is a function of antenna gain and receiver system sensitivity.

where R_I is the intercept range in kilometers, ERP_T is the effective radiated power of the target emitter in dBm, h_T is the height of the transmitting antenna in meters, h_R is the height of the receiving antenna in meters, G_R is the gain of the receiving antenna in the direction of the target emitter, and S is the sensitivity of the receiver system in dBm.

As you can see from these formulas, the intercept range is always impacted by the receiving antenna gain and the sensitivity of the receiving system. Note that the sensitivity is the required signal strength for a successful intercept. The more sensitivity the receiving system has, the lower this number will be. For example, a high-sensitivity receiver might have a sensitivity of −120 dBm, while a low-sensitivity receiver could have a sensitivity of −50 dBm.

The effective radiated power (ERP) of the target emitter is the amount of power it transmits in the direction of the intercepting receiver. Tactical communication threats will usually have 360° antennas with fairly constant gain versus azimuth; the ERP is the sum of the transmitter power (in dBm) and the antenna gain (in decibels). However, radar threats are expected to have narrow beam antennas. As shown in Figure 11.9, the narrow beam antenna has a main lobe and side lobes. The side lobes are shown simplified in that they are all the same strength; actual antenna side lobes vary. However, the drawing is realistic in that the nulls between the lobes are much narrower than the lobes. This means that an intercept receiver pointed at the radar threat emitter away from the direction of its main beam can be expected to encounter an ERP at the average side-lobe level. This level is usually stated as: $S/L = -N$ dB where N is the number of decibels that the average side lobe level is below the boresight gain.

Although not always true, it is fairly common for an ES system to be specified to receive the main lobe of a radar threat, while an ELINT system would be specified to intercept side-lobe transmissions from target radar emit-

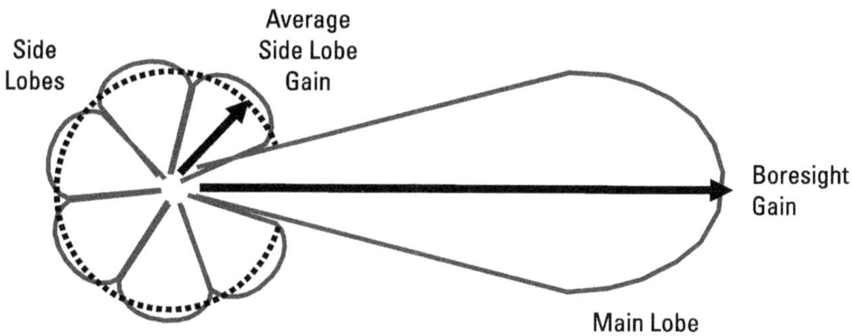

Figure 11.9 Radar ESM systems are often characterized as receiving signals from the boresight of threat radar antennas while ELINT systems are often characterized as receiving average side lobe level signals.

ters. This means that an ES system will often require less sensitivity and/or receiving antenna gain than an ELINT system.

SIGINT systems are generally assumed to require greater intercept range than ES systems; however, as with all generalities, this depends on the specific mission and situation. If we accept that SIGINT systems require greater intercept range, the receiving antenna gain and/or the sensitivity must be greater than required for ES systems. Narrow beam antennas have higher gain, but provide reduced probability of intercept (in a short time period). Thus, they are more appropriate for SIGINT applications. Full coverage antennas, while they provide less gain, can provide significantly better probability of intercept in a short time period, so are generally most appropriate for ES systems.

11.6 Receiver Considerations

Receiver issues can differentiate ES and SIGINT system requirements. Like the previously discussed issues, these differences have to do with the anticipated intercept geometry, the different types of information taken from intercepted hostile signals, and time criticality of intercepts.

There are a number of different types of receivers that can be used in either ES or SIGINT systems. Table 11.2 lists the most common types along with their characteristics that make them useful in ES or SIGINT applications. Each of these types of receivers is discussed in detail in other books, for example, Chapter 4 of [1].

The crystal video receiver is primarily used in radar warning receiver systems. It is ideal for this ES application because it covers a wide instantaneous frequency range, typically 4 GHz. This gives it the ability to receive any signal

Table 11.2
Receiver Types and Features

Receiver Type	Sensitivity	Dynamic Range	Bandwidth	Number of Signals	Features and Limitations
Crystal video	Low	High	Wide	1 in range	AM only
IFM	Low	High	Wide	1 in range	Frequency only
Superheterodyne	High	High	Narrow	1 of many	Recovers modulation
Channelized	High	High	Wide	Multiple simultaneous	Recovers modulation
Bragg cell	Low	Very Low	Wide	Multiple simultaneous	Frequency only
Compressive	High	High	Wide	Multiple simultaneous	Frequency only
Digital	High	High	Flexible	Multiple simultaneous	Recovers modulation, unique analysis capabilities

in a very short time. It typically has a wide enough bandwidth to receive very short pulses. However, it has the disadvantages of relatively poor sensitivity, an inability to determine the frequency of a received signal, and an inability to receive multiple simultaneous signals within its entire bandwidth. Although crystal video receivers have been used in reconnaissance systems under special circumstances, they are almost always radar ES receivers.

The instantaneous frequency measurement (IFM) receiver very quickly (typically 50 ns) determines the frequency (and only the frequency) of any received signal over an octave of bandwidth. It has approximately the same sensitivity as the crystal video receiver. Its big disadvantage is that it has an invalid output any time multiple, approximately equal power signals are present in its band coverage (i.e., during the same 50 ns). Because of its relatively low sensitivity, it is primarily used in radar ES systems.

The superhetrodyne receiver is very widely used in all communications applications. It is almost always found in any SIGINT communications ES system, and is sometimes used in radar ES systems. The primary advantages of superhetrodyne receivers are:

- Good sensitivity;
- The ability to receive one signal in a dense signal environment;
- The ability to recover any type of modulation;
- Measurement of the frequencies of received signals.

The main disadvantage of the superhetrodyne receiver is that it receives only a limited frequency range at a time and thus must be swept to search for threat signals. As explained next, there are trade-offs of sensitivity and the time required to find a signal at an unknown frequency versus bandwidth.

Channelized receivers allow the simultaneous recovery of multiple simultaneous signals. Their main disadvantage is the complexity (i.e., size, power, and weight) if there are a significant number of channels.

The electro-optical or Bragg cell receiver determines the frequency (and only the frequency) of multiple simultaneous signals in a dense environment. It has reasonable sensitivity, but it has the great disadvantage for both ES and SIGINT applications that it has extremely limited dynamic range. It is useful only in very limited applications.

Compressive receivers provide the frequency (and only the frequency) of multiple simultaneous signals in a dense environment. They have good sensitivity and are useful in both ES and SIGINT systems when used with superhetrodyne receivers set to identified signals of interest.

Digital receivers are used in many ES and SIGINT applications. They provide good sensitivity and dynamic range, although there are trade-offs relating to the state of the art in analog to digital converters. A digital receiver supports unique analysis capabilities. For example:

- A fast Fourier transform (FFT) circuit can perform a very fast spectrum analysis.
- Time compression algorithms can be implemented to detect noise-like signals.
- It can be configured to receive any type of modulation.

Sensitivity Versus Bandwidth

The following formula gives receiver system sensitivity (in dBm).

$$S = kTB + NF + \text{Required } RFSNR$$

where S is sensitivity in dBm, kTB is the thermal noise in the system in dBm, NF is the amount of noise added by the system above kTB in decibels, and Required $RFSNR$ is the required predetection SNR in decibels.

This is the definition of the amount of received power for the receiver to receive a signal with the required output quality. Sensitivity is also stated in terms of minimum discernable signal (MDS), which is determined as above, but with the RFSNR at 0 dB (i.e., signal equals noise at the receiver system input).

Because kTB is a function of the effective receiver system bandwidth, the graph in Figure 11.10 can be used to determine the MDS sensitivity from the receiver effective bandwidth and its noise figure. To use the figure, start at the bandwidth on the abscissa and draw directly up to the noise figure and then left to the ordinate, which gives the MDS sensitivity in dBm. To determine the sensitivity for full specified output performance, just add the required RFSNR to the MDS sensitivity.

11.7 Frequency Search Issues

Section 11.3 deals with the search issues associated with narrow beam antennas. Figure 11.6 is a graph allowing calculation of dwell time on a signal as a function of antenna beamwidth and scan rate. Now we deal with the other search issue, which is to find the unknown threat signal in frequency. A good rule of thumb is that to detect the presence of a signal, it must remain in our receiver bandwidth for a time equal to the inverse of the effective receiver bandwidth.

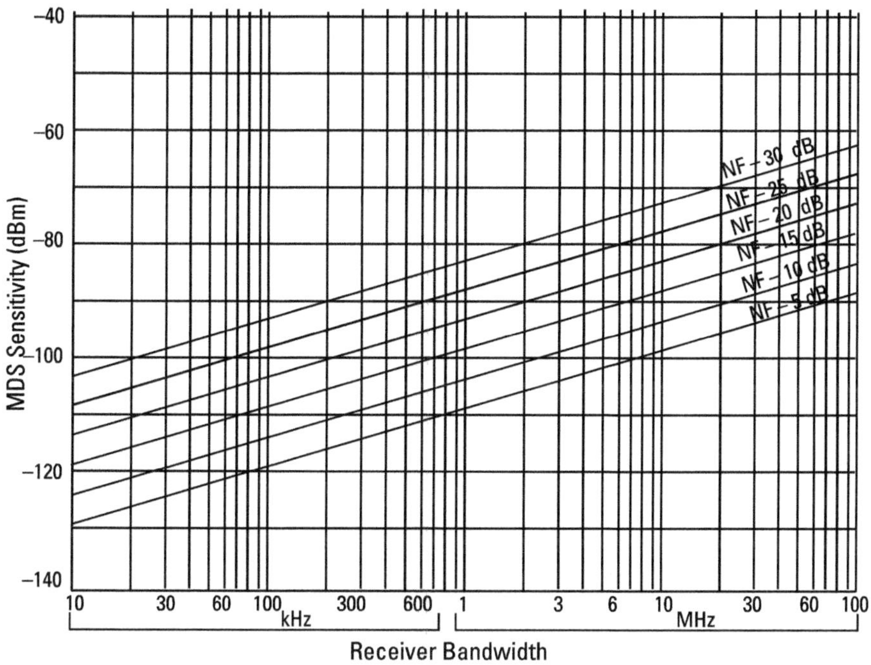

Figure 11.10 The MDS sensitivity of a receiver system is a function of its effective bandwidth and its noise figure.

For example, a receiver with a 1-MHz bandwidth must dwell at one frequency for 1 μs before it can be stepped to a new frequency.

The graph in Figure 11.11 allows the determination of the time that is required to cover a given frequency range (with adequate dwell time in the bandwidth) as a function of bandwidth and the range to be swept. To use the figure, draw up from the receiver bandwidth on the abscissa to the frequency range to be swept, then left to the total time required to find the signal.

11.8 Processing Issues

Now consider processing issues that can differentiate ES and SIGINT system requirements. These differences have to do with the nature of the information that must be collected from signals of interest and the time criticality of output reporting.

Perhaps the most important issue separating ES and SIGINT system missions is the nature and amount of data that must be collected on threat signals encountered.

Figure 11.12 summarizes the data requirements for radar and communications ES versus SIGINT systems.

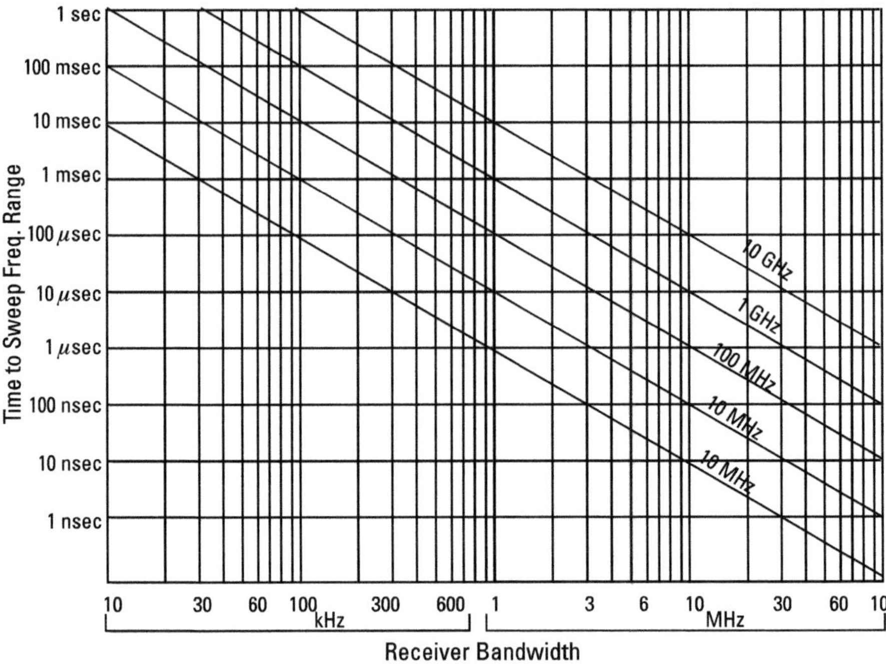

Figure 11.11 The time required to sweep a frequency range with adequate in-band dwell time is a function of the receiver bandwidth the range swept.

	ES Data Collection	SIGINT Data Collection
Radar Threat Signals	Collect only enough data to Determine radar type & operating mode Data range of known threats Adequate parametric resolution to resolve identification ambiguities	Collect enough data to support detailed analysis Data range limited by practical range of any future threat Adequate parametric resolution to determine capabilities of any future threat
Comm Threat Signals	Collect only externals data (frequency, Modulation type & level, Received signal strength, emitter location, intercept time) Parametric range and resolution only adequate to support threat ID, EOB, and Jamming activity	Collect externals and internals data. Externals data similar to ES requirement Internals data adequate to recover required military information from demodulated signals

Figure 11.12 Data collection requirements vary significantly between ES and SIGINT systems.

In general, radar ES systems collect only enough data to determine which of the enemy's weapons are being used and to allow selection of the correct countermeasures. All of this takes place within a single digit number of seconds. The collection and use of received data are shown in Figure 11.13. Note that the threat parameters stored in the receiver system's threat identification table (TID) are the result of extensive analysis of data previously collected by ELINT systems.

ELINT systems (i.e., SIGINT systems operating against radar threats) must collect much more complete data over the whole anticipated parametric range. This detailed data, which can be collected over extended time and/or multiple intercepts, supports the detailed analysis necessary to determine that the radar ES systems must identify immediately. A summary of the data that might be required for a typical pulse radar threat signal collected by and ELINT system is shown in Table 11.3.

Communication ES systems deal with the externals of threat signals to support electronic order of battle development, employment of countermeasures, and choice of fire and maneuver tactics. In general, this must be done very quickly because of the dynamic nature of tactical operations. The amount of data (typically digital) is determined by the number of parameters which must be collected and the required resolution to support tactical analysis. Table 11.4 shows a typical array of parameters that could be required for Comm ES collection on each threat.

If there are 250 signals present and the environment is collected 10 times per second, the required data bandwidth might be:

$$250 \text{ signals} \times 27 \text{ bits/signals} \times 10 \text{ collections per second} = 67{,}500 \text{ bits/sec}$$

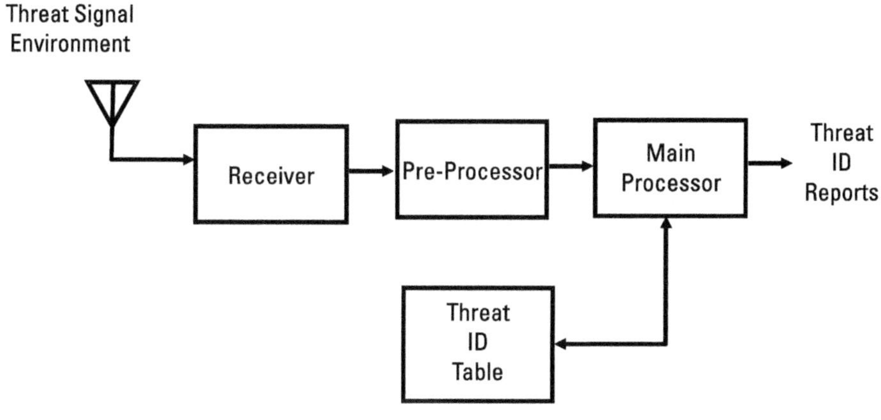

Figure 11.13 In a radar ES system, signal parameters are determined from each received signal and signal parameter files are compared to a threat ID table, and threat ID reports are output.

Table 11.3
ELINT Data for a Typical Pulse Radar Threat

Parameter	Range	Resolution	Number of Bits
Pulse width	0.1 to 20 μs	0.1 μs	9
Pulse repetition interval	3 to 3,000 μs	3 μs	11
Radio frequency	0.5 to 40 GHz	100 MHz	10
Scan period	0 to 30 seconds	0.1 second	9
BPSK clock rate	0 to 50 Mpps	100 pps	16
BPSK bits on pulse	0 to 1,000 {AU: Include unit of measurement}	1 bit	10
FM on pulse range	0 to 10 MHz	100 kHz	7
Total bits per signal			78

COMINT (i.e., SIGINT against communications threats) is normally assumed to extract militarily useful information carried by communication signals. However, this information must typically be tied to the location and type of emitter to be useful. Therefore, a COMINT system will, in most cases, be required to capture both the external and internal signal data as shown in Figure 11.14. In addition to the bits required for the externals data, the modulation must be captured. This will require some number of resolution bits (3 to 6) multiplied by twice the audio output bandwidth (or IF bandwidth) multiplied by the number of channels that are assumed to be active at any one time. For example, if 6-bit digitizing is used and there are twenty 25-kHz-wide channels of interest, the total bit rate could be:

$$20 \text{ channels} \times 2 \times 25{,}000 \text{samples/sec} \times 6 \text{ bits per sample} = 6 \text{ Mbps}$$

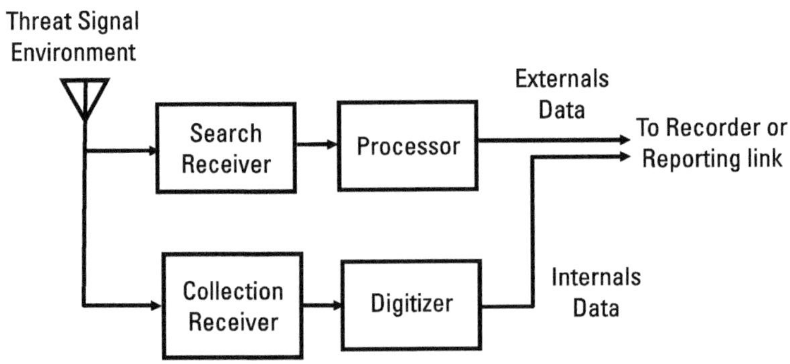

Figure 11.14 COMINT systems capture both the externals and internals of received signals.

Table 11.4
Typical Parameters Captured for Communications ES Threat Signal

Parameter	Range	Resolution	Number of Bits
Radio frequency	10 to 1,000 MHz	1 MHz	12
Modulation type			3
Encryption type			3
Direction of arrival	0° to 360°	1°	9
Total bits per signal			27

It is important to remember the age-old statement about situations in which EW and SIGINT systems are applied: "There is one and only one correct answer to any tactical problem: It depends on the situation and the terrain." More specifically, the correct answer depends on the threat signal modulations, threat operating characteristics, the environmental density, the geometric deployment and motion of threat and receiving assets, and the tactical situation. Thus, there is no single correct answer, so the main goal of this chapter is to help you make trade-offs to optimize results.

11.9 Just Add a Recorder

There are radar ES systems which include digital recorders to capture the characteristics of any new types of signals that may be encountered in the course of normal ES operations as shown in Figure 11.15. This has led some individuals to argue that such a system eliminates the need for SIGINT systems. This is possible: it depends on the situation and the terrain. In general, it is wise to consider the perhaps different circumstances in which a methodical search for and analysis of new threat signal types and the type of data which needs to be collected before making that kind of decision.

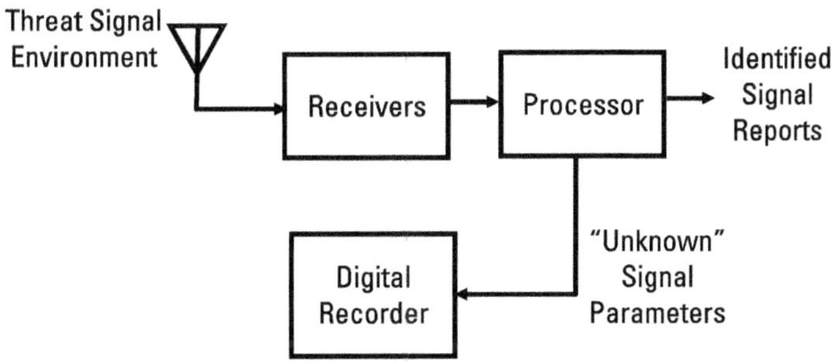

Figure 11.15 A digital recorder can be included in a radar ES system to capture the parameters of new types of signals encountered.

Reference

[1] Adamy, D., *EW 101: A First Course in Electronic Warfare*, Norwood, MA: Artech House, 2001.

About the Author

David L. Adamy is an internationally recognized expert in electronic warfare (EW), probably mainly because he has been writing the EW 101 columns for many years. In addition to writing these columns, he has been an EW professional (proudly calling himself a "Crow") in and out of uniform for more than 50 years. As a systems engineer, project leader, program technical director, program manager, and line manager, Mr. Adamy has directly participated in EW programs from just above DC to just above light. Those programs have produced systems that were deployed on platforms from submarines to space and met requirements from quick and dirty to high reliability.

He holds B.S.E.E. and M.S.E.E. degrees, both with majors in communication theory. In addition to the EW 101 columns, Mr. Adamy has published many technical articles in EW, reconnaissance, and related fields and has 14 books in print. He teaches EW-related courses all over the world and consults for military agencies and EW companies. He is a long-time member of the National Board of Directors and a past president of the Association of Old Crows.

Index

The Artech House Electronic Warfare Library

Advances in Statistical Multisource-Multitarget Information Fusion, Ronald P. S. Mahler

Antenna Systems and Electronic Warfare Applications, Richard A. Poisel

Electronic Intelligence: The Analysis of Radar Signals, Second Edition, Richard G. Wiley

Electronic Warfare for the Digitized Battlefield, Michael R. Frater and Michael Ryan

Electronic Warfare in the Information Age, D. Curtis Schleher

Electronic Warfare Receivers and Receiving Systems, Richard A. Poisel

*Electronic Warfare Target Location Methods,*Richard A. Poisel

EW 101: A First Course in Electronic Warfare, David L. Adamy

EW 103: Tactical Battlefield Communications Electronic Warfare, David L. Adamy

EW 104: EW Against a New Generation of Threats, David L. Adamy

Foundations of Communications Electronic Warfare, Richard A. Poisel

High-Level Data Fusion, Subrata Das

Human-Centered Information Fusion, David L. Hall and John M. Jordan

Information Warfare and Organizational Decision-Making, Alexander Kott, editor

Information Warfare and Electronic Warfare Systems, Richard A. Poisel

Information Warfare Principles and Operations, Edward Waltz

Introduction to Communication Electronic Warfare Systems, Richard A. Poisel

Knowledge Management in the Intelligence Enterprise, Edward Waltz

Mathematical Techniques in Multisensor Data Fusion, Second Edition, David L. Hall and Sonya A. H. McMullen

Modern Communications Jamming Principles and Techniques, Richard A. Poisel

Principles of Data Fusion Automation, Richard T. Antony

Stratagem: Deception and Surprise in War, Barton Whaley

Statistical Multisource-Multitarget Information Fusion, Ronald P. S. Mahler

Tactical Communications for the Digitized Battlefield, Michael Ryan and Michael R. Frater

Target Acquisition in Communication Electronic Warfare Systems, Richard A. Poisel

For further information on these and other Artech House titles, including previously considered out-of-print books now available through our In-Print-Forever® (IPF®) program, contact:

Artech House
685 Canton Street
Norwood, MA 02062
Phone: 781-769-9750
Fax: 781-769-6334
e-mail: artech@artechhouse.com

Artech House
16 Sussex Street
London SW1V 4RW UK
Phone: +44 (0)20-7596-8750
Fax: +44 (0)20-7630-0166
e-mail: artech-uk@artechhouse.com

Find us on the World Wide Web at: www.artechhouse.com